Microarrays

SECOND EDITION

METHODS IN MOLECULAR BIOLOGY™

John M. Walker, SERIES EDITOR

METHODS IN MOLECULAR BIOLOGY™

Microarrays

Volume 2: Applications and Data Analysis

SECOND EDITION

Edited by

Jang B. Rampal

Beckman Coulter, Inc.
Brea, CA

HUMANA PRESS ✳ TOTOWA, NEW JERSEY

© 2007 Humana Press Inc.
999 Riverview Drive, Suite 208
Totowa, New Jersey 07512

www.humanapress.com

This publication is printed on acid-free paper. ∞
ANSI Z39.48-1984 (American Standards Institute) Permanence of Paper for Printed Library Materials.

Production Editor: Jennifer Hackworth

Cover design by Karen Schulz

Cover illustration: Fig. 2, Chapter 4; *see* complete caption and discussion on p. 63.

For additional copies, pricing for bulk purchases, and/or information about other Humana titles, contact Humana at the above address or at any of the following numbers: Tel.: 973-256-1699; Fax: 973-256-8341; E-mail: orders@humanapr.com; or visit our Website: www.humanapress.com

Printed in the United States of America. 10 9 8 7 6 5 4 3 2 1

ISSN 1064-3745
E-ISBN 978-1-59745-304-2
Library of Congress Control Number: 2007924172

Dedication

To my parents and gurus

Preface

To meet the emerging needs of genomics, proteomics, and the other omics, microarrays have become unique and important tools for high-throughput analysis of biomolecules. Microarray technology provides a highly sensitive and precise technique for obtaining information from biological samples. It can simultaneously handle a large number of analytes that may be processed rapidly. Scientists are applying microarray technology to understand gene expression, to analyze mutations and single-nucleotide polymorphisms, to sequence genes, and to study antibody–antigen interactions, aptamers, carbohydrates, and cell functions, among many other research subjects.

The objective of *Microarrays* is to enable the researcher to design and fabricate arrays and binding studies with biological analytes. An additional goal is to provide the reader with a broader description of microarray technology tools and their potential applications. In this edition, *Microarrays* is divided in two parts: *Volume 1* deals with methods for preparation of microarrays, and *Volume 2* with applications and data analysis. Various methods and applications of microarrays are described and accompanied by exemplary protocols. *Volume 2* also covers topics related to bioinformatics, an important aspect of microarray technologies because of the enormous amount of data coming out of microarray experiments. Together, the two volumes provide useful information to the novice and expert alike.

Volume 2: Applications and Data Analysis is dedicated to describing applications of microarrays in DNA and protein studies, and for other biomolecules. Several chapters also focus on data analysis and bioinformatics. Chapter 1 covers the applications of microarrays in nonmammalian vertebrate systems and also provides some of the general steps involved in understanding the microarray process. Chapters 2–4 explain the process for selecting the appropriately coated glass slides for coupling of biomolecules and hybridization protocols. Chapter 5 describes the detection of bacterial pathogens using an oligonucleotide microarray format generated from the hydrolysis of PCR probe sequences. Information about genomic copy number changes in a human cancer cell line and analysis by microarray technology is illustrated in Chapter 6. Chapter 7 explains the interactions of biological sample parameters during microarray experiments. Chapter 8 describes the preparation of clinical samples for microarray study, e.g., preparation of nucleic acids from frozen and formalin-fixed paraffin-embedded human tissues using macro- and microdissection. Understanding microRNA gene expression in tissues, organs, and cell lines in eukaryotes is discussed in Chapter 9. Genotyping using minisequencing by arrayed primer extension (APEX) and printed arrays extended by a single-nucleotide base is discussed in Chapter 10. A protein chip method for single-base mutation determination in rpoB gene (from *Mycobacterium tuberculosis*) with high specificity is described in Chapter 11. Chapter 12 relates cDNA library construction by suppression subtraction hybridization. A simple data analysis pipeline using "linear models for microarray data" is discussed, which may be useful for studies of non-model organisms for which there is little genome sequence information. Target preparation methods using a fully integrated ArrayPlex® system based on Biomek FX and cDNA array hybridization using Affymetrix GeneChip® is discussed in Chapter 13. Chapter 14 describes the application of Affymetrix GeneChip® to extraction of mRNA from stimulated and unsitimulated neoplastic and fibroblastic stromal cells for cDNA array hybridization. Application of protein array (ProtoArray™) for profiling small molecules is discussed in Chapter 15. Monitoring of clinically relevant markers and regulatory pesticides by microarray is described in Chapter 16. Nanoengineered three-dimensional polyelectrolyte thin films coated glass slides used for the preparation of protein microarrays for detection of cytokine analytes is described in Chapter 17. In Overprinting, Chapter 18, printing is performed by contact and non-contact; it is demonstrated by microarray-based immunoassays without the need for wells or other fluid barriers. It represents about a 1000-fold reduction in consumption of reagents from that for conventional 96-well microtiter plate assay. A microarray based on general principal of microfluidic technology is presented in Chapter 19. In Chapter 20, Ciphergen ProteinChip® Array in combination with Protein Biological System 11C mass spectrometer was applied in analyzing SARS patient samples. Analysis of the data was performed by using Ciphergen ProteinChip® Software. Mass spectrometry procedure is applied in Chapter 21 for high-throughput analysis of affinity protein captured analytes. Linder reviews neural networks, including multiple ANN, in Chapter 22. De Bruyne has explained the typical workflow, error handling, PCA, SOM, and K-nearest neighbor related to microarray data analysis in Chapter 23. The *in situ* array hybridization thermodynamics, surface density of probes for predicting melting curve parameters is discussed in Chapter 24. Application of Cluster 3.0 and Java Treeview for microarray data clustering, and BGSSJ for functional classification is summarized in Chapter 25. Application of Perl script for designing various probes is describes in Chapter 26. Chapter 27 deals with the integration of array data with sequence, interaction, localization, and other parameters.

I believe this volume, *Applications and Data Analysis*, will provide valuable information to scientists at all levels, from novice to those intimately familiar with array technology. I would like to thank all the contributing authors for providing manuscripts. My thanks are also due to colleagues for their help in completing this volume. I thank John Walker for editorial guidance and the staff of Humana Press for making it possible to include a large body of available microarray technologies in this volume. Finally, my thanks to my family, especially to my sweet wife Sushma Rampal, for providing all sorts of incentives to complete this project successfully.

Jang B. Rampal

Contents

Contents of the Companion Volume

Volume 1: Synthesis Methods

Contributors

FAHD AL-MULLA • *Department of Pathology, Molecular Pathology Division, Faculty of Medicine, Kuwait University, Safat, Kuwait*

RABA AL-TAMIMI • *Department of Pathology, Molecular Pathology Division, Faculty of Medicine, Kuwait University, Safat, Kuwait*

CATHARINE AQUINO • *Functional Genomics Center Zurich, Zurich, Switzerland*

DAVE K. BERGER • *Department of Botany, Forestry and Agricultural Biotechnology Institute, University of Pretoria, Pretoria, South Africa*

LI-JUN BI • *Institute of Biophysics, Chinese Academy of Sciences, Beijing, China*

JOHNNY W. M. CHAN • *Department of Medicine, Queen Elizabeth Hospital, Kowloon, Hong Kong*

TOM S. CHAN • *Beckman Coulter, Inc., Fullerton, CA*

WAI-WAI CHENG • *Department of Clinical Oncology, Queen Elizabeth Hospital, Kowloon, Hong Kong*

WILLIAM C. S. CHO • *Department of Clinical Oncology, Queen Elizabeth Hospital, Kowloon, Hong Kong*

CLAUS B. V. CHRISTENSEN • *Department of Micro and Nanotechnology, Technical University of Denmark, Lyngby, Denmark*

BRIDGET G. CRAMPTON • *CSIR Biosciences and African Centre for Gene Technologies, Pretoria, South Africa*

MICHAEL C. CRESS • *Beckman Coulter, Inc., Fullerton, CA*

PHILIP J. R. DAY • *Manchester Interdisciplinary Centre, University of Manchester, Manchester, UK; Analytical Sciences, ISAS, Dortmund, Germany*

VERONIQUE DE BRUYNE • *Applied-Maths BVBA, Sint-Martens-Latem*

NORHA DELUGE • *Department of Biology and Biochemistry, University of Houston, Houston, TX*

VIJAY R. DONDETI • *Department of Cellular and Molecular Biology, University of Pennsylvania, Philadelphia, PA*

MARTIN DUFVA • *Department of Micro and Nanotechnology, Technical University of Denmark, Lyngby, Denmark*

XIAOLIAN GAO • *Department of Biology and Biochemistry, University of Houston, Houston, TX*

ERDOGAN GULARI • *Department of Chemical Engineering, University of Michigan, Ann Arbor, MI*

EL-MUSTAPHA HADDOUTI • *Institute of Pathology, University of Bonn, Bonn, Germany*

JENS CLAUS HAHNE • *Institute of Pathology, University of Bonn, Bonn, Germany*

INGO HEIN • *Scottish Crop Research Institute, Dundee, Scotland, UK*

AILING HONG • *Atactic Technologies Inc., Houston, TX*

HSUAN-CHENG HUANG • *Institute of Bioinformatics, National Yang-Ming University, Taipei, Taiwan*

HSUEH-FEN JUAN • *Department of Life Science, Institute of Molecular and Cellular Biology, National Taiwan University, Taipei, Taiwan*

KYU-YEON JUN • *Department of Biology and Biochemistry, University of Houston, Houston, TX*

ANNETTE KAMINSKI • *Institute of Pathology, University of Bonn, Bonn, Germany*

URBAN A. KIERNAN • *Intrinsic Bioprobes Inc., Tempe, AZ*

CHRISTINE N. B. LAU YIP • *Ciphergen Biosystems Incorporation, Fremont, CA*

STEPHEN C. K. LAW • *Department of Clinical Oncology, Queen Elizabeth Hospital, Kowloon, Hong Kong*

ROLAND LINDER • *Institute of Medical Informatics, University of Lübeck, Lübeck, Germany*

VICTOR W. S. MA • *Department of Clinical Oncology, Queen Elizabeth Hospital, Kowloon, Hong Kong*

ANNA MATEJKO • *Atactic Technologies Inc., Houston, TX*

ROBERT S. MATSON • *Beckman Coulter, Inc., Fullerton, CA*

GREGORY A. MICHAUD • *Invitrogen Corporation, Branford, CT*

RAYMOND C. MILTON • *Beckman Coulter, Inc., Fullerton, CA*

FAHD AL-MULLA • *Department of Pathology, Molecular Pathology Division, Faculty of Medicine, Kuwait University, Safat, Kuwait*

DOBRIN NEDELKOV • *Intrinsic Bioprobes Inc., Tempe, AZ*

RANDALL W. NELSON • *Intrinsic Bioprobes Inc., Tempe, AZ*

ROGER K. C. NGAN • *Department of Clinical Oncology, Queen Elizabeth Hospital, Kowloon, Hong Kong*

ERIC E. NIEDERKOFLER • *Intrinsic Bioprobes Inc., Tempe, AZ*

RANJAN J. PERERA • *Keck Graduate Institute, Claremont, CA*

BRUNO POT • *Applied-Maths BVBA, and Bacteriology of Ecosystems, Institut Pasteur de Lille (IBL), Lille Cedex, France*

PAUL F. PREDKI • *Invitrogen Corporation, Branford, CT*

JANG B. RAMPAL • *Beckman Coulter, Inc., Brea, CA*

TEREZA RICHARDS • *The Library, University of the West Indies, Mona, Kingston, Jamaica, West Indies*

MARGARET S. SAHA • *Department of Biology, College of William and Mary, Williamsburg, VA*

MICHAEL SALCIUS • *Invitrogen Corporation, Branford, CT*

RALPH SCHLAPBACH • *Functional Genomics Center Zurich, Zurich, Switzerland*

BARRY SCHWEITZER • *Invitrogen Corporation, Branford, CT*

NIJING SHENG • *Atactic Technologies Inc., Houston, TX*

CONOR W. SIPE • *Department of Biology, College of William and Mary, Williamsburg, VA*

JENS SOBEK • *Functional Genomics Center Zurich, Zurich, Switzerland*

ONNOP SRIVANNAVIT • *Atactic Technologies Inc., Houston, TX, and Department of Chemical Engineering, University of Michigan, Ann Arbor, MI*

RABA AL-TAMIMI • *Department of Pathology, Molecular Pathology Division, Faculty of Medicine, Kuwait University, Safat, Kuwait*

SCOTT J. TEBBUTT • *James Hogg iCAPTURE Center for Cardiovascular and Pulmonary Research, University of British Columbia, Vancouver, BC, Canada*

OLGA G. TROYANSKAYA • *Lewis-Sigler Institute for Integrative Genomics, Carl Icahn Laboratory, Princeton University, Princeton, NJ*

KEMMONS A. TUBBS • *Intrinsic Bioprobes Inc., Tempe, AZ*

ARNOLD VAINRUB • *College of Veterinary Medicine, Auburn University, Auburn, AL*

SUZANNE D. VERNON • *Division of Viral and Rickettsial Diseases, National Centers for Infectious Diseases, Center for Disease Control and Prevention, Atlanta, GA*

WIESNER VOS • *Department of Statistics, Oxford University, Oxford, UK*

MATHIAS WAGNER • *Department of Pathology, Saarland University, Homburg-Saar, Germany*

NICOLAS WERNERT • *Institute of Pathology, University of Bonn, Bonn, Germany*

TONI WHISTLER • *Division of Viral and Rickettsial Diseases, National Centers for Infectious Diseases, Center for Disease Control and Prevention, Atlanta, GA*

TAI-TUNG YIP • *Ciphergen Biosystems Incorporation, Fremont, CA*

TIMOTHY T. C. YIP • *Department of Clinical Oncology, Queen Elizabeth Hospital, Kowloon, Hong Kong*

HANDY YOWANTO • *Beckman Coulter, Inc., Fullerton, CA*

XIAN-EN ZHANG • *Institute of Biophysics, Chinese Academy of Sciences, Beijing, China*

XIAOLIN ZHANG • *Atactic Technologies Inc., Houston, TX*

JIZHONG ZHOU • *Environmental Science Division, Oak Ridge National Lab, Oak Ridge, TN*

XIAOCHUAN ZHOU • *Atactic Technologies Inc., Houston, TX*

XICHUN ZHOU • *Environmental Science Division, Oak Ridge National Lab, Oak Ridge, TN*

QI ZHU • *Department of Biology and Biochemistry, University of Houston, Houston, TX*

1

The Use of Microarray Technology in Nonmammalian Vertebrate Systems

Conor W. Sipe and Margaret S. Saha

Summary

Among vertebrates, the mammalian systems that are frequently used to investigate questions related to human health have gained the most benefit from microarray technology to date. However, it is clear that biological investigations and the generalized conclusions drawn from them, can only be enhanced by including organisms in which specific processes can be readily studied because of their genetic, physiological, or developmental disposition. As a result, the field of functional genomics has recently begun to embrace a number of other vertebrate species. This review summarizes the current state of microarray technology in a subset of these vertebrate organisms, including *Xenopus*, *Rana*, zebrafish, killifish *(Fundulus* sp.), medaka *(Oryzias latipes)*, Atlantic salmon, and rainbow trout. A summary of various applications of microarray technology and a brief introduction to the steps involved in carrying out a microarray experiment are also presented.

Key Words: Atlantic salmon; *Fundulus*; medaka; microarray; rainbow trout; *Xenopus*.

1. Introduction

The use of microarray technology has resulted in substantial progress in many areas of biology where standard experimental organisms (e.g., mouse, yeast, *Caenorhabditis elegans*, *Drosophila*, *Arabidopsis*) are employed. Among vertebrate species, the mammalian models (predominantly mouse and rat) used to study processes implicated in human health have reaped the most benefit from microarrays to date. A number of other vertebrate models traditionally utilized in developmental and toxicological investigations have been somewhat slower to take advantage of array technology, in large part because of having significantly fewer genomic resources available.

From: *Methods in Molecular Biology, vol. 382: Microarrays, Second Edition: Volume 2*
Edited by: J. B. Rampal © Humana Press Inc., Totowa, NJ

Biological investigations and specifically the validity of generalized conclusions, can only be enhanced by using a diversity of species in which specific physiological, developmental, or biochemical traits are readily studied *(1)*. Moreover, a better understanding of the genes effecting adaptive changes and leading to phenotypic differences can be gained by exploring gene expression profiles in other vertebrates for which a wealth of ecological, phenotypic, and genetic data are available *(2)*. Along these lines, functional genomics has recently branched out to a number of other vertebrate systems that might be more suitable for addressing particular questions in a variety of fields. This review will endeavor to: (1) summarize the current state of microarray technology in a subset of these vertebrate organisms that includes *Xenopus, Rana*, zebrafish, killifish (*Fundulus* sp.), medaka (*Oryzias latipes*), Atlantic salmon, and rainbow trout and (2) provide a brief introduction to the steps involved in carrying out a microarray experiment for investigators interested in introducing this technique into their laboratories.

2. Overview of Applications

2.1. Development

A primary goal of developmental biology is to understand the underlying networks of gene and protein interactions that control the processes leading to the final body plan of an organism *(3)*. However, this goal is technically challenging when using traditional molecular techniques, which rely on examining a limited number of genes in a given experiment (e.g., *in situ* hybridization), although extensive and informative analyses have been performed using these methods *(4–6)*. The advent of microarray technology has allowed the profiling of thousands of transcripts simultaneously and has given investigators the ability to examine large-scale changes in gene expression over the course of the development. The utility of this technique is demonstrated by the marked increase in the number of microarray investigations in two nonmammalian vertebrates, *Xenopus* and zebrafish, over the past year (a total of 19 at the time of writing).

These microarray experiments can be roughly divided into two categories, the first of which seeks to determine gene expression profiles at different points in normal development. The expression information obtained from such general surveys can suggest coexpressed gene clusters, identify region- or tissue-specific molecular markers, and make predictions regarding gene function. Recently, the first large-scale analysis of gene expression in the *Xenopus* embryo was reported *(7)*. Likewise, studies in other systems have established expression profiles for embryos at different developmental stages in zebrafish *(8–10)* and medaka *(11)*. More specific developmental processes have also been examined using microarrays, such as the maturation and development of trout ovary *(12)*, the differences between maternal and zygotic contribution to the embryonic transcriptome in *Xenopus (13)*, and the identification of *Xenopus* oocyte-specific transcripts *(14)*.

Another class of studies examines the changes in gene expression brought about by experimental perturbation of a specific developmental pathway. Following perturbation of normal gene expression, genes that are coordinately up- or downregulated can be identified and used to construct molecular pathways that aid in the prediction of the roles of genes with unknown function. In *Xenopus* and zebrafish, the two best-characterized organisms in the vertebrates considered in this review, this has been achieved using a number of molecular biology techniques. Most commonly, a specific protein or pathway is inhibited with chemical inhibitors or agonists, as in the case of the *Xenopus* FGF *(15)*, retinoic acid *(16)*, and BMP2 pathways *(17)*. Likewise, microinjection into embryos can be used to manipulate a given pathway through a dominant negative receptor mRNA *(18)*, injection of a misexpressed transcript *(19,20)*, or through morpholino-based knockdown of gene function *(21)*. The less technically demanding, yet equally effective, approach of physical amputation has been successfully applied in a large-scale analysis of fin regeneration in the adult medaka *(22)*. The previously described studies demonstrate the power of the microarray when combined with the molecular perturbations and/or other classical embryological techniques (i.e., explants, rotation operations, lineage tracing, and so on) available in these nonmammalian vertebrates.

2.2. Physiology

A number of studies in fish systems have investigated the transcriptional response to various physiological challenges. Microarray technology has proven a useful tool for dissecting the molecular foundations of host–microbial interactions. Molecular biomarkers for infection in the Atlantic salmon have been proposed based on the profile of genes upregulated in response to two common pathogens of the Atlantic salmon *(23–25)*. Similarly, zebrafish reared in a sterile environment have implicated genes regulated by microbiota of the gut, including those involved in epithelial proliferation, promotion of nutrient metabolism, and innate immune responses *(26)*.

The fish also provides an excellent system to answer more traditional physiological questions using microarrays. Combining a hypothesis-driven and discovery-based approach, examination of the transcriptional response was reported as a result of the large variations in temperature the killifish experiences on both daily and seasonal bases *(27)*. Unsurprisingly, the control of cell growth and proliferation are an important part of the response to temperature change; however, they report these pathways are regulated by different batteries of genes in constant against the fluctuating temperatures. Chronic temperature reduction in skeletal muscle from adult zebrafish results in a similar pattern of gene regulation when compared with *Fundulus (28)*. In addition, the transcriptional responses to hypoxia and physical stress, two physiological states having clear implications for human health and disease have been characterized in zebrafish *(8,29)* and trout *(30)*.

2.3. Toxicology

To date, classical model organisms have gained the most benefit from microarray research, owing mostly to the immense knowledge already accumulated in these systems. However, these organisms are rarely the most relevant for use in environmental biology and toxicology *(31)*. Fish species, on the other hand, are important biomonitoring tools for toxicologists, as aquatic ecosystems are the major recipient of human-produced pollutants *(32)*. As a result, many groups have begun using newly developed microarrays to measure the effects of contaminants on these species. Koskinen *(33)* endeavored to construct a rainbow trout cDNA microarray for use in toxicological studies and tested its efficacy by exposing trout fry to four model aquatic contaminants. Likewise, the effects of exposure to sublethal concentrations of zinc were studied in trout using a cross-species *Fugu ruprides* gill cDNA filter array *(31)*.

In addition to those present in fish models, a standardized test to determine the relative developmental toxicity hazard of chemical agents is already in place in *Xenopus* (frog embryo teratogenesis assay). This system has been used extensively to minimize the cost and time constraints associated with the use of conventional mammalian test organisms *(34)*. The power of the frog embryo teratogenesis assay toxicological testing system could be increased dramatically by using recently designed *Xenopus* microarrays in conjunction with purely morphological assays.

2.4. Evolution, Comparative Genomics, and Ecology

Spurred on by the success of functional genomics in the areas of development and physiology, investigators have sought to apply microarrays in the investigation other diverse biological processes *(35,36)*. Oleksiak et al. *(1)* have attempted to estimate the degree of natural variation in gene expression within and between natural populations of *Fundulus*. An initial study showed a relatively large number of genes in members of the same population often varied by a factor of 1.5 or more *(37)*. A more recent investigation found an association between cardiac metabolism and mRNA expression profiles, again suggesting that subtle variations in gene expression exist between individuals *(38)*. Other groups have attempted to examine the conservation of molecular pathways between species by using heterologous mRNA in microarray hybridizations. This approach has proved feasible in *Xenopus (39)*, zebrafish and cichlid fishes *(2)*, and salmonids *(31,40)*. Microarray technology has also been used to elucidate the molecular underpinnings of observed ecological phenomena For instance, Mori *(41)* recently examined the molecular basis of predator-induced morphological changes used as a defensive strategy in *Rana* tadpoles.

3. Microarray Design

The two most widely used microarray systems are long oligonucleotide and cDNA microarrays. Long oligonucleotide microarrays are generated by chemically synthesizing oligonucleotides (usually 40–70 bp in length) and affixing them to slides, while cDNA microarrays are created by spotting long strands of amplified cDNA sequences.

3.1. cDNA Arrays

Many of the earlier-mentioned references have successfully utilized cDNA microarrays in experiments. This strategy typically relies on randomly generated libraries of expressed sequence tags (ESTs) to dictate the number and makeup of probe sequences included on a given chip. While subtractive methods have been applied to enrich EST libraries to select for transcripts from specific structures *(12,14,40)*, the final sequence representation on most chips remains relatively indiscriminate. A different approach to selecting ESTs for inclusion on a microarray is to base it on membership in gene-oriented sequence clusters, such as the National Center for Biotechnology Information (NCBI) unigene database *(42,43)*. However, this method is hampered by an unsatisfactory level of sequence annotation, as numerous ESTs from these projects have not been assigned an identity based on homology *(7,18)*. Although, some of these unannotated sequences might represent species-specific genes, they are more likely unidentifiable owing to a lack of extensive genomic resources in these organisms. These issues will certainly be resolved in the future as more sequencing resources are applied to these nonmammalian vertebrates. One advantage that cDNA arrays offer over oligonucleotide-based chips is the ready access they provide to cDNAs with which to perform whole-mount *in situ* hybridization analysis for independent evaluation of data.

Microarrays are available for some of the species under consideration in this review from various research groups: two different salmon microarrays (4000 and 16,000 spots; also used for rainbow trout hybridizations) are available from the Genomic Research on Atlantic Salmon project (http://web.uvic.ca/cbr/grasp/) and a 5000 feature *Xenopus* cDNA from the *Xenopus* Microarray Project (http://silico.ucsd.edu/xenopus/). Other cDNA arrays can be acquired by contacting the corresponding authors in individual papers, who are usually willing to distribute arrays for a nominal fee to cover expenses.

3.2. Long Oligonucleotide

As a result of the effort associated with constructing cDNA arrays, the use of long oligonucleotide arrays has become increasingly popular as an alternative platform. It has even been suggested that oligonucleotide microarrays are more reliable for determining changes in gene expression than their cDNA counterparts *(44)*. A major drawback of these arrays is their associated costs, as the chemically synthesized oligonucleotide probes are usually acquired from a commercial source. Currently, there are pre-made oligonucleotide sets and microarrays available for *Xenopus* (Operon, www.operon.com), zebrafish (Operon; Agilent, www.agilent.com; Sigma-Genosys, www.sigma-genosys.com; Ocimum Biosolutions, www.ocimumbio.com), and medaka (Kimura et al. report their array will be available from Agilent in the future).

With the increased prevalence of custom microarray fabrication services, it has also become possible for investigators to design microarrays tailored to answer specific questions or investigate gene expression patterns in organisms for which commercial arrays do not exist. However, the high probe densities of oligonucleotide arrays make a manual selection of the genes included on a given chip, which is otherwise an impractical task. To overcome this problem, *in silico* probe selection method has recently been described that is applicable for any organism having sequence data *(45)*. This strategy relies on using publicly available microarray information to interrogate existing sequence data and identify a set of homologous genes in the organism of interest, and is particularly suitable for designing microarrays to investigate processes that are conserved among species. The large amounts of data made available by the large EST sequencing consortiums and also the rapid progress of various sequencing initiatives, make these nonmammalian vertebrates ideal candidates for the application of this microarray design method.

4. Experimental Design and Methods

The following is a general survey of methods involved in the major steps in a microarray experiment with no single topic treated in exhaustive detail. It is intended to be an introduction to those investigators who wish to begin the use of microarrays in their laboratories.

4.1. General Considerations

Ultimately, the specific question being asked is the most important consideration when designing an effective microarray experiment. This will generally dictate the type and amount of material available with which to work, as well as the most important comparisons to address the question being posed. A major decision is whether to use a direct (i.e., experimental vs control) or indirect (i.e., experimental vs reference, control vs reference) comparison of expression values. Pairwise comparisons between all samples will yield the most precise estimates of transcriptional differences *(46)*, and for many types of experiments, such as evaluating the effect of a toxin *(33)*, the perturbation of a specific pathway *(16,18)*, or comparing the disease state with the normal *(23,25)*, a direct comparison is the logical choice. When performing direct comparisons, most investigators have opted for dye-swap replications, where the hybridization conditions are repeated with dye assignments reversed in the second hybridization in order to minimize systematic bias introduced by differences in the intensities of the two dyes. A major drawback of this design is that it becomes increasingly less feasible in terms of cost and materials as more samples are added to an experiment.

Consequently, for experiments that measure differences in a large number of samples (e.g., over a developmental time-course) *(7,29)* or repeated measurements of fluctuating phenomena *(27)*, a reference design, where each experimental sample is compared with a common reference, is often used. The indirect expression values obtained from such a reference design have been shown to be robust when directly compared with those from all-pair experimental design *(47)*. Ideally, a reference sample should represent the genes expressed in all samples and with intermediate expression levels *(7)*. The abundance of amphibian and fish embryos that can be obtained and their discrete staging lends itself to the creation of such a pool. Additionally, the use of a universal reference sample for a given organism facilitates the comparison of microarray data sets generated in different experiments, which is otherwise an impractical task. Initial steps to define a common reference pool of RNA have been taken in *Xenopus*, where RNA from eight embryonic samples has been used *(7,47)*. In order to correct for subtle variation between different preparations of a reference RNA pool, it is necessary to fluorescently label each batch and compare the resulting signal intensities in a pairwise hybridization.

4.2. RNA Isolation and Amplification

Pure RNA preparations are critical to ensure reliable gene expression results. Partially degraded RNA can bias cDNA labeling toward sequences that lie near the 3′ terminus and therefore, might distort the relative proportions of various RNA species when hybridized. To preclude this, many investigators use the Agilent BioAnalyzer (Agilent Technologies, Palo Alto, CA), an instrument capable of visualizing picogram amounts of material, to measure RNA quality. Total RNA isolation techniques can roughly be divided into two categories represented in approximately equal proportions in the literature: (1) those that use phase separation to sequester RNA from proteins and DNA (e.g., Invitrogen's TRIzol [Invitrogen, Carlsbad, CA]) or (2) column-based chromatography strategies (e.g., Qiagen RNeasy, [Qiagen, Valencia, CA] Ambion MegaClear [Ambion, Austin, TX]). The use of both techniques, which fractionate RNA on a different physical basis, in conjunction provides significantly better purification than either method performed independently *(46)*.

In some experimental situations, it is impossible to isolate the substantial amounts of total RNA (40–60 µg) needed for a typical labeling reaction. Such is the case when working with tissue from embryonic explants, portions of organs, or other small populations of cells. To obtain the necessary template for target labeling, several RNA amplification methods have been developed. The predominant procedure relies on using a bacteriophage RNA polymerase to transcribe a cDNA template into aRNA (antisense RNA). This transcription is accomplished by synthesizing single-stranded cDNA from experimental RNA using primers containing a synthetic SP6, T3, or T7 RNA polymerase promoter. The final aRNA target has been demonstrated to reflect accurately the size, complexity, and abundance of mRNAs in the original RNA population *(48,49)*. Several commercial kits are available that implement this method (e.g., MessageAmp II, Ambion; RiboAmp, Arcturus [Sunnyvale, CA]; Superscript system, Invitrogen; BioArray system, Enzo Life Sciences [Farmingdale, NY]).

4.3. cDNA Labeling

Investigators are again presented with multiple options when preparing a labeled target solution for hybridization to a microarray. To date, most investigators have chosen to use the cyanine-based dyes Cy3 and Cy5 (GE Healthcare, Piscataway, NJ) to label cDNA; however, a range of alternative dyes (e.g., Alexa Fluor dyes, Invitrogen) are available that are reported to be stronger, more resistant to photobleaching, and less sensitive to pH changes *(50)*. Regardless of fluorophore choice, labels can be incorporated into cDNA directly through the inclusion of dye-conjugated nucleotides or through a coupling reaction between a reactive dye and amine-substituted nucleotides.

Direct labeling techniques are hampered by the ability of reverse transcriptase to incorporate bulky dye-conjugated nucleotides, resulting in labeled cDNA of low fluorescent intensity and a subsequent reduction in assay sensitivity. Nevertheless, many commercial enzymes have been engineered to improve incorporation of fluorophores (e.g., SuperScript III, Invitrogen; CyScribe, GE Healthcare; Omniscript, Qiagen), and many investigators have chosen to utilize this method as it is cost efficient and requires fewer steps. Indirect labeling relies on amino-allyl nucleotides (which are incorporated as efficiently as unmodified nucleotides) that are conjugated to an ester-linked dye molecule following cDNA synthesis. Indirect labeling generally results in a higher overall level of cDNA fluorescence by greatly reducing enzymatic bias, though is a more involved process.

The efficiency and yield of a dye labeling reaction is measured using standard spectroscopy, where the absorbance of the nucleic acid at 260 nm and the absorbance of the dye at its emission maximum are measured. Because the volumes involved are generally quite small, these measurements are usually taken in microcuvettes (10–200 µL) or spectrophotometers specifically designed for small sample volumes (e.g., the Nanodrop, Nanodrop Technologies).

4.4. Hybridization Conditions

Individual hybridization conditions will vary depending on the type of microarray employed in a given assay. The majority of commercial microarrays are distributed with detailed protocols containing a suggested hybridization regime. In general, custom arrays are hybridized under the same conditions of high stringency to deter nonspecific cross-hybridizations *(46)*. Besides labeled target cDNA, typical hybridization solutions contain buffers that are of high ionic strength to reduce electrostatic repulsion and promote complementary base pairing, as well as blocking agents and/or detergents to minimize background. The majority of hybridizations in the references reviewed earlier have also included deionized formamide (30–50%) in hybridization solutions. Formamide lowers the melting temperature of hybrids, thus allowing the temperature of hybridization to be lowered. This reduces the opportunity for target solution evaporation, which can occur at the higher temperatures needed for aqueous hybridizations.

4.5. Washing

Posthybridization washing is another critical step that has the potential to affect the quality of expression data obtained. Inadequate wash conditions will lead to splotches of high background fluorescence in the final scanned image, making detection of genuine signal impossible. Although many commercial arrays come with standard washing protocols, the optimal wash conditions for a particular set of targets should be determined empirically. Generally, these will consist of successively weaker SSC washes (e.g., from 2X to 0.1X) to remove all traces of the target solution, which contains unhybridized cDNA molecules and other substances (e.g., SDS) that can contribute to background fluorescence *(46)*. A common error to be avoided is allowing the slide to dry (even partially) during the washing procedure, as this will lead to a deposition of salt and SDS on the surface of the array that is difficult to remove. Liberal agitation and repeated dunking of slides during these washes is often required to avoid high amounts of irregular background fluorescence. If, after scanning an array, high background fluorescence interferes with signal, the washing steps can be repeated at a higher stringency (i.e., higher temperature, longer incubation times, or more agitation) to remove all traces of hybridization solution.

4.6. Image Acquisition

The importance of the image acquisition step cannot be overstated, as all subsequent data analyses follow from it. Scanning presents the challenge of optimizing a number of parameters to obtain the best possible image in the fewest number of scans. Some of these parameters vary depending on both array type and scanner manufacturer and they need to be set before the final image acquisition begins; these can include focal length, scan resolution, and scan speed. Common to all arrays, however, is the need to maximize spot intensity over that of background. In an ideal microarray image, the intensity of spots corresponding to the lowest-expressed genes should lie just above background pixel intensities, and the intensities of the highest-expressed genes just below saturation levels (i.e., the point at which pixel intensity falls outside the instrument's dynamic range of detection).

The majority of microarray scanners presently available (e.g., Perkin-Elmer ScanArray, Axon GenePix scanners) rely on confocal optics with one or more lasers as excitation sources to scan spots in a pixel-by-pixel manner. In these systems, photomultiplier tubes (PMT) detect photons emitted by labeled cDNA hybridized on the microarray and change the signal into an electrical current that is converted, in turn, to discrete digital values. Thus, two main scanning parameters increase or decrease the spot signal intensities obtained in a given scan: the power of the excitation source (i.e., laser power) and the gain of the PMT. As a rule, PMT gain should be increased before raising laser power, as higher laser energies will often lead to rapid photobleaching of the array. On the other hand, PMT gain indiscriminately raises the intensity of all pixels and magnifies background noise as well as true signal. Frequently, a coordinate adjustment of both parameters is required to ensure that as many spots as possible lie in between the two extremes of saturation and background.

4.7. Data Analysis

Spot intensities in microarray image files are quantified using any one of a multitude of commercially available software packages (e.g., GenePix Pro, Axon; ScanArray Express, Perkin-Elmer; Imagene, Biodiscovery). A computer file (most commonly a .gal file, provided with commercial arrays or created by spotting software) provides a link between the scanned image and the identities of the genes or sequences at a given position on the microarray. Generally, in the first step of data analysis, background-corrected median or mean spot intensity values are calculated, and features of poor quality are filtered out. Individual software packages use different parameters to flag poor quality spots, but all rely on some combination of morphology, number of saturated pixels, or overall intensity. Using a *Xenopus* array, a simple and robust method is developed for determining true spot signal using mean to median correlation *(51)*. Those spots that survive this initial filtering step are normalized to remove intensity- and position-dependent bias in the quantification of each feature (most commonly using the locally weighted scatterplot smoothing [LOESS] method), and a final ratio of intensities (experimental/reference or experimental/control) is calculated for each spot.

4.7.1. Pattern Discovery

Numerous techniques have been developed to aid researchers in discovering and visualizing patterns present in gene expression data generated by microarray experiments *(52)*. In general, these techniques are excellent choices for initial data analysis, as they can suggest gene interactions or members of functional pathways *(46)*. Probably the most prevalent of these computational methods is cluster analysis, which constructs relationships based on expression patterns observed in data sets as a whole *(53,54)*. Both treatment groups and genes can be clustered by some measure of similarity (which vary by method) to organize samples together into familiar tree-like dendograms. Those genes (or groups having genes), which are up- or downregulated in a roughly similar manner lie closest to one another in the tree. Another family of pattern discovery techniques, which includes principal component analysis and multidimensional scaling, strives to reduce the number of variables needed to represent a data set while retaining a maximum amount of its variability *(55)*. The most informative (i.e., those contributing the most variability to the dataset as a whole) variables are then projected into two- or three-dimensional space, and can separate treatments or gene clusters into revealing groups.

4.7.2. Statistical Analysis

The statistical treatment of microarray data has been extensively treated in the literature (for detailed reviews, *see* **refs. 52,56** and **57**). Until relatively recently, fold changes in gene expression were widely used to select candidate genes from a data set for further study (as in **ref. 13**). However, this approach has been the object of criticism, as it does not give a measure of how significant such a difference is likely to be *(57)*. As a result, a number of different statistical tests have been utilized in the previously listed references to detect significantly differentially expressed genes in microarray data sets, including the Students' *t*-test *(21,29,30,33)*, *F*-test *(10)*, analysis of variance *(37,38,58)*, and Bayesian analyses *(16)*. Another statistical method, Significance analysis of microarrays, was developed by Tusher *(59)* specifically for microarrays and has gained widespread acceptance in the literature *(14,25,39)*.

4.7.3. Independent Confirmation of Results

It is important to note that the data obtained from a single microarray experiment will only implicate genes as candidates for involvement in a given network or pathway. It is up to the investigator to decide whether the results are accurate for the particular biological system under study *(60)*. Consequently, other experimental methods are used to validate independently the gene expression levels measured using a microarray. The exact technique used will vary based on the scientific question, but commonly used techniques include semi-quantitative reverse transcription polymerase chain reaction (RT–PCR), quantitative real-time PCR, Northern blot, ribonuclease protection assay, *in situ* hybridization, or immunohistochemistry. Real-time PCR is the choice of many for acquiring precise measurements of candidate gene expression levels, as the method is rapid and relatively simple to perform once established in a laboratory.

5. Online Information Sources

5.1. Genome Sequencing

1. *Danio rerio* Sequencing Project at the Sanger Institute, http://www.sanger.ac.uk/ Projects/D_rerio/.
2. JGI *Xenopus tropicalis* genome project, http://genome.jgi-psf.org/Xentr4/Xentr4. home.html.
3. Salmon Genome Project, http://www.salmongenome.no/cgi-bin/sgp.cgi.
4. Medaka Genome Project, http://dolphin.lab.nig.ac.jp/medaka/.
5. NBRP Medakafish Genome Project, http://shigen.lab.nig.ac.jp/medaka/genome/ top.jsp.

5.2. EST Sequencing

1. *Xenopus*, zebrafish, salmon, trout, and killifish Gene Indices at TIGR, http://www.tigr.org/tdb/tgi/index.shtml.
2. Washington University Zebrafish EST sequencing project, http://www.genetics. wustl.edu/fish_lab/frank/cgi-bin/fish/.
3. *Xenopus tropicalis* EST project at the Sanger Institute, http://www.sanger.ac.uk/ Projects/X_tropicalis/.
4. BLAST interface to search the GRASP EST database, http://snoopy.ceh.uvic.ca/.
5. FunnyBase, a database of functional information for Fundulus ESTs, http://genomics. rsmas.miami.edu/funnybase/super_craw4/.
6. MeBase, a database of medaka ESTs, http://mbase.bioweb.ne.jp/~dclust/ me_base.html.

5.3. Other Sites of Interest

1. The Zebrafish Information Network, http://www.zfin.org.
2. A web-based *Xenopus* resource, http://www.xenbase.org.
3. Genomic Research on Atlantic Salmon Project, http://web.uvic.ca/cbr/grasp/.
4. The main medaka web resource site, http://biol1.bio.nagoya-u.ac.jp:8000/.
5. The official NCBI Handbook, http://www.ncbi.nlm.nih.gov/books/bv.fcgi?call=bv. View.ShowTOC&rid=handbook.TOC&depth=2.
6. Y.F. Leung's huge database of microarray links, http://ihome.cuhk.edu.hk/~b400559/ array.html.

References

1. Oleksiak, M. F., Kolell, K. J., and Crawford, D. L. (2001) Utility of natural populations for microarray analyses: isolation of genes necessary for functional genomic studies. *Mar. Biotechnol. (NY)* **3,** S203–S211.
2. Renn, S. C., Aubin-Horth, N., and Hofmann, H. A. (2004) Biologically meaningful expression profiling across species using heterologous hybridization to a cDNA microarray. *BMC Genomics* **5,** 42.
3. Levine, M. and Davidson, E. H. (2005) Gene regulatory networks for development. *Proc. Natl. Acad. Sci. USA* **102,** 4936–4942.
4. Pollet, N., Muncke, N., Verbeek, B., et al. (2005) An atlas of differential gene expression during early Xenopus embryogenesis. *Mech. Dev.* **122,** 365–439.
5. Kudoh, T., Tsang, M., Hukriede, N. A., et al. (2001) A gene expression screen in zebrafish embryogenesis. *Genome Res.* **11,** 1979–1987.
6. Quiring, R., Wittbrodt, B., Henrich, T., et al. (2004) Large-scale expression screening by automated whole-mount in situ hybridization. *Mech. Dev.* **121,** 971–976.
7. Baldessari, D., Shin, Y., Krebs, O., et al. (2005) Global gene expression profiling and cluster analysis in Xenopus laevis. *Mech. Dev.* **122,** 441–475.
8. Ton, C., Stamatiou, D., Dzau, V. J., and Liew, C. C. (2002) Construction of a zebrafish cDNA microarray: gene expression profiling of the zebrafish during development. *Biochem. Biophys. Res. Commun.* **296,** 1134–1142.
9. Lo, J., Lee, S., Xu, M., et al. (2003) 15000 unique zebrafish EST clusters and their future use in microarray for profiling gene expression patterns during embryogenesis. *Genome Res.* **13,** 455–466.
10. Linney, E., Dobbs-McAuliffe, B., Sajadi, H., and Malek, R. L. (2004) Microarray gene expression profiling during the segmentation phase of zebrafish development. *Comp. Biochem. Physiol. C Toxicol. Pharmacol.* **138,** 351–362.
11. Kimura, T., Jindo, T., Narita, T., et al. (2004) Large-scale isolation of ESTs from medaka embryos and its application to medaka developmental genetics. *Mech. Dev.* **121,** 915–932.
12. von Schalburg, K. R., Rise, M. L., Brown, G. D., Davidson, W. S., and Koop, B. F. (2005) A comprehensive survey of the genes involved in maturation and development of the rainbow trout ovary. *Biol. Reprod.* **72,** 687–699.
13. Altmann, C. R., Bell, E., Sczyrba, A., et al. (2001) Microarray-based analysis of early development in *Xenopus laevis. Dev. Biol.* **236,** 64–75.
14. Vallee, M., Gravel, C., Palin, M. F., et al. (2005) Identification of novel and known oocyte-specific genes using complementary DNA subtraction and microarray analysis in three different species. *Biol. Reprod.* **73,** 63–71.
15. Chung, H. A., Hyodo-Miura, J., Kitayama, A., Terasaka, C., Nagamune, T., and Ueno, N. (2004) Screening of FGF target genes in *Xenopus* by microarray: temporal dissection of the signalling pathway using a chemical inhibitor. *Genes Cells* **9,** 749–761.
16. Arima, K., Shiotsugu, J., Niu, R., et al. (2005) Global analysis of RAR-responsive genes in the Xenopus neurula using cDNA microarrays. *Dev. Dyn.* **232,** 414–431.

17. Peiffer, D. A., Von Bubnoff, A., Shin, Y., et al. (2005) A *Xenopus* DNA microarray approach to identify novel direct BMP target genes involved in early embryonic development. *Dev. Dyn.* **232,** 445–456.
18. Shin, Y., Kitayama, A., Koide, T., et al. (2005) Identification of neural genes using Xenopus DNA microarrays. *Dev. Dyn.* **233,** 248.
19. Taverner, N. V., Kofron, M., Shin, Y., et al. (2005) Microarray-based identification of VegT targets in Xenopus. *Mech. Dev.* **122,** 333–354.
20. Munoz-Sanjuan, I., Bell, E., Altmann, C. R., Vonica, A., and Brivanlou, A. H. (2002) Gene profiling during neural induction in Xenopus laevis: regulation of BMP signaling by post-transcriptional mechanisms and TAB3, a novel TAK1-binding protein. *Development* **129,** 5529–5540.
21. Leung, A. Y., Mendenhall, E. M., Kwan, T. T., et al. (2005) Characterization of expanded intermediate cell mass in zebrafish chordin morphant embryos. *Dev. Biol.* **277,** 235–254.
22. Katogi, R., Nakatani, Y., Shin-i, T., Kohara, Y., Inohaya, K., and Kudo, A. (2004) Large-scale analysis of the genes involved in fin regeneration and blastema formation in the medaka, Oryzias latipes. *Mech. Dev.* **121,** 861–872.
23. Rise, M. L., Jones, S. R., Brown, G. D., von Schalburg, K. R., Davidson, W. S., and Koop, B. F. (2004) Microarray analyses identify molecular biomarkers of Atlantic salmon macrophage and hematopoietic kidney response to *Piscirickettsia salmonis* infection. *Physiol. Genomics* **20,** 21–35.
24. Tsoi, S. C., Cale, J. M., Bird, I. M., Ewart, V., Brown, L. L., and Douglas, S. (2003) Use of human cDNA microarrays for identification of differentially expressed genes in Atlantic salmon liver during *Aeromonas salmonicida* infection. *Mar. Biotechnol. (NY)* **5,** 545–554.
25. Ewart, K. V., Belanger, J. C., Williams, J., et al. (2005) Identification of genes differentially expressed in Atlantic salmon (Salmo salar) in response to infection by Aeromonas salmonicida using cDNA microarray technology. *Dev. Comp. Immunol.* **29,** 333–347.
26. Rawls, J. F., Samuel, B. S., and Gordon, J. I. (2004) Gnotobiotic zebrafish reveal evolutionarily conserved responses to the gut microbiota. *Proc. Natl. Acad. Sci. USA* **101,** 4596–4601.
27. Podrabsky, J. E. and Somero, G. N. (2004) Changes in gene expression associated with acclimation to constant temperatures and fluctuating daily temperatures in an annual killifish Austrofundulus limnaeus. *J. Exp. Biol.* **207,** 2237–2254.
28. Malek, R. L., Sajadi, H., Abraham, J., Grundy, M. A., and Gerhard, G. S. (2004) The effects of temperature reduction on gene expression and oxidative stress in skeletal muscle from adult zebrafish. *Comp. Biochem. Physiol. C Toxicol. Pharmacol.* **138,** 363–373.
29. Ton, C., Stamatiou, D., and Liew, C. C. (2003) Gene expression profile of zebrafish exposed to hypoxia during development. *Physiol. Genomics* **13,** 97–106.
30. Krasnov, A., Koskinen, H., Pehkonen, P., Rexroad, C. E., 3rd, Afanasyev, S., and Molsa, H. (2005) Gene expression in the brain and kidney of rainbow trout in response to handling stress. *BMC Genomics* **6,** 3.

31. Hogstrand, C., Balesaria, S., and Glover, C. N. (2002) Application of genomics and proteomics for study of the integrated response to zinc exposure in a non-model fish species, the rainbow trout. *Comp. Biochem. Physiol. B Biochem. Mol. Biol.* **133,** 523–535.

32. Wester, P. W., van der Ven, L. T. M., and Vos, J. G. (2004) Comparative toxicological pathology in mammals and fish: some examples with endocrine disrupters. *Toxicology* **205,** 27–32.

33. Koskinen, H., Pehkonen, P., Vehniainen, E., et al. (2004) Response of rainbow trout transcriptome to model chemical contaminants. *Biochem. Biophys. Res. Commun.* **320,** 745–753.

34. Fort, D. J., Rogers, R. L., Thomas, J. H., Buzzard, B. O., Noll, A. M., and Spaulding, C. D. (2004) Comparative sensitivity of Xenopus tropicalis and Xenopus laevis as test species for the FETAX model. *J. Appl. Toxicol.* **24,** 443–457.

35. Feder, M. E. and Mitchell-Olds, T. (2003) Evolutionary and ecological functional genomics. *Nat. Rev. Genet.* **4,** 651–657.

36. Stearns, S. C. and Magwene, P. (2003) The naturalist in a world of genomics. *Am. Nat.* **161,** 171–180.

37. Oleksiak, M. F., Churchill, G. A., and Crawford, D. L. (2002) Variation in gene expression within and among natural populations. *Nat. Genet.* **32,** 261–266.

38. Oleksiak, M. F., Roach, J. L., and Crawford, D. L. (2005) Natural variation in cardiac metabolism and gene expression in Fundulus heteroclitus. *Nat. Genet.* **37,** 67–72.

39. Chalmers, A. D., Goldstone, K., Smith, J. C., Gilchrist, M., Amaya, E., and Papalopulu, N. (2005) A Xenopus tropicalis oligonucleotide microarray works across species using RNA from Xenopus laevis. *Mech. Dev.* **122,** 355–363.

40. Rise, M. L., von Schalburg, K. R., Brown, G. D., et al. (2004) Development and application of a salmonid EST database and cDNA microarray: data mining and interspecific hybridization characteristics. *Genome Res.* **14,** 478–490.

41. Mori, T., Hiraka, I., Kurata, Y., Kawachi, H., Kishida, O., and Nishimura, K. (2005) Genetic basis of phenotypic plasticity for predator-induced morphological defenses in anuran tadpole, Rana pirica, using cDNA subtraction and microarray analysis. *Biochem. Biophys. Res. Commun.* **330,** 1138–1145.

42. Chen, Y. A., McKillen, D. J., Wu, S., et al. (2004) Optimal cDNA microarray design using expressed sequence tags for organisms with limited genomic information. *BMC Bioinformatics* **5,** 191.

43. Lorenz, M. G., Cortes, L. M., Lorenz, J. J., and Liu, E. T. (2003) Strategy for the design of custom cDNA microarrays. *Biotechniques* **34,** 1264–1270.

44. Li, J., Pankratz, M., and Johnson, J. A. (2002) Differential gene expression patterns revealed by oligonucleotide versus long cDNA arrays. *Toxicol. Sci.* **69,** 383–390.

45. Dondeti, V. R., Sipe, C. W., and Saha, M. S. (2004) In silico gene selection strategy for custom microarray design. *Biotechniques* **37,** 768–770, **72,** 74–76.

46. Bowtell, D. and Sambrook, J. (2003) DNA microarrays: a molecular cloning manual, Cold Spring Harbor Laboratory Press, Cold Spring Harbor, NY.

47. Konig, R., Baldessari, D., Pollet, N., Niehrs, C., and Eils, R. (2004) Reliability of gene expression ratios for cDNA microarrays in multiconditional experiments with a reference design. *Nucleic Acids Res.* **32,** E29.

48. Van Gelder, R. N., von Zastrow, M. E., Yool, A., Dement, W. C., Barchas, J. D., and Eberwine, J. H. (1990) Amplified RNA synthesized from limited quantities of heterogeneous cDNA. *Proc. Natl. Acad. Sci. USA* **87,** 1663–1667.

49. Eberwine, J., Yeh, H., Miyashiro, K., et al. (1992) Analysis of gene expression in single live neurons. *Proc. Natl. Acad. Sci. USA* **89,** 3010–3014.

50. Cox, W. G., Beaudet, M. P., Agnew, J. Y., and Ruth, J. L. (2004) Possible sources of dye-related signal correlation bias in two-color DNA microarray assays. *Anal. Biochem.* **331,** 243–254.

51. Tran, P. H., Peiffer, D. A., Shin, Y., Meek, L. M., Brody, J. P., and Cho, K. W. (2002) Microarray optimizations: increasing spot accuracy and automated identification of true microarray signals. *Nucleic Acids Res.* **30,** E54.

52. Slonim, D. K. (2002) From patterns to pathways: gene expression data analysis comes of age. *Nat. Genet.* **32,** 502–508.

53. Tavazoie, S., Hughes, J. D., Campbell, M. J., Cho, R. J., and Church, G. M. (1999) Systematic determination of genetic network architecture. *Nat. Genet.* **22,** 281–285.

54. Eisen, M. B., Spellman, P. T., Brown, P. O., and Botstein, D. (1998) Cluster analysis and display of genome-wide expression patterns. *Proc. Natl. Acad. Sci. USA* **95,** 14,863–14,868.

55. Landgrebe, J., Wurst, W., and Welzl, G. (2002) Permutation-validated principal components analysis of microarray data. *Genome Biol.* **3,** RESEARCH0019.

56. McShane, L. M., Shih, J. H., and Michalowska, A. M. (2003) Statistical issues in the design and analysis of gene expression microarray studies of animal models. *J. Mammary Gland Biol. Neoplasia* **8,** 359–374.

57. Cui, X. and Churchill, G. A. (2003) Statistical tests for differential expression in cDNA microarray experiments. *Genome Biol.* **4,** 210.

58. Whitehead, A. and Crawford, D. L. (2005) Variation in tissue-specific gene expression among natural populations. *Genome Biol.* **6,** R13.

59. Tusher, V. G., Tibshirani, R., and Chu, G. (2001) Significance analysis of microarrays applied to the ionizing radiation response. *Proc. Natl. Acad. Sci. USA* **98,** 5116–5121.

60. Chuaqui, R. F., Bonner, R. F., Best, C. J., et al. (2002) Post-analysis follow-up and validation of microarray experiments. *Nat. Genet.* **32,** 509–514.

2

Quality Considerations and Selection of Surface Chemistry for Glass-Based DNA, Peptide, Antibody, Carbohydrate, and Small Molecule Microarrays

Jens Sobek, Catharine Aquino, and Ralph Schlapbach

Summary

The complexity of workflows for the production of high quality microarrays asks for the careful evaluation and implementation of materials and methods. As a cornerstone of the whole microarray process, the microarray substrate has to be chosen appropriately and a number of crucial considerations in respect to matching the research question with the technical requirements and possibilities have to be taken into account. In the following, how to lay the fundamental for high performance microarray experiments by evaluating basic quality requirements and the selection of suitable slide surface architectures for a variety of applications was concentrated.

Key Words: Microarrays; quality; surface chemistry; spot morphology; immobilization.

1. Introduction

Microarrays have become an indispensable and highly efficient tool for the investigation of genome alterations and large-scale gene expression patterns in basic and applied research in the academic and industrial world. Emerging uses of microarrays for the elucidation of protein-binding activity, antibody–epitope specificity, and functional protein enzyme assays, can be foreseen for the near future. The fact that microarray technology is not limited to DNA and protein applications only is well-illustrated by latest developments in the generation and use of carbohydrate arrays for binding studies or cellular on-the-chip assays.

In order to cope with the various technical issues connected to the fabrication and use of microarrays in these diverse areas, expertize in many fields of physical and organic chemistry, biochemistry, and molecular biology has to be combined with latest technologies in engineering and bioinformatics. In this article, guidelines on the selection of a suitable slide surface that is usually dictated by the

From: *Methods in Molecular Biology, vol. 382: Microarrays: Second Edition: Volume 2*
Edited by: J. B. Rampal © Humana Press Inc., Totowa, NJ

planned application are provided. In turn, the chosen surface chemistry determines the experimental conditions of immobilization, blocking, and hybridization, which have to be carefully optimized. Stating this, the principle production parameters have been optimized in a large number of experiments for a variety of slides with different slide chemistries and for diverse applications, such as long and short DNA, peptide, protein, antibody, carbohydrate, and small molecule microarrays. As a result, general recommendations have been established on how experimental conditions for custom made microarrays can be optimized. This workflow and optimized protocols will be presented in the subsequent Chapters 3 and 4.

2. Methods

2.1. General Quality Considerations

Glass quality and the quality of the coatings determine the optical properties of the slide such as probe immobilization efficiency and spot morphology. Slide quality is a crucial factor in the production of microarrays and is determined by a combination of chemical composition, flatness, autofluorescence, and glass homogeneity. Scratches, deposits, or other artifacts might arise from the production process of the glass and the subsequent chemical coating. Additionally, abrasion from the plastic packings might leave particles on the slide surface (*see* **Note 1**). In order to evaluate the quality of a slide, a quick glance across the surface of the slide held into the light under a small angle reveals any visible chemical depositions, detectable as gray shadows or small islands of speckles. Irrespective of the type of damage, such slides should not be used for microarray applications. A second simple but highly efficient quality test is scanning the slide at highest laser intensity and PMT amplification (**Fig. 1**). A very low-signal intensity of homogeneous distribution should be the basis for rejection of the slide. It is important to note that most of the slides on the market are based on a wet chemical silanation reaction, which is very difficult to perform under controlled conditions. Manufacturers should be selected carefully for such type of slide in order to obtain the best possible results in a microarray experiment regarding reliability, reproducibility, and detection limit. All slide errors must be kept as small as possible. Low quality slides should be rejected as they bear the risk of failing an experiment (*see* **Note 2**).

The next important parameter in the production of microarrays is the spotting process, which often leads to imperfect spots with an unsatifactory spot morphology and large spot-to-spot variations. Spots might deviate from the perfect round shape especially when a contact printer is used, and more significantly, the deposited compounds often display a nonuniform distribution within the spots. Imperfect spots might introduce a large experimental error that compromises the absolute accuracy of the individual measurement.

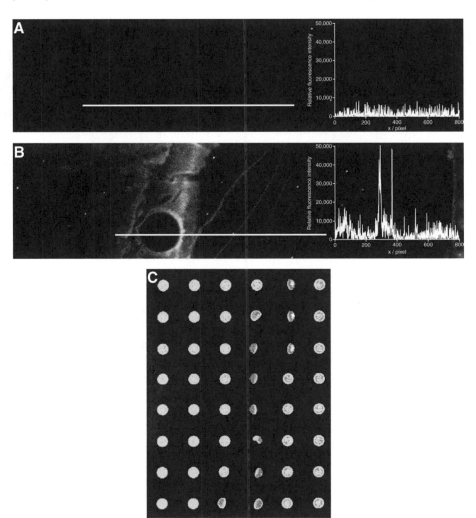

Fig. 1. Slide quality test and effect of deposits on the slide surface on spot morphology. (**A,B**) Laser scan images (ScanArray5000, Perkin Elmer, Downers Grove, IL;
details 17×4.3 mm^2) of commercial slides obtained from different manufacturers.
Excitation of fluorescence at 543 nm (laser intensity 100%) and detection at 570 nm
(photomultiplier tube setting 100%) shows significant differences in the quality of the
slide coatings. In this comparison, only slide (**A**) should be considered for the production of a microarray. Structures on slide (**B**) are presumably caused by drying effects
during slide production. Inserts show a line profile of the marked region. (**C**) Deposits
on the slide surface result in bad spot quality as illustrated for spots of a Cy3-labeled
13-mer oligonucleotide on a commercial epoxy silane slide of insufficient quality
(scan obtained with LS400, Tecan, Salzburg, Austria).

The sensitivity of spot formation to surface artifacts can be used for a simple slide quality test and can serve to assess the overall quality of whole batches of slides from manufacturers, as according to our experience, a single slide most often represents a batch of 10–25 slides. Using the advantage of piezo printers that are able to deliver pL drops in a highly accurate and reproducible manner, printing a dye-labeled oligonucleotide in some thousand replicates results in an array of spots showing—in an ideal case—identical fluorescence intensities and spot morphology (**Fig. 2A**). Deviations from a uniform spot intensity across the whole slide surface such as gradients (large-range artifacts, **Fig. 2C**) or deformed spots (small-range imperfections, **Fig. 2B**) suggest a bad overall slide quality (*see* **Notes 3** and **4**).

2.2. Overview of Slide Surface Chemistry and Immobilization Procedures in the Production of Microarrays

2.2.1. General Considerations on Surface Chemistry and Immobilization Procedures

The concept of surface chemistry is directly related to the immobilization of probe molecules. A precondition for the production of microarrays is a surface with a suitable chemical coating. Uncoated clean glass cannot be used because it is too hydrophilic and strongly adsorbs components from the air that change the surface in an uncontrolled way. Moreover, because of lacking binding functionalities, a specific immobilization of the probe molecules is impossible. The large spectrum of surface chemistries used for microarray applications can be represented by their simplified molecular architecture as shown in **Fig. 3**. The most simple chemical coatings consist of a layer of silane with a reactive terminal group (**Fig. 3A**). Additionally, spacer groups that can be introduced by the reaction with a bifunctional crosslinker increase the distance from the glass surface and ensure a more solution-like behaviour of the binding reaction between probe and target molecules (**Fig. 3B**). More sophisticated coatings consist of a layer of different types of polymers (**Fig. 3C**) or dendrimers (**Fig. 3D**). In all cases, the coating on the slide surface determines the properties of hydrophilicity, long-term chemical stability, binding properties, such as loading capacity and chemical reactivity, the degree of adsorption, and spot size and shape. In addition, the surface chemistry provides the chemical environment for the sample compound and influences stability and accessibility of the immobilized probe molecules. In respect to the latter, there are three general methods for immobilization: the formation of strong covalent bonds (1) by a thermal chemical reaction, and (2) ultraviolet (UV) crosslinking, and (3) physical adsorption. It is obvious that the immobilization of different types of probe molecules require a matching surface chemistry in order to obtain optimal experimental results. Immobilization conditions are extensively discussed in the subsequent Chapter 3.

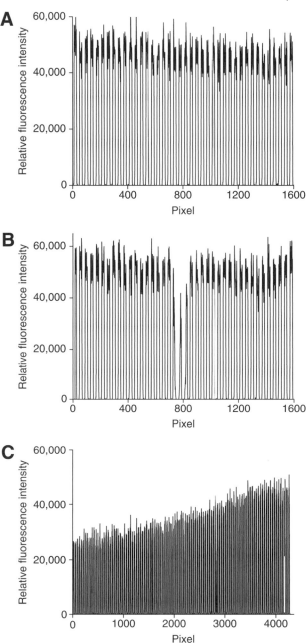

Fig. 2. Line scan through a number of replicate spots. (**A**) Uniform spots on a high-quality slide surface. The standard deviation of fluorescence intensity is 5% (40 spots). (**B**) Because of deposits on the slide surface spot shape and intensity is strongly distorted in the middle part of the scan. (**C**) Inhomogeneous coating results in a strong gradient in spot intensity. (For verification, the slide was turned by 180° and scanned again to exclude effects arising from scanning out of focal plane.)

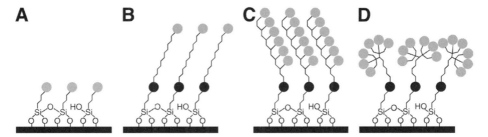

Fig. 3. Schematic representation of slide surface architecture. (**A**) Simple silane-based slides. (**B**) Linker-modified slides. (**C**) Polymer-coated slides. (**D**) Dendrimer-coated slides. The representation is simplified such that no structural effects are taken into account (molecular orientational effects, silane multilayer formation, cross-linking of polymers, and gel formation). Black spheres are linking groups introduced by the chemical synthesis of the slide. Gray spheres are reactive groups. On polymer slides of type 3C these groups might be either reactive functional groups or nonreactive groups, for example, amide. Structures are not drawn to scale.

The cost of the slides might increase from simple silane slides to linker-modified and then more from polymer to dendrimer slides. The following paragraphs provide recommendations on the selection of suitable surfaces for various applications *(1)*.

2.2.2. Oligonucleotide Microarrays

Now-a-days oligonucleotide arrays can be considered a standard application featuring a number of advantages over cDNA arrays. Because oligonucleotides modified with a nucleophilic linker react under mild conditions with electrophilic groups including glycidyloxy (epoxy), aldehyde, isothiocyanate, and activated ester (e.g., N-hydroxysuccinimide [NHS]), slides presenting these functional groups are preferably used. Aminosilane-coated slides does not provide a suitable functionality to covalently immobilize oligonucleotides in a mild and specific manner. On these slides, immobilization can only be achieved by applying harsh conditions, such as heating to 80°C or UV crosslinking *(2)*. In many articles, epoxysilane- and aldehyde-modified slides (prepared by the reaction of aminosilane slides with glutaraldehyde) are successfully used for oligonucleotide applications (refer to publications). However, when comparing processing protocols for epoxysilane- and aldehyde-modified slides, the former has clear advantage regarding simplicity, processing time, flexibility with regard to the blocking reagent, no harmful and toxic chemicals (sodium borohydride and sensitive) and hence less cost for waste disposal. It has been found that there is no need for a (rather expensive and sensitive) nucleophilic linker group to immobilize long oligonucleotides at surfaces presenting epoxy groups (*see* Chapter 3).

To achieve functional immobilization of oligonucleotides smaller than about 20 nt requires a more careful selection of a matching surface chemistry than for commonly used longer oligonucleotides. Becuase of the short length of the molecules, the accessibility of the immobilized probe by the target molecule can be severely hindered *(3)*, longer reaction times are required when hybridized with large target molecules such as cDNA. This is caused by sterical hindrances and by the particular conditions in close proximity to surfaces, where diffusion rate constants are significantly reduced *(4)*. Moreover, surface charges change the properties and reactivity of the immobilized molecules, leading to substantial differences in hybridization thermodynamics *(5)*. In order to avoid these effects and to accelerate binding reactions with a target molecule, slides of type 3B–D (according to **Fig. 3**) are required to ensure a more solution-like interaction of probe and target molecules.

2.2.3. cDNA Microarrays

DNA of high-molecular weight, such as cDNA or bacterial artificial chromosomes (BACs) are usually immobilized on two-dimensional (2D) slide surfaces of type 3A (according to **Fig. 3**) *(6)*. BACs consist of a mixture of different DNA molecules of different size that show heterogeneous properties on a slide surface regarding the spot morphology. Traditionally, aminosilane- and epoxysilane-coated slides are used for these applications. Comparing the two types of surfaces, aminosilane-coated slides offer some advantages with respect to flexibility in the choice of spotting solutions and the quality of the resulting spot morphology. Frequently used spotting solutions, such as dimethyl sulfoxide (DMSO) and a saline-citrate buffer (3X SSC) can serve as examples. DMSO is a very useful solvent for many different organic molecules and a 50% aqueous solution is often used for spotting DNA. This solution denatures DNA so that single strands are deposited and immobilized. Thereby, no protocol step for strand separation on the slide surface is necessary and spares the slides to be boiled at 95°C, which is a harsh treatment (*see* experiments and **Fig. 1** in Chapter 3). In addition, spotting cDNA in 50% DMSO to aminosilane-coated slides results in a good spot morphology, which is not the case for epoxysilane-coated slides that are incompatible with DMSO in water and produce spots of undefined shape and morphology.

In many cases, 3X SSC is a very good spotting buffer in combination with surfaces coated with epoxysilane or aminosilane and results in spots of excellent morphology for many types of probe molecules. However, in the case of BACs printed in 3X SSC the spot morphology on epoxysilane slides is unsatisfactory. The same solution printed to aminosilane slides results in spots of a very good morphology, which illustrates the sensitvity of the spot morphology to the combined properties of the spotting solution and the surface coating. As outlined next, spotting dye-labeled model compounds in recommended solutions to the slide of choice shows the resulting spot morphology right away (*see* Chapter 3, **Subheadings 2.1.2.** and **2.1.4.**). This helps to choose a functional combination of spotting solution and slide surface without the need of hybridization.

2.2.4. Peptide Microarrays

For applications that require the immobilization of molecules of low-molecular weight linker-modified slides of type 3B, polymer slides of type 3C, and dendrimer slides of type 3D (according to **Fig. 3**) should be used for the reasons mentioned earlier (*see* **Subheading 2.2.2.**). If the probe molecules contain a sufficiently long linker, simple silane-based slides of type 3A can be used. In any case, a matching surface chemistry is indispensable. There are many examples published, offering a variety of sophisticated immobilization strategies (*7–14*). In most cases, home-made slide surfaces were used including linker-modified aminosilane (*12,15–17*) and epoxysilane slides (*12*), epoxy-activated polyethylene glycol (PEG)-coated surfaces (*18*), aldehyde-modified slides (*11,19,20*), semicarbazide slides (*13,21,22*), and alkanethiolates forming self-assembled monolayers on gold (*9,10,23*), among others. Our laboratory have successfully used in-house made PEG-coated slides for spotting peptide arrays for epitope mapping of corresponding antibodies (*24*). There are only a limited selection of type 3B and 3C slides on the market. As a consequence, many microarray groups prepare slides of their own.

2.2.5. Carbohydrate Microarrays

Carbohydrate microarrays are a relatively new application and not yet well established. The biggest problem in producing a carbohydrate microarray is the limited availability of defined carbohydrate samples (*25*). For carbohydrates of low-molecular mass, the best method for immobilization is covalent (*9,26–31*) or noncovalent coupling (*32–34*) using a linker with a specific coupling chemistry. However, this requires sophisticated chemical modifications as carbohydrates from natural sources usually lack such a suitable linker needed for a site-directed immobilization to the surface.

An alternative is immobilization in the absence of a modification of the probe molecules. Microbial polysaccharides of high-molecular mass (*35*) and neoglycolipids (*36*) were adsorbed to nitrocellulose-coated slides. Bryan and Wong immobilized unmodified di- and trisaccharides to nonpolar polystyrene microtiter plates by adsorption (*34,37*). Because to their high content of nucleophilic hydroxyl groups carbohydrates easily immobilize at surfaces presenting electrophilic groups (e.g., epoxy or NHS ester groups). Fluorescein-labeled dextrane was immobilized to epoxysilane of type 3B and other electrophilic-activated polymer surfaces of type 3C (*see* **Fig. 3**) in model experiments to determine processing parameter (*see* Chapter 3). However, without site-specific immobilization chemistry the carbohydrates are bound in a random distribution at the surface. In turn, this undefined orientation might lead to blocking of molecular recognition sites, a situation similar to that observed for protein and antibody microarrays *(38–40)*. Feizi et al. *(27)* presented an overview of immobilization methods for carbohydrates. Slide surface coatings used so far include thioles on gold *(28)* or glass *(31,41)*, nitrocellulose *(35,36,42)*, and (oxidized black) polystyrene *(37,43)*.

2.2.6. Small Molecule Microarrays

The term "small molecule" refers to chemical compounds of low-molecular mass that can be extracted from natural sources or produced synthetically *(44–46)*. For an overview on the use of small molecules in microarray applications *see* **refs. *16,47–51***. The molecular structures of these probes can be highly diverse. A large variety of chemical reactions leading to a stable chemical bond and efficiently coupling the probe to the surface may be applied for immobilization. Usually, in the course of the chemical synthesis of probes, a building block consisting of a linker and a reactive group are introduced for which a slide with matching surface chemistry is available. As discussed for short oligonucleotide, carbohydrate, and peptide microarrays, two main principles must be followed: First, a site-directed immobilization in terms of matching chemical groups for coupling the probe to the surface must be applied. Second, a suitable linker to gain distance from the surface is needed to obtain a good accessibility of the probes by the target molecules.

Examples for immobilization of small molecules include the extensive work of the Schreiber group coupling probes containing a hydroxyl group to a chlorinated surface *(52–55)*, and the reaction of azide groups with a phosphane-coated surface in a Staudinger ligation reaction *(56)*. Aminosilane-coated slides were modified with a bifunctional crosslinker including maleimide derivatives *(15)*, a PEG derivative *(18)*, or a diazobenzylidene *(57)*. Hoff et al. *(15)* used a chemolabile linker for a slow release of immobilized probes for uptake by cells incubated on the microarray. An alternative surface chemistry applied to small molecule microarrays includes photoaffinity labeling of the surface with a light sensitive 3H-diazirinyl presented by Kanoh et al. *(58)*. At UV irradiation, this chemical group decomposes and forms a reactive electrophilic carbene that efficiently reacts with nucleophilic groups in steroids. The authors state that because of the high reactivity of the carbene, structurally distinct molecules can be immobilized.

A completely different approach for the production and processing of small molecule microarrays was invented by Gosalia and Diamond *(59)*. Their probes were dissolved in a nonvolatile glycerol/DMSO mixture and were spotted to cleaned blank glass slides. These nanoliter droplets did not evaporate and were used as stable liquid microreactors. Without the need for immobilization and washing procedures, target compounds were subsequently applied by aerosol deposition. This method allows for the transfer of the 96-well plate assay format to a microarray while requiring much smaller volumes for a screen in liquid solution.

Note that some authors use the term small molecule microarray in the context of hybridization of small molecule-tag conjugates to, for example, an oligonucleotide array *(60)*. In this type of application, PNA or oligonucleotides *(61)* are used as decoding tag.

2.2.7. Antibody Microarrays

Unlike the examples described earlier, antibodies and other proteins are large molecules that immobilize at nearly all surfaces by passive adsorption. In addition, covalent bonds are formed if the slide surface provides reactive electrophilic groups. Most importantly, as the biological function of proteins depends on their three-dimensional structure, conformational changes must be avoided. In practice, this is rarely possible to the full extent and explains why certain proteins immobilized at surfaces do not show the expected function observed in solution *(62,63)*. The degree of denaturation of a protein on a surface strongly depends on the chemical nature of the coating.

Although many surfaces are prone to denature proteins, PEG-coated surfaces and hydrogels are used to reduce the risk of protein denaturation *(64,65)*. Immobilization of antibodies on 2D surfaces was optimized by Peluso et al. *(66)*. Angenendt et al. *(67)* described an overview on methods of specific immobilization of antibodies and proteins. Some commercial antibody and protein microarrays are based on nitrocellulose membrane slides. Unfortunately, application of these slides is strongly limited owing to the generation of a large background signal upon laser excitation of fluorophores caused by extensive stray light arising from the some 10-μm thick nitrocellulose membrane *(68)*.

A variety of phosphorylation specific antibodies have been successfully used on all types of slides recommended earlier, which illustrates that antibodies are more stable than many other proteins and that a variety of surfaces can be used for antibody applications *(69)*.

3. Notes

1. Never touch the surface of a microarray or wipe off dust and other particles. Deposits should be removed exclusively by clean compressed air or inert gases.
2. Coated but unprocessed slides are sensitive to air (amino groups) and humidity (nearly all reactive groups used on slides). Keep in mind that there is only a monolayer of molecules with reactive groups immobilized at a 2D slide surface. Once a package is opened, the slides should be used immediately or sealed appropriately. The best way to store slides is under nitrogen in a desiccated box. Slides can usually be used past the manufacturer recommended date if properly stored.

Some reactive groups (NHS ester-activated slides) degrade even when stored under optimal condition but can be reactivated *(70)*.

3. Spotted slides should be stored without further processing (as long as there is no specific reason for processing, for example, as a result of a change in spot morphology with time caused by properties of the spotting solution). There is a large amount of probe in the spot covering and thereby protecting those molecules immobilized at the surface. An exception are slide coatings based on nitrocellulose and nylon, where the three-dimensional matrix protects the spotted probe.

4. Processed slides stored desiccated under nitrogen in the dark can be used for months for later scanning. Experiments performed during the summer (2005) have shown that dyes degrade on the surface of microarrays upon exposed to air even when stored under exclusion of light. Results of dye stability tests clearly correlate with the air ozone level *(71)*.

Acknowledgment

JS likes to thank Prof. Dr. Joe Jiricny (University of Zurich, ETH Zurich), Dr. Orlando Schärer (Stony Brook University Hospital), and Dr. Philip Day (University of Manchester) for their extensive help getting started with slide development, microarray production, and hybridization experiments.

References

1. For availability of microarray slides, *see* Sobek, J. and Schlapbach, R. (2004) Substrate architecture and functionality define the properties and performance of DNA, peptide, protein, and carbohydrate microarrays. *Pharmagenomics* **Sept.15,** 32–44.

2. Wang, H. -Y., Malek, R. L., Kwitek, A. E., et al. (2003) Assessing unmodified 70-mer oligonucleotide probe performance on glass-slide microarrays. *Genome Biol.* **4,** R5.

3. Butler, J. E. (2000) Solid supports in enzyme-linked immunosorbent assay and other solid-phase immunoassays. *Methods* **22,** 4–23.

4. Chan, V., Graves, D. J., Fortina, P., and McKenzie, S. E. (1997) Adsorption and surface diffusion of DNA oligonucleotides at liquid/solid interfaces. *Langmuir* **13,** 320–329.

5. Vainrub, A. and Pettitt, B. M. (2003) Surface electrostatic effects in oligonucleotide microarrays: control and optimization of binding thermodynamics. *Biopolymers* **68,** 265–270.

6. Taylor, S., Smith, S., Windle, B., and Guiseppi-Elie, A. (2003) Impact of surface chemistry and blocking strategies on DNA microarrays. *Nucleic Acids Res.* **31,** E87.

7. Yeo, D. S. Y., Srinivasan, R., Chen, G. Y. J., and Yao, S. Q. (2004) Expanded utility of the native chemical ligation reaction. *Chem. Eur. J.* **10,** 4664–4672.

8. Melnyk, O., Duburcq, X., Olivier, C., Urbès, F., Auriault, C., and Gras-Masse, H. (2002) Peptide arrays for highly sensitive and specific antibody-binding fluorescence assays. *Bioconjugate Chem.* **13,** 713–720.

9. Houseman, B. T., Gawalt, E. S., and Mrksich, M. (2003) Maleimide-functionalized self-assembled monolayers for the preparation of peptide and carbohydrate biochips. *Langmuir* **19,** 1522–1531.
10. Houseman, B. T. and Mrksich, M. (2002) Towards quantitative assays with peptide chips: a surface engineering approach. *Trends Biotechnol.* **20,** 279–281.
11. Falsey, J. R., Renil, M., Park, S., Li, S., and Lam, K. S. (2001) Peptide and small molecule microarray for high throughput cell adhesion and functional assays. *Bioconjugate Chem.* **12,** 346–353.
12. Lesaicherre, M. -L., Uttamchandani, M., Chena, G. Y. J., and Yao, S. Q. (2002) Developing site-specific immobilization strategies of peptides in a microarray. *Bioorg. Med. Chem. Lett.* **12,** 2079–2083.
13. Olivier, C., Hot, D., Huot, L., et al. (2003) α-Oxo semicarbazone peptide or oligodeoxynucleotide microarrays. *Bioconjugate Chem.* **14,** 430–439.
14. Chelius, D. and Shaler, T. A. (2003) Capture of peptides with N-terminal serine and threonine: a sequence-specific chemical method for peptide mixture simplification. *Bioconjugate Chem.* **14,** 205–211.
15. Hoff, A., André, T., Fischer, R., et al. (2004) Chemolabile cellular microarrays for screening small molecules and peptides. *Molec. Diversity* **8,** 311–320.
16. Uttamchandani, M., Walsh, D. P., Khersonsky, S. M., Huang, X., Yao, S. Q., and Chang, Y. -T. (2004) Microarrays of tagged combinatorial triazine libraries in the discovery of small-molecule ligands of human IgG. *J. Comb. Chem.* **6,** 862–868.
17. Xiao, S. -J., Textor, M., Spencer, N. D., and Sigrist, H. (1998) Covalent attachment of cell-adhesive, (Arg-Gly-Asp)-containing peptides to titanium surfaces. *Langmuir* **14,** 5507–5516.
18. Lee, M. -R. and Shin, I. (2005) Fabrication of chemical microarrays by efficient immobilization of hydrazide-linked substances on epoxide-coated glass surfaces. *Angew. Chem.* **117,** 2941–2944.
19. This type of slide is commercially available.
20. Salisbury, C. M., Maly, D. J., and Ellman, J. A. (2002) Peptide microarrays for the determination of protease substrate specificity. *J. Am. Chem. Soc.* **124,** 14,868–14,870.
21. Duburcq, X., Olivier, C., Malingue, F., et al. (2004) Peptide protein microarrays for the simultaneous detection of pathogen infections. *Bioconjugate Chem.* **15,** 307–316.
22. Duburcq, X., Olivier, C., Desmet, R., et al. (2004) Polypeptide semicarbazide glass slide microarrays: characterization and comparison with amine slides in serodetection studies. *Bioconjugate Chem.* **15,** 317–325.
23. Houseman, B. T., Huh, J. H., Kron, S. J., and Mrksich, M. (2002) Peptide chips for the quantitative evaluation of protein kinase activity. *Nat. Biotech.* **20,** 270–274.
24. Polymenidou, M. and Sobek, J., to be published.
25. Plante, O. J., Palmacci, E. R., and Seeberger, P. H. (2001) Automated solid-phase synthesis of oligosaccharides. *Science* **291,** 1523–1527.
26. Ortiz Mellet, C. and Garcia Fernandez, J. M. (2002) Carbohydrate microarrays. *Chem. Bio. Chem.* **3,** 819–822.

27. Feizi, T., Fazio, F., Chai, W., and Wong, C. -H. (2003) Carbohydrate microarrays—a new set of technologies at the frontiers of glycomics. *Curr. Opin. Struct. Biol.* **13,** 637–645.

28. Houseman, B. T. and Mrksich, M. (2002) Carbohydrate arrays for the evaluation of protein binding and enzymatic modification. *Chem. Biol.* **9,** 443–454.

29. Bryan, M. C., Lee, L. V., and Wong, C. -H. (2004) High-throughput identification of fucosyltransferase inhibitors using carbohydrate microarrays. *Bioorg. Med. Chem. Lett.* **14,** 3185–3188.

30. Daines, A. M., Maltman, B. A., and Flitsch, S. L. (2004) Synthesis and modifications of carbohydrates, using biotransformations. *Curr. Opin. Chem. Biol.* **8,** 106–113.

31. Park, S., Lee, M. -R., Pyo, S. -J., and Shin, I. (2004) Carbohydrate chips for studying high-throughput carbohydrate-protein interactions. *J. Am. Chem. Soc.* **126,** 4812–4819.

32. Fazio, F., Bryan, M. C., Blixt, O., Paulson, J. C., and Wong, C. -H. (2002) Synthesis of sugar arrays in microtiter plate. *J. Am. Chem. Soc.* **124,** 14,397–14,402.

33. Fazio, F., Bryan, M. C., Lee, H. -K., Chang, A., and Wong, C. -H. (2004) Assembly of sugars on polystyrene plates: a new facile microarray fabrication technique. *Tetrahedron Lett.* **45,** 2689–2692.

34. Bryan, M. C., Lee, L. V., and Wong, C. -H. (2004) High throughput identification of fucosyltransferase inhibitors using carbohydrate microarrays. *Bioinorg. Med. Chem. Lett.* **14,** 3185–3188.

35. Wang, D., Liu, S., Trummer, B. J., Deng, C., and Wang, A. (2002) Carbohydrate microarrays for the recognition of cross-reactive molecular markers of microbes and host cells. *Nat. Biotech.* **20,** 275–281.

36. Fukui, S., Feizi, T., Galustian, C., Lawson, A. M., and Chai, W. (2002) Oligosaccharide microarrays for high-throughput detection and specificity assignments of carbohydrate-protein interactions. *Nat. Biotech.* **20,** 1011–1017.

37. Bryan, M. C. and Wong, C. -H. (2004) Aminoglycoside array for the high-throughput analysis of small molecule–RNA interactions. *Tetrahedron Lett.* **45,** 3639–3642.

38. MacBeath, G. and Schreiber, S. L. (2000) Printing proteins as microarrays for high-throughput function determination. *Science* **289,** 1760–1763.

39. Sakanyana, V. (2005) High-throughput and multiplexed protein array technology: protein–DNA and protein–protein interactions. *J. Chromatogr. B,* **815,** 77–95.

40. Cha, T., Guo, A., and Zhu, X. Y. (2005) Enzymatic activity on a chip: the critical role of protein orientation. *Proteomics* **5,** 416–419.

41. Park, S. and Shin, I. (2002) Fabrication of carbohydrate chips for studying protein-carbohydrate interactions. *Angew. Chem. Int. Ed.* **41,** 3180–3182.

42. Wang, D. (2003) Carbohydrate microarrays. *Proteomics* **3,** 2167–2175.

43. Willats, W. G. T., Rasmussen, S. E., Kristensen, T., Mikkelsen, J. D., and Knox, J. P. (2002) Sugar-coated microarrays: a novel slide surface for the high-throughput analysis of glycans. *Proteomics* **2,** 1666–1671.

44. Khandurina, J. and Guttman, A. (2002) Microchip-based high-throughput screening analysis of combinatorial libraries. *Curr. Opin. Chem. Biol.* **6,** 359–366.

45. Lam, K. S., Liu, R., Miyamoto, S., Lehman, A. L., and Tuscano, J. M. (2003) Applications of one-bead one-compound combinatorial libraries and chemical microarrays in signal transduction research. *Acc. Chem. Res.* **36,** 370–377.

46. He, X. G., Gerona-Navarro, G., and Jaffrey, S. R. (2005) Ligand discovery using small molecule microarrays. *J. Pharmacol. Exp. Ther.* **313,** 1–7.

47. Uttamchandani, M., Walsh, D. P., Yao, S. Q., and Chang, Y. -T. (2005) Small molecule microarrays: recent advances and applications. *Curr. Opin. Chem. Biol.* **9,** 1–10.

48. Lam, K. S. and Renil, M. (2002) From combinatorial chemistry to chemical microarray. *Curr. Opin. Chem. Biol.* **6,** 353–358.

49. Spring, D. R. (2005) Chemical genetics to chemical genomics: small molecules offer big insights. *Chem. Soc. Rev.* **34,** 472–482.

50. Butcher, R. A. and Schreiber, S. L. (2005) Using genome-wide transcriptional profiling to elucidate small-molecule mechanism. *Curr. Opin. Chem. Biol.* **9,** 25–30.

51. Stegmaier, K., Ross, K. N., Colavito, S. A., O'Malley, S. O., Stockwell, B. R., and Golub, T. R. (2004) Gene expression-based high-throughput screening (GE-HTS) and application to leulemia differentiation. *Nat. Genet.* **36,** 257–263.

52. MacBeath, G., Koehler, A. N., and Schreiber, S. L. (1999) Printing small molecules as microarrays and detecting protein-ligand interactions en masse. *J. Am. Chem. Soc.* **121,** 7967–7968.

53. Hergenrother, P. J., Depew, K. M., Schreiber, S. L. (2000) Small-molecule microarrays: covalent attachment and screening of alcohol-containing small molecules on glass slides. *J. Am. Chem. Soc.* **122,** 7849–7850.

54. Koehler, A. N., Shamji, A. F., and Schreiber, S. L. (2003) Discovery of an inhibitor of a transcription factor using small molecule microarrays and diversity-oriented synthesis. *J. Am. Chem. Soc.* **125,** 8420–8421.

55. Kuruvilla, F. G., Shamji, A. F., Sternson, S. M., Hergenrother, P. J., and Schreiber, S. L. (2002) Dissecting glucose signalling with diversity-oriented synthesis and small-molecule microarrays. *Nature* **416,** 653–657.

56. Köhn, M., Wacker, R., Peters, C., et al. (2003). Staudinger ligation: a new immobilization strategy for the preparation of small-molecule arrays. *Angew. Chem. Int. Ed.* **42,** 5830–5834.

57. Barnes-Seeman, D., Park, S. B., Koehler, A. N., and Schreiber, S. L. (2003) Expanding the functional group compatibility of small-molecule microarrays: discovery of novel calmodulin ligands *Angew. Chem. Int. Ed.* **42,** 2376–2379.

58. Kanoh, N., Kumashiro, S., Simizu, S., et al. (2003) Immobilization of natural products on glass slides by using a photoaffinity reaction and the detection of protein–small-molecule interactions. *Angew. Chem. Int. Ed.* **42,** 5584–5587.

59. Gosalia, D. N. and Diamond, S. L. (2003) Printing chemical libraries on microarrays for fluid phase nanoliter reactions. *Proc. Natl. Acad. Sci.* **100,** 8721–8726.

60. Winssinger, N., Ficarro, S., Schultz, P. G. and Harris, J. L. (2002) Profiling protein function with small molecule microarrays. *Proc. Natl. Acad. Sci.* **99,** 11,139–11,144.

61. Melkko, S., Scheuermann, J., Dumelin, C. E., and Neri, D. (2004) Encoded self-assembling chemical libraries. *Nat. Biotechnol.* **22,** 568–574.
62. Butler, J. E., Ni, L., Brown, W. R., et al. (1993) The immunochemistry of sandwich ELISAs. VI. Greater than 90% of monoclonal and 75% of polyclonal anti-fluorescyl capture antibodies (CAbs) are denatured by passive adsorption. *Mol. Immunol.* **30,** 1165–1175.
63. Zhu, H. and Snyder, M. (2003) Protein chip technology. *Curr. Opin. Chem. Biol.* **7,** 55–63.
64. Kingshott, P. and Griesser, H. J. (1999) Surfaces that resist bioadhesion. *Curr. Opin. Solid State and Mat. Sci.* **4,** 403–412.
65. Arenkov, P., Kukhtin, A., Gemmell, A., Voloshchuk, S., Chupeeva, V., and Mirzabekov, A. (2000) Protein microchips: use for immunoassay and enzymatic reactions. *Anal. Biochem.* **278,** 123–131.
66. Peluso, P., Wilson, D. S., Do, D., et al. (2003) Optimizing antibody immobilization strategies for the construction of protein microarrays. *Anal. Biochem.* **312,** 113–124.
67. Sobek, J., Bartscherer, K., Jacob, A., Hoheisel, J. D., and Angenendt, P. (2006) Microarray technology as a universal tool for high-throughput analysis of biological systems. *Comb. Chem. High Throughput Screen* **9,** 365–380.
68. Kusnezow, W., Jacob, A., Walijew, A., Diehl, F., and Hoheisel, J. D. (2003) Antibody microarrays: an evaluation of production parameters. *Proteomics* **3,** 254–264.
69. Kusnezow, W. and Hoheisel, J. D. (2003) Solid supports for microarray immunoassays. *J. Mol. Recognit.* **16,** 165–176.
70. Gong, P. and Grainger, D. W. (2004) Comparison of DNA immobilization efficiency on new and regenerated commercial amine-reactive polymer microarray surfaces *Surface Sci.* **570,** 67–77.
71. Sobek, J. to be published. We thank Dr. J. Brunner, Environment and Health Protection of the City of Zurich, for providing these data.

3

Optimization Workflow for the Processing of High Quality Glass-Based Microarrays

Applications in DNA, Peptide, Antibody, and Carbohydrate Microarraying

Jens Sobek, Catharine Aquino, and Ralph Schlapbach

Summary

As the performance of microarray experiments is directly dependent on the quality of the materials, the suitability of the protocols, and the accuracy of the work performed, optimization of existing microarray workflows is needed in almost every experiment to achieve higher quality and meaningfulness of the generated data. In the following, we describe a workflow for the optimization of microarray processing parameters, based on the previously selected surface structure. Four simple model experiments with dye-labeled compounds is used to determine crucial experimental parameters including spotting concentration, spotting solution, immobilization efficiency, and blocking conditions even in cases where recommendations from the slide manufacturer or from the literature are missing. In this article, processing parameters for DNA, peptide, antibody, and carbohydrate microarrays are outlined. The applicability of the model experiments is demonstrated and described in detail on the example of short oligonucleotides.

Key Words: Immobilization; microarrays; optimization; quality; surface chemistry; workflow.

1. Introduction

The increasing use of microarrays for the highly parallel analysis of a multitude of molecular interactions and binding events parallels the advances that the technology has seen in recent years in respect to sensitivity, flexibility, and reliability. Although, these criteria are largely fulfilled for standard applications of microarrays, new applications featuring nonnucleic acid probe substances, are still in their infancy and can be optimized in most cases. In this article, guidelines are provided on how production and processing parameters of microarrays can be optimized in order to obtain high-quality results. We restrict ourselves to

From: *Methods in Molecular Biology, vol. 382: Microarrays: Second Edition: Volume 2*
Edited by: J. B. Rampal © Humana Press Inc., Totowa, NJ

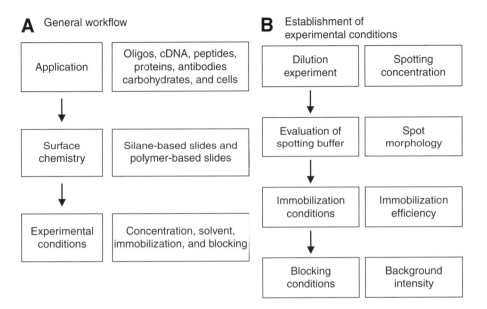

Fig. 1. Workflow of the optimization process. **(A)** General workflow. **(B)** Establishment of experimental conditions.

the optimization of array production and primary processing, from choosing a suitable microarray slide to the point of the analytical experiment on its surface. We do not elaborate on questions of sample preparation, labeling, and only briefly touch subjects of hybridization and incubation protocols.

The overall process of the article is illustrated by the workflow scheme shown in **Fig. 1**. The selection of a suitable slide surface chemistry is often dictated by the planned application. In turn, the chosen surface chemistry determines the experimental conditions of immobilization, blocking, and hybridization, which have to be carefully optimized. Stating this, the principle production parameters in a large number of experiments have been optimized for a variety of slides with different slide chemistries and for diverse applications, such as long and short DNA, peptide, protein, antibody, and carbohydrate microarrays. As a result, we have established general recommendations on how experimental conditions for custom made microarrays can be optimized. In the following, the usefulness of this method has been demonstrated by means of some simple and well-known model systems.

2. Materials

1. Deionized and filtered water using a Milli-Q Synthesis A10 purification system (Millipore, Billerica, MA) (*see* **Notes 1** and **2**).
2. 3′-Aminohexyl-modified 5′-Cy3-labeled oligonucleotides and the 5′-Cy5-labeled complementary sequence from Microsynth (Balgach, CH).

3. The sequence of the 13-mer used for model experiments was 5′-TGC CTG AAG CTA T.

4. All oligonucleotides purified by high-performance liquid chromatography or polyacrylamide gel electrophoresis (48-mers and longer), respectively.

5. Dye-labeled bacterial artificial chromosomes (BACs) and complementary deoxyribonucleic acid (cDNA) prepared by Dr. R. Schaub and D. Bausch (University Children's Hospital, Zurich).

6. TRITC-(tetramethylrhodamine isothiocyanate) labeled antibodies were obtained from Jackson ImmunoResearch Laboratories (West Grove, PA).

7. Unlabeled antibodies from Santa Cruz (Santa Cruz, CA) and from Cell Signaling (Danvers, MA) (mouse monoclonal antibodies, rabbit and goat were polyclonal antibodies).

8. Anti-fluorescein antibody prepared by Dr. S. Melkko (ETH, Zurich).

9. Model peptide 5(6)-Carboxyfluorescein-KTKESLGRKIQIQRSG-NH$_2$, gift from Dr. M. Schutkowski (JPT Peptide Technologies, Berlin).

10. Sodium chloride (molecular biology grade) from Sigma-Aldrich (Buchs, CH).

11. (N-[2-hydroxyethyl]piperazine-N′-[2-ethanesulfonic acid]) (HEPES, >99.5%) from Sigma-Aldrich (Buchs, CH).

12. Sodium mono- and diphosphate (both ≥99.0%) from Sigma-Aldrich (Buchs, CH).

13. Trisodium citrate (>99%) from Sigma-Aldrich (Buchs, CH).

14. Fluorescein isothiocyanate dextrane conjugate (molar mass 4000 g/mol) from Fluka, (Buchs, CH).

15. Hydrochloric acid (p.a.) from Fluka (Buchs, CH).

16. Ethanolamine (p.a.) from Fluka (Buchs, CH).

17. Betaine hydrochloride (>99%) from Fluka (Buchs, CH).

18. Formamide (p.a., ≥99.5%) from Fluka (Buchs, CH).

19. Glycerol (p.a., ≥99.5%) from Fluka (Buchs, CH).

20. Triton X-100 (molecular biology grade) from Fluka (Buchs, CH).

21. Tween-20 from Fluka (Buchs, CH).

22. D-(+)-Trehalose dihydrate (≥99.0%) from Fluka (Buchs, CH).

23. Sodium dodecyl sulfonate (SDS, >98%) was from Fluka (Buchs, CH).

24. Dimethyl sulfoxide (DMSO) was from Fluka (Buchs, CH).

25. Buffers: PBS-T (phosphate buffered saline, Tween-20), TBS-T (TRIS buffered saline, Tween-20), SSC (saline sodium citrate), SSC/SDS.

26. 2-Amino-2-(hydroxymethyl)-1,3-propanediol, hydrochloride (Tris-HCl, molecular biology grade) from VWR International (Dietikon, CH).

27. For microarray spotting non-contact arrayer were used (sciFLEXARRAYER, Scienion, Berlin, Germany; Piezorray, PerkinElmer, Downers Grove, IL).

28. Washing, blocking, hybridization, and drying of the slides in an automated hybridization station (HS4800, Tecan, Salzburg, Austria).

29. GeneMachines HybStation (Genomic Solutions, Ann Arbor, MI).

30. For read-out 4 color confocal laser scanner (LS400, Tecan; Salzburg, Austria; ScanArray5000, PerkinElmer, Boston, MA, USA).

31. ArrayPro Analyzer 4.5.1 for quantitative data evaluation (Media Cybernetics, Silver Spring, MD, USA).

32. Genespotter 2.4.2 for quantitative data evaluation (Microdiscovery, Berlin, Germany).
33. Polyethylene glycol coated slides (sciChip-PEG, E-PEG), epoxysilane coated slides (sciChip 2D, E-2D), a polymer coated slide (sciChip 3D, E-3D) from Scienion (Berlin, Germany).
34. CodeLink™ hydrogel slide (CDL, GE Healthcare, Piscataway, NJ) (*see* **Notes 3** and **4**).

3. Methods

3.1. General Workflow for the Optimization of Microarrays

Although the recommendations given in Chapter 2 are helpful to identify a suitable substrate, the achievement of perfect results usually needs considerable fine tuning. The advantage of the workflow presented in the following is that every single optimization step can be observed experimentally. The basis of our experimental scheme is the use of dye-labeled model compounds and the judgment of effects resulting from protocol optimization by means of diagrams of signal intensity and background against sample concentration. It is assumed that one or a few model compounds represent the diversity of samples to be printed which can be easily achieved for a library of 70-mer oligonucleotides or PCR products. More difficult is the situation with proteins and peptides as their physicochemical properties, chemical composition, and three-dimensional (3D) structure variations, which leads to specific requirements in order to achieve optimal conditions for array production, particularly for spotting. However, even in these cases the outlined approach have found to be sufficiently accurate to determine the crucial parameters, even more so if one uses more than one model compound to increase precision and specificity of the optimization process.

The recommended procedure focuses on four parameters: spotting concentration, spotting buffer composition, and the conditions for immobilization and blocking. The results of the corresponding model experiments are easy to evaluate in a fluorescence intensity–concentration diagram reflecting signal intensity and the standard deviation, background intensity and noise. The steps of the optimization workflow are independent from each other. However, changes made in one step of the protocol often require the adaptation of subsequent protocol steps.

As the very first step preceding the optimization, the selection of a slide type determines the experimental parameters and protocols, including sample concentration, spotting buffer, immobilization conditions, and blocking reagents. If there are no specific recommendations available, one should use slides of the different types shown in **Fig. 3** in Chapter 2. In our experiments, we have performed the optimization of oligonucleotide, peptide, and antibody arrays using epoxysilane-coated slides (E-2D, type 3A), polyethylene glycol (PEG) slides (E-PEG, type 3B), a polymer-coated epoxy-activated 3D slide

(E-3D, type 3C) available from Scienion, Berlin, and a 3D hydrogel slide with NHS ester activation (CDL, type 3C, crosslinked structure not shown) from GE Healthcare. For all commercially obtained slides it is recommended that processing protocols given by the manufacturer are followed consistently, as only these are based on the full understanding of the respective surface chemistry (*see* **Note 4**). In respect to finding the perfect remaining experimental parameters, only scarce information is usually provided by the manufacturer. For the dilution experiment outlined next, it is suggested that the recommended spotting buffer and immobilization conditions should be used in order to stay consistent with the proceedings in the text.

3.1.1. Spotting Concentration

The spotting concentration is an important factor and it depends on the molecular size of the compound to be spotted. As a rule of thumb, the higher the molar mass the lower the optimal spotting concentration needs to be used. Optimal spotting concentrations are in the range of some millmolar solutions for carbohydrates or small molecules with molar masses of <1 kDa, and about 1 μ*M* for antibodies of 150 kDa. In the example workflow below 13-mer, 25-mer, 48-mer, and 71-mer oligonucleotides have been chosen as frequently used probe substances. By concentrating in some steps on the use of 13-mer oligonucleotides, we focus on the most difficult of oligonucleotide applications as all critical parameters influence the outcome of an experiment with a small molecule more prominently.

As a first step, dilution series of dye-labeled model oligonucleotides modified with an amino linker for immobilization are prepared in 3X SSC and phosphate buffer in concentrations of 2.5–40 μ*M*. Three piezo-driven droplets (~1 nL total) of each solution are printed in blocks of 10–20 replicates onto three slides of each type using a noncontact spotter. Spotting a number of slides is necessary to increase the validity of the test experiments. The slides are scanned immediately after spotting and the probe molecules are immobilized in a humidity chamber overnight (*see* **Notes 5** and **6**), according to recommendations of the slide supplier. In the present case, the following processing steps are all performed in the chamber of an automated hybridization station (*see* **Note 7**). Following spotting and immobilization, the slide surface is blocked with a suitable blocking agent that can be optimized later on, if it is required. In detail, E-2D, E-PEG, and E-3D slides are wetted with 0.1% Triton X-100 followed by the reaction with a 5-m*M* solution of HCl to deactivate all remaining epoxy groups that are not used for binding oligonucleotide. For CDL, the slide is blocked in ethanolamine solution in Tris-HCl at pH 9.0 at 50°C (*see* **Note 8**). In all cases, blocking is followed by extensive washing with water and/or buffer. After drying in the hybridization station at 30°C with nitrogen (or manual drying using a nitrogen flow), the slides

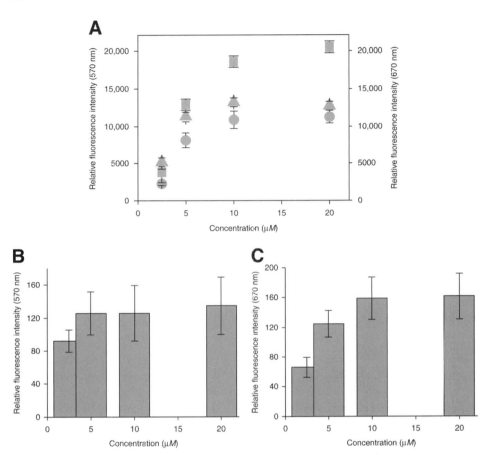

Fig. 2. Dependence of fluorescence intensity on concentration. **(A)** Dependence of fluorescence intensity on concentration. Mean fluorescence intensity of 40 replicates of a Cy3-labeled 13-mer oligonucleotide in 3X SSC on a E-2D slide is plotted against the concentration. Squares: after spotting; triangles: after washing and blocking (fluorescence detected at 570 nm for both, left axis); circles: after hybridization of a complementary Cy5-labeled oligonucleotide (detected at 670 nm, right axis). **(B,C)** Dependence of background fluorescence intensity on concentration. **(B)** Background after washing and blocking the slide and **(C)** after hybridization. Data are from the experiment in **2A**.

are scanned again using the same scanner settings as for the first scan. For data evaluation, the average of the mean spot fluorescence intensity and the background intensity, respectively, of all replicates are plotted against the spotting concentration. Examples for the resulting diagrams are shown in **Fig. 2A,B** for a 13-mer oligonucleotide spotted in 3X SSC to E-2D slides. The plots reveal information on the slide properties and the protocol performance and provide the quantitative basis of our workflow as effects of parameter changes in the optimization experiments can easily be determined and quantified.

Starting from the origin of the diagram in **Fig. 2A**, the expected linear increase of fluorescence intensity with increasing concentration is observed (squares). At higher concentrations a negative deviation from the linear behavior is found, which is supposedly caused by fluorescence quenching when molecules come into close contact. The scan after washing and blocking the slide results in a similar plot (triangles). The fluorescence intensity after washing is a measure of the binding capacity of the surface. On E-2D slides, a plateau is formed at concentrations above 10 µM reflecting the saturation of the surface with oligonucleotides. Similar maximal concentrations, illustrated by the plateau, are observed for E-PEG and E-3D. Because of their extended 3D structure, hydrogel slides such as CDL have a higher loading capacity. In an intensity–concentration diagram this results in a linear increase in fluorescence intensity for spotting solutions of concentrations of up to 30 µM. As a result of these experiments a maximum spotting concentration of 20 µM for the following steps have been chosen. It is a concentration that matches recommendations given in the literature and by slide manufacturers. At this concentration the spot quality is found to be excellent, which can be illustrated indirectly by the very small error bars in the diagrams in **Fig. 2A** (*see* **Notes 9** and **10**).

Spotting of higher concentrations is not recommended as it can cause serious smearing effects in proximity of the spots or across the whole slide surface (comet tails). Such artifacts are always the result of too slow blocking reaction compared with the immobilization of probe molecules at the slide surface. The effect can often be observed with proteins or other compounds that can immobilize very fast to surfaces. An extreme example is biotinylated compounds of low-molecular weight spotted to streptavidin-coated slides. In principle, besides reducing the spotting concentration, smearing effects can be avoided by using a blocking solution in which the spotted probe is not soluble (*see* **Subheading 3.2.**, in the next Chapter 4 of this volume). However, the general applicability of these substances is critical as they may harm sensitive probes of other compound classes. We have found that smearing artifacts can strongly be reduced by extensive washing at 4°C with following solvent or buffer solutions even when extremely high concentrations of probes are spotted.

3.1.2. Spotting Solution

With the exception of rare cases, spot quality obtained using standard parameters is unsatisfactory. Spot morphology is a crucial issue, especially in one-color experiments, as it determines the extent of experimental error (e.g., spot-to-spot variation) and the validity of the experiment. For this reason, the second step in the optimization workflow is dedicated to the optimization of the spotting solution in order to obtain perfect spot morphology. For a given surface and with the knowledge of the optimal spotting concentration, only the properties of the spotting solution are left to influence the spot morphology when a noncontact piezo spotter is used at room temperature and the recommended relative humidity. In addition to influencing the spot morphology, the spotting solution should stabilize printing solution system, lead to consistent and stable spots on the surface, and allow for a high efficiency of the immobilization. Procedures on how to optimize a solvent for spotting cDNA are described *(1,2)*. However, as surface properties strongly depend on the specific experimental conditions during the surface-coating process, substrates that in principle have the same surface architecture but are obtained from a variety of manufacturers can feature different slide properties. These properties affect the quality of spot morphology and reproducibility of the spot size on slides from different sources, even when being processed in parallel and theoretically featuring the same surface chemistry. Therefore, it might be necessary to optimize printing conditions for a given type of slide *(2)*.

3.1.3. Spot Size

As specific applications might ask for spots of a given size, production, and processing parameters have to be adjusted accordingly. Aqueous solutions delivered in amounts of 300–400 pL (calculated diameter for a spherical droplet of 300 pL is 83 µm) to surfaces by a noncontact piezo printer results in spots of some 10–100 µm in diameter (ranging from about 85 µm in a betaine-formamide-based solution on E-2D with a contact angle of 50–55°, to 390 µm in DMSO 50% on E-PEG slides with a contact angle of 40–45°). The spot size depends on the specific composition of the solution determining the surface tension in combination with the slide surface energy. The more hydrophobic a surface is (large contact angles), the smaller the resulting spot will be. The spot size can also strongly depend on the nature and concentration of the probe and on the spotted volume. As an example, with oligonucleotides an increase in spot size correlates with an increasing number of nucleotides. Because of the hydrophobic nature of most dyes used for labeling, the spotting of oligonucleotide-dye conjugates might result in larger spots as compared with unlabeled oligonucleotides. This effect is most pronounced for short oligonucleotides. As shown in **Fig. 3** for unlabeled 13-mer oligonucleotides spotted in 3X SSC to E-2D slides (1 nL) the spot diameter is about 100 µm whereas, for a 71-mer the diameter is 175 µm *(3)*. Labeled with Cy3, the diameter of 13-mer changes to 170 µm. The spot size of the 71-mer remains largely the same (180 µm). The 13-mer oligonucleotide labeled with Cy5, FAM, and ROX, gives spot sizes of 180 µm, 190 and 200 µm, respectively. The concentration dependence of the spot size for a Cy3-labeled 13-mer is shown in

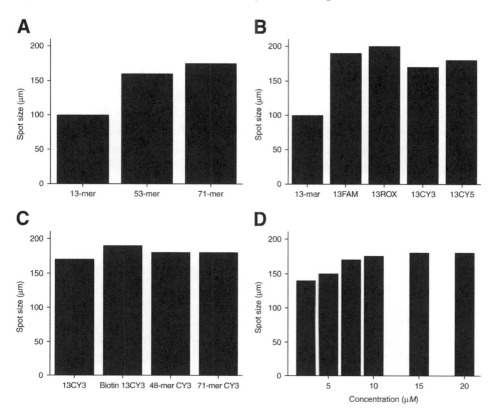

Fig. 3. Dependence of spot size on molecular properties for 13- to 71-mer 5'- dye-labeled and unlabeled oligonucleotides spotted to E-2D (1 nL, 20 µM, in 3X SSC). Biotin 13CY3 is 3'- biotinylated. (**A**) For unlabeled oligonucleotides of different lenghts. (**B**) Influence of dye-labeling. (**C**) Influence of molecular size and additional substitution. (**D**) Concentration dependence for a Cy3-labeled 13-mer. Results are obtained from one representative slide.

Fig. 3D. As mentioned earlier, the spot diameter might vary significantly depending on slide hydrophobicity with the relative sizes between the mentioned spot compositions remaining about the same.

3.1.4. Spot Morphology

Spot morphology, that is the combination of the outer shape of the spot and, more importantly, the distribution of probe molecules within the spot, strongly depends on surface properties, solvent, and sample concentration. Spots are formed owing to spreading of the liquid on the surface and its evaporation *(4)*. Additionally, the quality of the glass surface in terms of pureness and flatness, the homogeneity and structure of the chemical coating, as well as potential residues left from the coating process on the surface have a strong influence on spot morphology (*see* **Subheading 2.1.**, in Chapter 2). Imperfect spot morphologies are the result of a combination of these effects. In most instances it is impossible to separate the contributing effects in order to determine the factors responsible for sample distribution in a spot. Therefore, to optimize spot morphology, it is essential to select a suitable spotting solution through a systematic strategy instead of relying on an empirical trial-and-error approach.

In the course of our model experiments we have investigated the spot morphology of oligonucleotides and peptides dissolved in various solvents and spotted to microarray slides of different surface chemistries. In most cases, spot morphologies and their concentration dependencies, which cannot be explained easily have been observed. However, using a polymer-coated surface (E-3D) and a Cy3-labeled 71-mer oligonucleotide spotted in phosphate buffer, we have found a strikingly clear concentration dependence of the spot morphology as shown in **Fig. 4A–D**. The spot morphology clearly changes with concentration, from perfect doughnut structures at low concentrations (1–3 μ*M*) to spots with a nearly perfect sample distribution at high concentration (20 μ*M*). To explain the origin of this effect, mixtures of two different dye-labeled oligonucleotides of different length (a 13-mer and a 71-mer) were spotted at different concentrations to a variety of surfaces. It is found that the distribution of the two components in a spot can be very different. It is suggested that a chromatographical separation of the compounds induced by interactions with the surface takes place. This effect is comparable to processes which cause a sample separation in thin-layer chromatography with the surface coating acting as stationary phase and the spreading droplet as mobile phase.

Addition of glycerol to the spotting solution often improves spot morphology particular at low probe concentrations (**Fig. 4E–G**). However, besides helping spot morphology, glycerol influences the immobilization efficiency and produces a change of the spot morphology over a few days, caused by a partial evaporation of water up to the point where the composition of the azeotrope (with 87% glycerol) is reached (*see* **Note 11**). This results in an inhomogeneous distribution of sample in the spot, making it an urgent need to process these slides within a few days or at least to wash the slides immediately after immobilization. As an alternative, spots obtained from solutions containing formamide or betaine do not change their morphology but show the benefit of improved spot morphology.

As already mentioned previously, the quality of the spotted compound has also significant impact on spot morphology. Nonpurified or degraded substances might lead to a nonuniform compound distribution especially when residues of detergents are left in the spotting solution. Usually, the spot morphology does not change after washing or other processings of the slide.

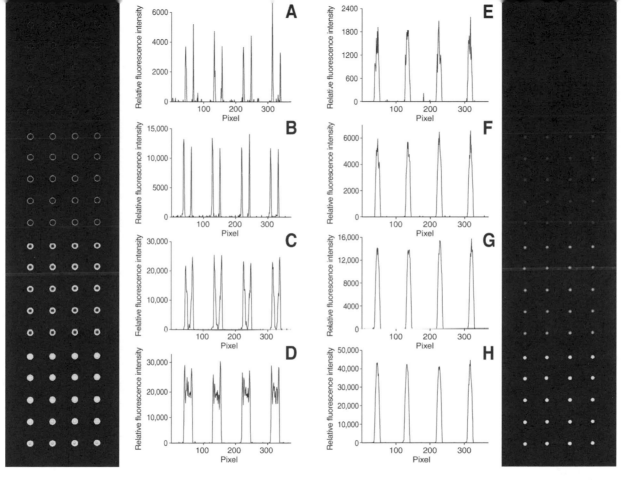

Fig. 4. Concentration dependence of spot morphology. 2.5 μM, 5 μM, 10 μM, and 20 μM solutions of a 71-mer oligonucleotide in 150 mM phosphate, pH 8.5 (**4A–D**) and 150 mM phosphate/5% glycerol (**4E–H**), respectively, were spotted to E-3D slides. The line scans show a cross-section through the spots.

3.1.5. Immobilization Conditions

In next step of the optimization process, suitable immobilization conditions must be found. It is a very critical step as it determines the amount of substance available for the following analytical experiment. Immobilization efficiency strongly depends on the combination of the chemical nature of the probe molecules and the slide surface chemistry, as well as on experimental conditions such as the spotting solution, air humidity, temperature, and reaction time. As a rule of thumb, the smaller the probe molecules to be immobilized are the more successful is an immobilization strategy using a specific chemical reaction. In principle, a very large spectrum of chemical reactions might be used for specific coupling of compounds to surfaces. This requires the respective reactive chemical modification at the probe molecule and a matching chemical group at the surface. Because biological probes are often equipped with nucleophilic groups such as amino, hydroxyl, and thiol, the most commonly used chemical reaction for immobilization is coupling to electrophilic groups at the slide surface including aldehyde, epoxy, isothiocyanate, and activated ester groups. In the presence of more than one reactive group, a more sophisticated and site-specific coupling chemistry might be needed. To give a few examples, Diels-Alder reaction *(5)*, Michael addition *(6)*, or Staudinger ligation *(7)* were used for this purpose. In contrast, large molecules are easier to immobilize because of their higher tendency for adsorption. Proteins, antibodies, and polysaccharides are adsorbed at coated surfaces slides such as nitrocellulose, polyamide, and polyacrylamide. This coupling process does not require any reactive group. Most of these slides are commercially available *(8)*. Immobilization by adsorption does not need any modification of the probe molecules. However, in this process the orientation of the probe on the surface is undefined.

In practical terms, for immobilization of oligonucleotides an amino linker is frequently used in combination with surfaces presenting electrophilic groups. It our experiments using 48-mer, 53-mer, and 71-mer oligonucleotides, that oligonucleotides with and without amino linker immobilize with the same efficiency to surfaces activated with epoxy groups were observed. For small molecules, peptides, and carbohydrates, a large variety of coupling strategies from the large repertoire of organic synthesis have been applied to avoid multiple immobilization at different molecular sites. Probes of low-molecular weight are usually chemically modified or synthesized starting from small building blocks in an elaborate procedure, providing the molecules with a suitable linker including a reactive group for coupling to a matching surface for a site-specific immobilization.

In all cases, where probe molecules are immobilized by a specific chemical reaction, immobilization in a humidity chamber for 4–72 h is recommended.

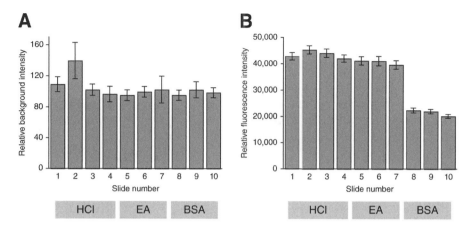

Fig. 5. Signal intensities and background measured at 570 nm for a Cy3-labeled 13-mer oligonucleotide spotted to 10 E-2D slides. The slides were blocked at room temperature with hydrochloric acid (HCl, S1–2: 1 mM, S3–4: 5 mM), 50 mM ethanolamine in Tris-HCl, pH 9.0 (EA, S5–7), and BSA 1% in PBS (S8–10). (**A**) Background intensity and noise. (**B**) Fluorescence intensity and standard deviation.

This is rather a long time period for a reaction of picomolar amounts of compounds with the surface, especially in view of the usually applied highly reactive coupling chemistry. And it results from the fact that picoliter to nanoliter amounts of water delivered to a surface at room temperature usually dry within seconds. Once the spot is dried, the mobility of molecules to rearrange for a chemical reaction with the surface is significantly reduced and an extended period of time for the reaction to take place is indispensable. In contrast, large molecules such as proteins and antibodies immobilize faster at surfaces because adsorption needs much less molecular rearrangement. Another reason for long reaction times comes from the reduced rate of immobilization that is observed in solutions containing glycerol, although, these spots do not dry in contrast to salt-based spotting solutions. In this case the probe molecules remain distributed in the spot volume with an accordingly reduced surface concentration and therefore, need an increased reaction time to interact with the binding groups at the surface.

Apart from thermal reactions, ultraviolet (UV) light can be applied for immobilization. This method is often used for immobilization of cDNA on aminosilane-coated surfaces. Radiation with UV light generates highly reactive radicals that form unspecific covalent bonds. In principle, any type of organic molecule can be immobilized by UV light (*9*). However, owing to the nonspecific nature of the process, it should be taken into account that uncontrolled (and presumably unwanted) side rections might lead to molecular changes including cycloadditions of nucleotides in DNA; and induction of S–S bond reduction, oxidation, and photocleavage in proteins (*10*). In addition, significant

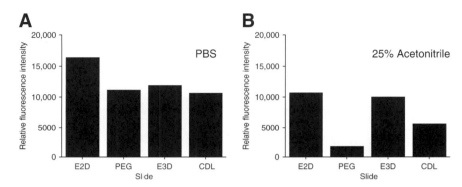

Fig. 6. Immobilization efficiency of a dye-labeled peptide. The model peptide 2.5 mM in PBS (**A**) and PBS/25% acetonitrile (**B**) was spotted to different types of slides. Immobilization was performed for 16 h in a humidity chamber (75% relative humidity, room temperature). Slides were scanned after washing with PBS and blocking with ethanolamine in Tris-HCl, pH 9.0.

structural disturbances have been described including crosslinking of sample molecules among themselves *(11)*. In experiments aiming to substitute cDNA by unmodified 70-mer oligonucleotides, Wang et al. immobilized samples by UV crosslinking to aminosilane-coated slides. As a result, distorted concentration dependencies were observed including a high degree of sequence dependence (*see* **Fig. 5A**). The authors state that this is presumably caused by crosslinking oligonucleotides in unwanted side-reactions preventing them from reacting with the target. As stated earlier, use of epoxysilane slides is recommended for applications with long (unmodified) oligonucleotides that can be immobilized in a humidity chamber overnight.

In contrast to recommendations given in the literature cDNA does not require an immobilization by UV light to poly-L-lysine (PLL)-coated surfaces *(12)*. This is explained by the surface structure of PLL that allows a high degree of molecular contact with the oppositely charged DNA backbone (at pH 7.0). PLL-coated slides do have some issues (*see* **Note 12**) *(8,13,14)*.

As previously discussed, the fluorescence intensity of the plateau observed after washing the slide is a good measure for comparing different immobilization protocols (triangles in **Fig. 2A**). Moreover, spotting suitable model compounds in different solvents helps to find suitable conditions, as shown in **Fig. 6** for the immobilization of a model peptide to different slide surfaces. The plot shows that the immobilization efficiency of the peptide spotted in PBS and immobilized for 16 h in a humidity chamber at room temparature is nearly independent from the slide used (**Fig. 6A**). However, when E-PEG or the hydrogel slide in combination with acetonitrile is used, the immobilization is not

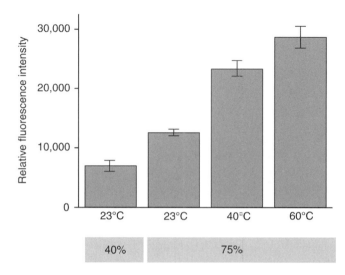

Fig. 7. Optimization of immobilization conditions. The model peptide was immobilized to E-PEG slides at room temperature (23°C, 40% relative humidity) and in a humidity chamber (23–60°C/75% relative humidity) for 16 h.

sufficient under the chosen conditions (**Fig. 6B**). The most simple solution to improve the immobilization efficiency is to extend the immobilization duration or process the slides at an elevated temperature. An instructive example is given in **Fig. 7**. The diagram shows that there is a strong dependence of immobilization efficiency on temperature and relative humidity. A similar effect is observed for the immobilization of oligonucleotides. From our experience, insufficient immobilization often arises from discrepancies between conditions for the chemical reaction of the sample with surface reactive groups and the composition or pH of the solvent.

3.1.6. Blocking Conditions

In order to take advantage of the so far increased quality and performance of the microarray, the last step in the process aiming to block reactive surface groups that have not reacted with the probe substance. This step needs optimization. Blocking depends on the surface chemistry and is restricted in respect to usable blocking agents by the class of immobilized compounds. In most cases, the blocking procedure deactivates reactive chemical groups in a specific reaction (e.g., with acids, bases, or succinimide), or covers the surface with an inert blocking agent such as a protein (bovine serum albumin [BSA], casein), a mixture of small proteins (TopBlock), or salmon sperm DNA (prehybridization).

In our example, the effect of different blocking solutions on E-2D is investigated by means of a Cy3-labeled 13-mer oligonucleotide. After immobilization, the slides are blocked with HCl of different concentrations (pH 3.0–4.0), with ethanolamine in Tris-HCl buffer at pH 9.0 for 15 min, or with 1% BSA in PBS for 1 h, at room temperature. In order to determine the resistance of the blocked surface, a hybridization can be performed using a high concentration of complementary Cy5-labeled 13-mer oligonucleotide (1 μM) at room temperature for 1 h. Fluorescence signal and background intensities and the standard deviations for a respective experiment are shown in **Fig. 5** for results obtained in 3X SSC. Whereas background and noise are comparable for the three blocking methods (**Fig. 5A**), the fluorescence intensity is decreased by a factor of 2 for blocking with BSA (**Fig. 5B**). This effect could be ascribed to blocking the accessibility of the comparably small oligonucleotide by the large protein. As a follow-up, it has to be ensured that the combination of blocking agent and solvent used does not lead to degradation or denaturation of the immobilized sample or to the destruction of the slide coating. As an example, many solvents and aqueous solutions at low pH lead to denaturation of proteins and antibodies. Reactive side groups can be modified with blocking agents, for example, with succinimide for blocking aminosilane slides. Polyethylene glycol-coating materials can be cleaved under acidic conditions (at very low pH values). However, using E-PEG slides for 3 yr we have not observed such effects to a measurable amount.

In the case of oligonucleotides as probe molecules, HCl is known to cause depurination at low pH values. In order to evaluate this effect, the hybridization is performed using the complementary Cy5-labeled 13-mer oligonucleotide under stringent conditions near T_m (calc. 38°C) after blocking as described earlier. If depurination occurs, one missing purine would cause a significant decrease in hybrid stability. From our data we did not observe a difference in signal intensities when slides are blocked under acidic (pH 3.0–4.0) and basic (pH 9.0) conditions (**Fig. 5B**), so that any depurination measurable effect can be excluded. In our example of oligonucleotide hybridization on epoxysilane-coated slides, background and noise are small and independent from the blocking solution. Therefore, the choice of blocking agents for epoxysilane-coated slides depends exclusively on the nature of the immobilized compound. The same is observed for other slides activated with epoxy groups including E-PEG and E-3D.

Ideally, blocking is performed under controlled and reproducible conditions in the chamber of an automated hybridization station just before the actual hybridization. This eliminates an additional drying step and reduces the risk of generating artifacts on the slide surface that could lead to a high-scattered background. As some of the reagents described earlier might have the potential to damage the hybridization chamber, it is necessary to check the compatibility of the hybridization chamber material and the blocking reagents.

3.1.7. Hybridization Test

After every optimization step in the workflow, the performance of the washed and blocked arrays are tested by a hybridization. For hybridization of immobilized Cy3-labeled oligonucleotides we have used complementary Cy5-labeled compounds in a HEPES- or SSC-based buffers at concentrations of 1–100 nM. The resulting plots of fluorescence signal and background intensity against concentration, respectively, follow the curve shapes for the spotted dye-labeled compounds after washing the slide (circles in **Fig. 2A**, bars in **Fig. 2C**). These results indicate that the immobilized probes are accessible to the target. However, problems with probe accessibility may arise with crosslinked polymer slides when reactions with large molecules including proteins and antibodies are performed.

3.2. Final Remarks

The execution of the workflow described takes about 1–2 wk and consumes about 25 substrates of every slide type that is tested. As only a limited number of model compounds are needed, the respective costs for the optimization are rather low. As shown by the scans in **Fig. 2** in Chapter 4, the workflow provides a good basis for the experimental optimization of microarray production parameters.

4. Notes

1. Chemicals used for microarray should be of highest quality. Keep in mind that additives (e.g., stabilizer) can be present in some chemicals, which might interfere with the experiment. Some solvents degrade and might cause undesired side reactions.
2. All solutions used for microarray experiments should be of high quality. Buffer solutions should be freshly prepared and filtered.
3. If coming in contact with organic solvents barcodes can be washed away and might cause artifacts on the slide surface.
4. For general recommendations on slide handling, *see* **Notes 1–4** in Chapter 2.
5. Extreme care must be taken when slides are processed in a laboratory oven. The slides can easily be contaminated by compounds which were placed in the oven long ago. They should also be protected from the ventilation system by a densely closed box. A glass box should be used (e.g., as humidity chamber), especially at elevated temperature. Plastic materials often contain monomers left from production that might contaminate the slide surface.
6. A saturated solution of NaCl generates an atmosphere with a relative humidity of 75% independent of temperature (measured between 23 and 60°C). If the salt solution is not saturated a higher relative humidity can be obtained depending on the temperature (according to thermodynamic basic laws). However, in this case immobilization conditions are less well defined.
7. This offers the advantage that reactions are performed under strictly controlled conditions without the risk of a partial slide drying or uncertainties introduced by

manual processing, which is the reason why the use of such instruments is strongly recommended. Nonetheless, the given recommendations are equally valid when using a manual hybridization chamber but special attention has to be paid to the mentioned risk of introducing irregularities.

8. In the final protocol, optimized for using CDL slides in the hybridization station, the slide was first of all wetted with 0.1% Triton X-100. For E-PEG slides it has been found that washing with an SSC/SDS based buffer at 4°C before blocking is optimal to avoid smearing effects. *See* Chapter 4.

9. Under optimal conditions we obtain a spot-to-spot variability of about 2–3% (using a needle printer these numbers are usually much higher. For comparison see values given in citation 3). Slide-to-slide variability in one-color experiment is much larger (about 8–10%) mainly depending on slide quality and the skill of the user. However, in a batch of slides deviations as large as 20–25% for single slides might occur as determined in model experiments. Manufacturer of high quality slides usually guarantee for small batch-to-batch variation.

10. Fluorescence intensities plotted in **Fig. 2A** should be used with some caution. Fluorescence dyes, especially cyanines including Cy3 and Cy5, tend to aggregate thus influencing their fluorescence properties. Moreover, the fluorescence intensity of dyes in solution strongly depends on molecular interactions with the environment and on the molecule the dye is interacting with. The latter effect is most pronounced with dye-labeled hydrophobic peptides. Fluorescence intensities measured with a confocal laser scanner are relative values and therefore, can only be compared, when the experimental conditions are comparable regarding environmental effects and laser scanner settings. For example, the fluorescence intensities of a dye-labeled oligonucleotide spotted in 3X SSC and in 3X SSC/10% glycerol solutions, cannot be compared when the slide is scanned without further processing. The latter will always show a higher fluorescence intensity at the same dye concentration as the spots do not dry and the fluorescence is measured in a liquid environment. Spotted in 3X SSC the spots dry within seconds and dye fluorescence is strongly quenched owing to interactions with molecules in close contact. However, after mild immobilization using a humidity chamber and washing off salts and glycerol the conditions are comparable as in both cases only a monolayer of immobilized oligonucleotide is left. Because of environmental effects, the ratio of fluorescence intensities of two scans before and after processing (e.g.,washing) the slide does not necessarily reflect the actual ratio of the number of molecules delivered to and immobilized at the surface.

11. Spots were found to be stable when DMSO is used as cosolvent.

12. The application of PLL is limited by aging effects of the coating. Moreover, the coating is much less stable than those based on silanes. PLL binds to surfaces through a multitude of electrostatic bonds, whereas, silanes form much stronger covalent bonds. In particular at high salt concentrations the electrostatic forces between PLL and the surface are diminished to an extent that the coating might be torn off.

Acknowledgments

JS likes to thank Prof. Dr. Joe Jiricny (University of Zurich, ETH Zurich), Dr. Orlando Schärer (Stony Brook University Hospital), and Dr. Philip Day

(University of Manchester) for their extensive help getting started with slide development, microarray production, and hybridization experiments. We would like to thank Dr. Rahel Schaub and Damaris Bausch (University Children's Hospital Zurich), Dr. Mike Schutkowski (JPT Peptide Technologies, Berlin), Dr. Gerald Radziwil (University of Zurich), and Dr. Samu Melkko (ETH Zurich) for kindly providing samples of model compounds.

References

1. Wrobel, G., Schlingemann, J., Hummerich, L., Kramer, H., Lichter, P., and Hahn, M. (2003) Optimization of high-density cDNA-microarray protocols by design of experiments'. *Nucl. Acids Res.* **31,** E67. Erratum (2003) **31,** 7057.
2. Rickman, D. S., Herbert, C. J., and Aggerbeck, L. P. (2003) Optimizing spotting solutions for increased reproducibility of cDNA microarrays. *Nucleic Acids Res.* **31,** E109.
3. Dawson, E. D., Reppert, A. E., Rowlen, K. L., and Kuck, L. R. (2005) Spotting optimization for oligo microarrays on aldehyde–glass. *Anal. Biochem.* **341,** 352–360.
4. Determined after hybridization with complementary Cy5-labeled oligonucleotides.
5. Deegan, R. D., Bakajin, O., Dupont, T. F., Huber, G., Nagel, S. R., and Witten, T. A. (1997) Capillary flow as the cause of ring stains from dried liquid drops. *Nature* **389,** 827–829.
6. Houseman, B. T. and Mrksich, M. (2002) Towards quantitative assays with peptide chips: a surface engineering approach. *Trends Biotechnol.* **20,** 279–281.
7. Park, S., Lee, M. R., Pyo, S. -J., and Shin, I. (2004) Carbohydrate chips for studying high-throughput carbohydrate-protein interactions. *J. Am. Chem. Soc.* **126,** 4812–4819.
8. Köhn, M., Wacker, R., Peters, C., et al. (2003) Staudinger ligation: a new immobilization strategy for the preparation of small-molecule arrays. *Angew. Chem. Int. Ed.* **42,** 5830–5834.
9. Sobek, J. and Schlapbach, R. (2004) Substrate architecture and functionality define the properties and performance of DNA, peptide, protein, and carbohydrate microarrays. *Pharmagenomics* **Sep. 15,** 32–44.
10. Caelen, I., Gao, H., and Sigrist, H. (2002) Protein density gradients on surfaces. *Langmuir* **18,** 2463–2467.
11. Steen, H. and Nørregaard Jensen, O. (2002) Analysis of protein-nucleic acid interactions by photochemical cross-linking and mass spectrometry. *Mass Spectrom. Rev.* **21,** 163–182.
12. Wang, H. -Y., Malek, R. L., Kwitek, A. E., et al. (2003) Assessing unmodified 70-mer oligonucleotide probe performance on glass-slide microarrays. *Genome Biol.* **4,** R5.
13. André Gerber, ETH Zurich, private communication.
14. Hessner, M. J., Meyer, L., Tackes, J., Muheisen, S., and Wang, X. (2004) Immobilized probe and glass surface chemistry as variables in microarray fabrication. *BMC Genomics* **5,** 53.

4

Processing Protocols for High Quality Glass-Based Microarrays

Applications in DNA, Peptide, Antibody, and Carbohydrate Microarraying

Jens Sobek, Catharine Aquino, and Ralph Schlapbach

Summary

Based on the selection of a suitable surface chemistry and bearing the option for optimization using a defined workflow, standard experimental protocols for the *processing* of microarrays can be used as starting points for a successful experiment. In Chapters 2 and 3, general *quality* considerations and the selection of *surface* chemistry have been discussed. A workflow for the selection of slide surface architectures and the optimization of microarray processing parameters also has been described. In the present article, *processing* parameters for DNA, peptide, antibody, and carbohydrate microarrays are outlined that serve as a first recommended step in the iterative establishment process. For a number of popular applications of microarray technology the outlined protocols can be applied to directly generate high-quality results.

Key Words: Microarrays; DNA; peptide; antibody; carbohydrate; processing protocols; optimization.

1. Introduction

Although there is vast and comprehensive resource of literature and protocols for the production and processing of microarrays, the challenge to find a suitable protocol for the implementation and use of a given microarray is all but trivial. On this background and building on Chapters 2 and 3, a collection of protocols have been provided that have proven useful in actual experiments and are supposed to serve as a starting point in the optimization process.

From: *Methods in Molecular Biology, vol. 382: Microarrays: Second Edition: Volume 2*
Edited by: J. B. Rampal © Humana Press Inc., Totowa, NJ

2. Materials

1. Deionized and filtered water using a Milli-Q Synthesis A10 purification system (Millipore, Billerica, MA, USA) (*see* **Notes 1** and **2**).
2. 3′-Aminohexyl-modified 5′-Cy3-labeled oligonucleotides and the 5′-Cy5-labeled complementary sequence from Microsynth (Balgach, CH).
3. The sequence of the 13-mer used for model experiments was 5′-TGC CTG AAG CTA T.
4. All oligonucleotides purified by high-performance liquid chromatography or polyacrylamide gel electrophoresis (48-mers and longer), resp.
5. Dye-labeled bacterial artificial chromosomes (BACs) and complementary deoxyribonucleic acid (cDNA) prepared by Dr. R. Schaub and D. Bausch (University Children's Hospital Zurich).
6. TRITC-(tetramethylrhodamine isothiocyanate) labeled antibodies were obtained from Jackson ImmunoResearch Laboratories (West Grove, PA).
7. Unlabeled antibodies from Santa Cruz (Santa Cruz, CA) and from Cell Signaling (Danvers, MA) (mouse monoclonal antibodies, rabbit and goat were polyclonal antibodies).
8. Anti-fluorescein antibody prepared by Dr. S. Melkko (ETH Zurich).
9. Model peptide 5(6)-Carboxyfluorescein-KTKESLGRKIQIQRSG-NH_2, gift from Dr. M. Schutkowski (JPT Peptide Technologies, Berlin).
10. Sodium chloride (molecular biology grade) from Sigma-Aldrich (Buchs, CH).
11. (N-[2-hydroxyethyl]piperazine-N′-[2-ethanesulfonic acid]) (HEPES, >99.5%) from Sigma-Aldrich (Buchs, CH).
12. Sodium mono- and diphosphate (both ≥99.0%) from Sigma-Aldrich (Buchs, CH).
13. Trisodium citrate (>99%) from Sigma-Aldrich (Buchs, CH).
14. Fluorescein isothiocyanate dextrane conjugate (molar mass 4000 g/mol) from Fluka, (Buchs, CH).
15. Hydrochloric acid (p.a.) from Fluka (Buchs, CH).
16. Ethanolamine (p.a.) from Fluka (Buchs, CH).
17. Betaine hydrochloride (>99%) from Fluka (Buchs, CH).
18. Formamide (p.a., ≥99.5%) from Fluka (Buchs, CH).
19. Glycerol (p.a., ≥99.5%) from Fluka (Buchs, CH).
20. Triton X-100 (molecular biology grade) from Fluka (Buchs, CH).
21. Tween-20 from Fluka (Buchs, CH).
22. D-(+)-Trehalose dihydrate (≥99.0%) from Fluka (Buchs, CH).
23. Sodium dodecyl sulfonate (SDS, >98%) was from Fluka (Buchs, CH).
24. Dimethyl sulfoxide (DMSO) was from Fluka (Buchs, CH).
25. Buffers: PBS-T (phosphate buffered saline, Tween-20), TBS-T (Tris buffered saline, Tween-20), SSC (saline sodium citrate), SSC/SDS.
26. 2-Amino-2-(hydroxymethyl)-1,3-propanediol, hydrochloride (Tris-HCl, molecular biology grade) from VWR International (Dietikon, CH).
27. For microarray spotting non-contact arrayer were used (sciFLEXARRAYER, Scienion, Berlin, Germany; Piezorray, PerkinElmer, Downers Grove, IL, USA).

28. Washing, blocking, hybridization, and drying of the slides in an automated hybridization station (HS4800, Tecan, Salzburg, Austria).
29. GeneMachines HybStation (Genomic Solutions, Ann Arbor, MI).
30. For read-out 4-color confocal laser scanner (LS400, Tecan, Salzburg, Austria; ScanArray5000, PerkinElmer, Boston, MA, USA).
31. ArrayPro Analyzer 4.5.1 for quantitative data evaluation (Media Cybernetics, Silver Spring, MD, USA).
32. Genespotter 2.4.2 for quantitative data evaluation (Microdiscovery, Berlin, Germany).
33. Polyethylene glycol-coated slides (sciChip-PEG, E-PEG), epoxysilane coated slides (sciChip 2D, E-2D), a polymer coated slide (sciChip 3D, E-3D) from Scienion (Berlin, Germany).
34. CodeLink™ hydrogel slide (CDL, GE Healthcare, Piscataway, NJ) (*see* **Notes 3** and **4**).

3. Methods

In the following, precise protocol suggestions are provided for testing the microarrays, which were optimized in Chapters 2 and 3. The protocols have been developed for a Tecan HS4800 automated hybridization station, but they can be used with any other hybridization machine. In a practical comparison of the performance of the HS4800 with a SlidePro as well as a GeneMachines HybStation (*see* **Note 5**), it was found that protocols for the processing of oligonucleotide microarrays can easily be adapted, even though the these machines significantly differ in the technical way of slide processing. The given protocols can be transcribed and adapted for manual hybridization as well (*see* **Note 6**). The most important change applies to washing steps. Because of a very efficient washing in the automated systems, the duration of manual washing should be increased by at least a factor of 2 depending on the efficiency of agitation (*see* **Note 7**). Slide drying in HS4800 system is performed by flushing the chamber with nitrogen, a unique feature, thus avoiding problems with ozone-sensitive dyes that have to be taken into account especially when processing the slides manually *(1)*.

3.1. Oligonucleotides

The optimization of oligonucleotide microarrays has been described in detail in Chapter 3. In principle, it has been shown that all the four slide types investigated (*see* **Subheading 2.**) can be used as substrates for this type of application. However, when comparing fluorescence intensities in our model experiments, the hydrogel slide significantly outperforms the other slides.

It is found that, the shorter the oligonucleotide the more important an extended molecular architecture on the slide surface is, to increase the accessibility of the immobilized sample and to avoid surface effects (*see* **Fig. 3**, in Chapter 2, types 3B-D) *(2,3)*. Comparing the fluorescence intensity of a dye-labeled 13-mer oligonucleotide immobilized on the four different types of slides it can be estimated that the loading density of E-2D and E-PEG slide (slide types 3A and 3B, respectively) are similar (data not shown). The polymer coatings of the E-3D slide and the hydrogel slide CDL (both of type 3C) are extended to the third dimension and allow for a three-dimensional immobilization of the oligonucleotides. The thickness of the coating is much higher for CDL than for the E-3D slide which is reflected by the signal intensities obtained in immobilization experiments with oligonucleotides of different lenghts (13- to 71-mers). For a 13-mer dye-labeled oligonucleotide, the signal intensity on CDL slides is about four times higher reflecting the higher number of immobilized probe molecules (*see* **Note 8**). However, the accessibility of the probe can be expected to be better for the E-3D slide because of its freely moving polymer chains in contrast to the highly crosslinked rigid polymer network of the hydrogel. Our experiments have shown that this type of hydrogel slide is especially suitable for immobilization and hybridization with short oligonucleotides (*see* **Note 9**). For experiments with large molecules, i.e., immobilized antibodies, these advantages could not be confirmed. Moreover, the CDL slide costs about twice as much as a high-quality epoxysilane-coated slide and therefore, will be used only for applications with the specific need for the respective functionality.

3.1.1. Validation Protocol for Hybridization of 13-mer Oligonucleotides to Oligonucleotide Arrays (for E-2D, E-PEG, E-3D, and CDL Slides)

1. Preparation of hybridization solution. A commercial buffer (Ambion no. 1) and a HEPES-based buffer with similar success. The sample concentration applied is 1–100 nM. For testing the ability of the slide to withstand nonspecific adsorption, we used concentrations as high as 1 μM *(4)*.
2. Some slides (E-PEG) require an additional washing step to eliminate probe-smearing effects. Washing is performed with wash buffer (0.2X SSC/0.2 %SDS) for 2–5 min at 4°C. The CDL slide is wet by 0.1% Triton X-100. Wetting the slide is a general requirement of the hybridization station to avoid the formation of air bubbles in the hybridization chamber.
3. Slide blocking is performed at highest agitation rate using 5 mM HCl (E-PEG), 5 mM HCl containing 0.1% Triton X-100 (E-2D, E-3D) for 15 min at room temperature, or with 50 mM ethanolamine in Tris-HCl, pH 9.0 at 50°C for 30 min (CDL). The concentration of HCl can be reduced to 1 mM without a noticable change in background intensity (*see* **Fig. 7** in Chapter 3).
4. Wash extensively with wash buffer (e.g., 0.2X SSC/0.2% SDS) for 3 min in order to remove the HCl. When washed manually, the solvent should be exchanged four to five times.
5. Introduce hybridization solution at the hybridization temperature. This avoids the formation of air bubbles in the hybridization chamber (especially when a hybridization is performed at high temperature). For immobilized 13-mer oligonucleotides (calculated $T_m = 38°C$), hybridized was performed at temperatures between 23 and 33°C.

6. For hybridization of complementary oligonucleotides in the model experiments (no mismatch discrimination needed), 30 min to 2 h are sufficient, depending on target concentration. A strong agitation is recommended.
7. For model experiments with matching oligonucleotides all washing steps are performed at room temperature:
   a. First wash: 2 × 2 min at 23°C, 2X SSC/0.2% SDS.
   b. Second wash: 2 × 2 min at 23°C, 0.2X SSC/0.2% SDS.
   c. Third wash: 2 × 1 min at 23°C, 0.2X SSC (*see* **Note 7**).
   If a mismatch discrimination is required, the temperature of the washing solutions can be increased to 36–40°C.
8. Slide drying is performed automatically with nitrogen for 5 min at 30°C.

3.2. cDNA and BACs

Protocols for cDNA microarray applications are well established and a number of protocols are available *(5,6)*. Aminosilane, epoxysilane and aldehyde activated slides (*see* **Fig. 3** in Chapter 3, types 3A and 3B, respectively), as well as poly-L-lysine (PLL)-coated slides *(7)* are the substrates mainly used.

For the workflow presented in Chapter 3 we used dye-labeled cDNA and BACs were used as model compounds *(8)*. By spotting dilution series in 3X SSC, the optimal spotting concentration is determined to be in the range of 200–400 mg/L. Besides 3X SSC and 50% DMSO there are a number of spotting solutions developed for cDNA applications including mixtures of formamide, betaine, and nitrocellulose *(9)*, and DMSO or detergent mixtures *(10)*. Strand separation of cDNA immobilized from a nondenaturing spotting solution is usually done by heating the slide for 2–3 min in water at 95°C. This harsh treatment causes considerable thermal stress to the immobilized DNA. The unfavorable nature of the latter widely used treatment can be illustrated by determination of the amount of dye-labeled cDNA that is removed from epoxysilane- and aminosilane-coated surfaces by this treatment. Depending on the immobilization conditions (ultraviolet [UV] light dose used for immobilization: 200, 300, and 600 mJ/cm^2), between 25 and 35% of the originally immobilized substance is torn off the surface (**Fig. 1**). For this reason, we recommend spotting from denaturing solutions instead of later denaturing steps, with the advantage of resulting in higher final signal intensities *(9)*.

Immobilization of cDNA is usually performed by radiation with UV light and/or heating the slide to 80°C for 2–4 h. Because of the large molecular size, unwanted effects caused by radiation as discussed in Chapter 3 obviously have no consequence on the experimental outcome, which is in contrast to comparable experiments with oligonucleotides *(11)*. cDNA immobilizes on PLL-coated slides without the need for UV crosslinking *(12)*. Limitations of using PLL slides are discussed (*see* **Note 12** in Chapter 3).

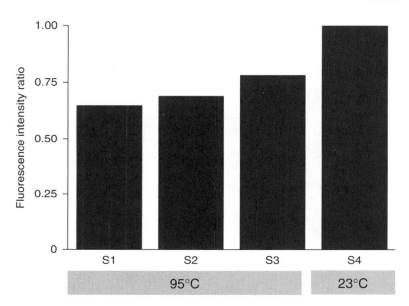

Fig. 1. Effect of hot water for strand separation on dye-labeled cDNA immobilized on E-2D. The fluorescence intensity ratio before and after the treatment is plotted. As a reference, the intensity ratio of an untreated slide, which was extensively washed at 23°C is set to 1. The amount of remaining cDNA correlates with the UV dose used for immobilization: S1 = 200 mJ/cm², S2 = 300 mJ/m², and S3 = 600 mJ/cm².

Blocking cDNA slides works well with protocols recommended in the literature using succinimide for aminosilane- and PLL-coated slides, acids or bases for epoxysilane coated slides, sodium borohydride for aldehyde modified slides, or using universal mixture of salmon sperm DNA and bovine serum albumin (BSA). An interesting feature is the use of organic solvents in which the DNA is not soluble including N-methylpyrrolidone *(13)*, N-methylimidazol and methylene chloride *(14)* to avoid smearing effects in situations, where DNA is spotted at too high concentrations.

3.3. Peptides

In contrast to oligonucleotides and cDNA, peptides show a much larger variety in their physical and chemical properties exhibiting a strong sequence dependence. Problems with array production often arise from the need of an organic cosolvent to dissolve hydrophobic peptides. A fluorescein-labeled model peptide with a free N-terminus containing 15 amino acids, soluble in 1X PBS at a concentration of 10 m*M* (for sequence information *see* **Subheading 2.**) *(15)* were used in the experiments. In view of solubility issues encountered with other peptides, experiments were performed in different solutions including

PBS, PBS containing DMSO (25%, 50%), and acetonitrile (25%). Additionally, a basic catalyst is added (diisopropylethylamine, Hünigs Base) to accelerate the coupling reaction and glycerol for improving the spot morphology. Some results are shown in **Figs. 5** and **6** in Chapter 3.

As outlined in Chapters 2 and 3, a suitable coupling chemistry is needed to immobilize peptides at only one molecular site. In our example, the model peptide contains additional lysine primary amino groups, which should react in parallel with the N-terminus. This results in immobilization at different molecular sites and presumably multiple binding to electrophilic surface groups. Despite these disadvantages, the peptide can be used for an array optimization as this would only affect a final incubation assay but not the general experimental parameters including spotting concentration, spotting solution, immobilization efficiency, and surface blocking.

Following our workflow, the optimal spotting concentration is determined in a dilution experiment and found to be 3–5 mM, independent from the spotting solutions previously listed. Immobilization is performed in a humidity chamber overnight. Especially in presence of glycerol, peptides, and other substrates spotted in organic solvents frequently show impaired immobilization efficiency on certain surfaces, even when a matching coupling chemistry is applied. To improve this and to accelerate the reaction with reactive surface groups, immobilization is performed at different temperatures. There is a strong dependence of the immobilization efficiency on temperature and relative humidity (*see* **Fig. 6** in Chapter 3).

Spotting peptides in PBS results in a very good spot morphology on all surfaces that we have investigated (*see* **Subheading 2.**). Especially on hydrophilic 3D-polymer-coated slides the presence of organic solvents in the spotting solution changes spot size and morphology considerably. Good spot morphology is obtained when using PBS or PBS containing 25% acetonitrile on E-PEG and epoxysilane-coated slides, respectively. Under certain conditions and depending on the immobilization chemistry, unpurified peptides can be spotted as well. Using this approach, peptides are synthesized and functionalized in a way that only full length products will immobilize at the surface and by-products can be washed off. As a limitation of this method, it might be difficult to obtain a good spot morphology as discussed in Chapter 3.

Blocking reactions of peptide arrays can be performed under acidic or basic conditions using hydrochloric acid or ethanolamine, depending on the slide surface chemistry. In order to determine the functionality of the immobilized peptide after all the processing steps we have performed a cleavage experiment with trypsin. The cleavage of the peptide, incubated with the enzyme at 37°C for 1 h, results in the loss of the dye and a decrease in fluorescence intensity. The ratio of fluorescence intensities before and after cleavage depends on the type of slide and is found to be between 1.5 (E-2D) and 2.2 (E-PEG). The variability between slide types as a result of different surface architectures influencing the cleavage ratio originates from details of the immobilization (i.e., multiple attachment) and reflects the accessibility of the peptide by the protease.

3.3.1. Validation Protocol for the Reaction of Proteins and Antibodies on a Peptide Array (for E-2D, E-PEG, E-3D, and CDL Slides)

1. Preparation of incubation solution. The reactions are performed in 1X PBS buffer containing the same amount of blocking agent as the solution used for blocking.
2. Blocking is performed 1–3% TopBlock in PBS at room temperature for 1 h or at 4°C overnight.
3. A washing step to remove the blocking solution is not necessary.
4. Introduce sample solution at the incubation temperature.
5. Reaction time depends on the target concentration. As a rule of thumb, the product of target concentration and reaction time is a measure of final signal intensity. A strong agitation is recommended.
6. Washing steps. First wash in a suitable buffer, for example, PBS(T) or TBS(T) for 5 × 2 min at room temperature. If detergent is used in the washing solution, a second wash is needed: 3 × 1 min at room temperature.
7. Slide drying with nitrogen for 5 min at 30°C.

3.4. Antibody Microarrays

In Chapters 2 and 3, we have already listed a variety of different slide surfaces that can be used for antibody microarrays, including epoxysilane-coated slides, aldehyde-modified slides, PLL slides, and polymer-coated slides *(16,17)*. For a review featuring many useful recommendations on the optimization and processing of antibody arrays *(see* **ref. 18***)*.

As an example, we applied our workflow to optimize an array of phosphorylation specific antibodies by means of FITC-labeled antimouse, antigoat, and antirabbit (model) antibodies, respectively, using E-2D, E-PEG, E-3D, and CDL slides. In dilution experiments spotting the antibodies in PBS the optimal concentration is determined, at 0.65–1.3 μM (100–200 mg/L). The curve shape found in the diagram is identical with the one for oligonucleotides and peptides *(see* **Fig. 2A** in Chapter 3) and reflects limitations of the maximal surface packing density.

The range of spotting solutions for antibodies (and proteins in general) is limited to aqueous solutions because of the sensitive nature of the compounds which may loose their 3D structure. Usually, PBS-based buffers are used. Additives are recommended that are known to improve the stabilty of antibodies and proteins on the slide surface like 10–40% PEG (molar mass 200 g/mol, *(19)*, up to 40% glycerol *(20)*, and disaccharides such as sucrose and trehalose. The first two prevent the spot from drying. Disachararides are thought to protect proteins by forming a matrix and act as a water substitute *(21)*. In our tests we obtained very good results with spotting from PBS containing 10% PEG resulting in a very good spot morphology and higher signal intensities compared to PBS alone.

Immobilization of antibodies at 4°C overnight is performed in a humidity chamber to avoid drying. In the literature different procedures are recommended including immobilization for a period of 2 h or overnight at room temperature or 4°C, respectively, depending on the stability of the compounds. In contrast to oligonucleotides and peptides, antibodies easily immobilize by adsorption to nearly all surfaces and form covalent bonds through electrophilc groups (epoxy, aldehyde, isothiocyanate, NHS ester) present at the slide surface.

The biggest challenge in optimizing antibody arrays is a high background signal. First, widespread smearing effects can result from too high spotting concentrations or when the slide is washed and blocked under inappropriate conditions. As stated earlier, smearing effects can be avoided by chosing a suitable spotting concentration in order not to overload the surface. Second, incubation of the slide with dye-labeled proteins or antibodies often results in nonspecific adsorption and a huge background signal, therefore, making blocking of the surface a most critical step. From our experience, following immobilization, wash the slides extensively for some minutes with flowing buffer at 4°C thus nearly completely reducing smearing of spotted probe. Incubating the slide with 1–3% BSA or inert protein mixtures (TopBlock, 1–3% in PBS) overnight at 4°C results in a low background. The following incubation with the target ideally would be performed in the presence of the same amount of blocking agent, which can further reduce background problems. Antibody arrays are stored at 4°C in a humidity chamber. For long-term storage the slide box is additionally sealed under nitrogen.

3.4.1. Validation Protocol for the Reaction of Proteins or Antibodies on an Antibody Array (for E-2D, E-PEG, E-3D, and CDL Slides)

1. Preparation of incubation solution. The reactions are performed in 1X PBS buffer containing the same amount of blocking agent as the solution used for blocking.
2. Washing the slide in PBS-T or TBS-T in order to remove unbound antibody for 2–5 min at 4°C using streaming buffer. A detergent should be used that is compatible with the antibodies spotted.
3. Slide blocking with 1–3% BSA or 1–3% TopBlock, both in PBS for 1–2 h at room temperature or at 4°C overnight.
4. A washing step to remove the blocking solution is not necessary.
5. Introduce sample solution at the temperature of incubation.
6. Reaction time depends on the concentration of antibodies in the sample. Strong agitation is recommended.

7. Washing steps. First wash with PBS-T (TBS-T) for 5 × 2 min at room temperature. If detergent used in the washing solution, a second wash is needed with PBS (TBS) 3 × 1 min at room temperature (*see* **Note 10**).
8. Slide drying with nitrogen for 5 min at 30°C.

3.5. Carbohydrates

As a model compound for the determination of processing conditions we used a fluorescein-labeled dextrane of relatively low-molar mass (4000 g/mol). Similar compounds are used for immobilizing dextrane derivatives of high-molecular mass (20 k–2000 kDa) to nitrocellulose-coated slides by nonspecific adsorption *(22)*.

Dilution series of the model compound are spotted to E-2D, E-PEG, and E-3D slides, which all have a matching surface chemistry for covalent immobilization. The optimal spotting concentration in 3X SSC (pH 7.4), phosphate (pH 8.5), and 50% DMSO is found to be 2–4 mM. Very good spot morphologies are obtained with all spotting solutions on all types of slides used. Even on E-2D (small) spots of good quality are obtained when using DMSO containing buffers, which indicates that the carbohydrate changes the properties of the spotting solution.

Immobilization is performed in a humidity chamber overnight. In accordance with the reactivity of hydroxyl groups, the immobilization of the dextrane is more efficient at higher pH values. Blocking the slides is performed in basic media (ethanolamine in Tris-HCl at pH 9.0). For testing the perfomance of the array an antifluorescein antibody is used *(23)* and proved the specific interaction with the immobilized fluorescein-dextrane conjugate.

3.5.1. Validation Protocol for the Reaction of Proteins or Antibodies on Carbohydrate Arrays (for E-2D, E-PEG and E-3D Slides)

1. Preparation of incubation solution. The antibody is dissolved in PBS/3% TopBlock at a concentration of 10 nM.
2. E-PEG slides are washed with wash buffer (PBS-T) for 2–5 min at 4°C. E-2D and E-3D are not sensitive to smearing effects and are wet with PBS-T for 2 min.
3. Slide blocking was performed at highest agitation rate using 50 mM ethanolamine in Tris-HCl, pH 9.0 at room temperature for 15 min. Alternatively, the slides can be blocked with PBS/3% TopBlock at room temperature for 1 h.
4. Wash extensively with flowing buffer (PBS-T) for 3 min to remove the ethanolamine. When washed manually, the solvent should be exchanged four to five times.
5. Introduce antibody solution at the incubation temperature (23°C).
6. Reaction time depends on antibody concentration. At 10 nM it was incubated for 1–2 h. A strong agitation is recommended.
7. Washing: 5 × 2 min at 23°C with PBS(T). A second wash is needed when a detergent is used: second wash for 3 × 1 min room temperature.
8. Slide drying is performed automatically with nitrogen for 5 min at 30°C.

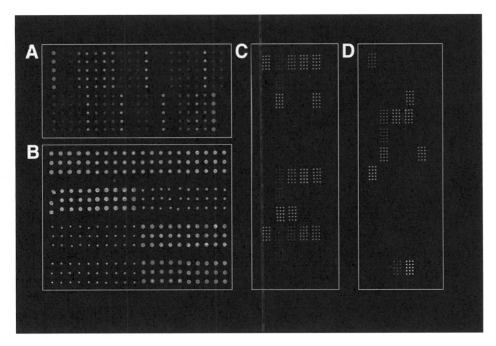

Fig. 2. Results of optimization applying the workflow shown in **Fig. 1** in Chapter 3. The pictures were not electronically processed. **(A)** 19-mer oligonucleotide array (detail) hybridized with 48-mer dye-labeled single-strand PCR products in experiments used for decoding a chemical library *(24)*. The red spots (upper left and lower right column) belong to dye-labeled oligonucleotides spotted as positive control. **(B)** A typical result of the optimization procedure. The aim was to find a solvent mixture resulting in a good spot morphology and, simultaneously, to determine the immobilization efficiency. Blocks of 5 × 3 replicates of the model peptide were spotted in different solvent mixtures and at two concentrations (optimal spotting concentration found in a preceding experiment, and as twice this concentration) to a E-PEG slides. Note the large differences in spot morphology (probe distribution), spot size, and fluorescence signal intensity. The latter is related to the immobilization efficiency. **(C,D)** Scans of two antibody microarrays. Twenty phosphorylation specific antirabbit **(C)** and antimouse antibodies **(D)** were spotted to E-3D **(C)** and E-PEG **(D)**. The arrays were incubated with matching dye-labeled target antibodies according to the protocol previously given.

3.6. Concluding Remarks

The execution of the workflow described in this and Chapters 2 and 3 takes about 1–2 wk and consumes about 25 specimens of every slide type that is tested. As only a limited number of model compounds are needed, the respective costs for the optimization are rather low. As shown in **Fig. 2**, the workflow provides a good basis for the experimental optimization of microarray production parameter.

4. Notes

1. Chemicals used for microarray applications should be of highest quality. Keep in mind that additives (e.g., stabilizer) can be present in some chemicals which might interfere with the experiment. Some solvents degrade and might cause undesired side reactions.
2. All solutions used for microarray experiments should be of high quality. Buffer solutions should be freshly prepared and filtered.
3. Organic solvents coming in contact with barcode areas can release some residues that can cause artifacts on the slide surface.
4. For general recommendations of slide handling, *see* **Notes 1–4** in Chapter 3.
5. This offers the advantage that reactions are performed under strictly controlled conditions without the risk of a partial slide drying or uncertainties introduced by manual processing, which is the reason why the usage of such instruments is strongly recommended. Nonetheless, the given recommendations are equally valid when using a manual hybridization chamber but special attention has to be paid to the mentioned risk of introducing irregularities.
6. If processed manually, slides must be transferred between solutions quickly without (partial) drying during processing steps. Otherwise residues will be left behind that cannot be removed and will cause background artifacts.
7. To obtain a high-washing efficiency it is necessary to wash with fresh washing solutions frequently. For example, if slides are washed after a chemical modification of the surface (silanation, reaction with a linker) or to completely remove an acidic or basic blocking solution or after hybridization step and so on, replacing of the wash solution four to five times is necessary.
8. Some the of commercial hydrogel slides do not possess reactive functional groups to allow an immobilization through covalent bonds. Compounds of low-molecular weight often do not immobilize efficiently by adsorption on these slides.
9. As a result of environmental effects on the fluorescence intensity, the number of immobilized molecules cannot be exactly determined by comparing the fluorescence intensities on both slides; *see* **Note 10** in Chapter 3.
10. Calcium phosphate precipitates when calcium containing incubation solutions are used in presence of PBS(T).

Acknowledgments

JS would like to thank Prof. Dr. Joe Jiricny (University of Zurich, ETH Zurich), Dr. Orlando Schärer (Stony Brook University Hospital), and Dr. Philip Day (University of Manchester) for their extensive help getting started with slide development, microarray production and hybridization experiments. We would like to thank Dr. Rahel Schaub and Damaris Bausch (University Children's Hospital Zurich), Dr. Mike Schutkowski (JPT Peptide Technologies, Berlin), Dr. Gerald Radziwil (University of Zurich), and Dr. Samu Melkko (ETH Zurich) for kindly providing samples of model compounds.

References

1. Fare, T. L., Coffey, E. M., Dai, H., et al. (2003) Effects of atmospheric ozone on microarray data quality. *Anal. Chem.* **75,** 4672–4675.
2. Chan, V., Graves, D. J., Fortina, P., and McKenzie, S. E. (1997) Adsorption and surface diffusion of DNA oligonucleotides at liquid/solid interfaces. *Langmuir* **13,** 320–329.
3. Vainrub, A. and Pettitt, B. M. (2003) Surface electrostatic effects in oligonucleotide microarrays: control and optimization of binding thermodynamics. *Biopolymers* **68,** 265–270.
4. The protocol was developed for hybridization of 14-mer FAM-labeled oligonucleotides to an array of 412 13-mer probes.
5. Hegde, P., Qi, R., Abernathy, K., et al. (2000) A concise guide to cDNA microarray analysis. *BioTechniques* **29,** 548–562.
6. http://cmgm.stanford.edu/pbrown/protocols/index.html.
7. Hessner, M. J., Meyer, L., Tackes, J., Muheisen, S., and Wang, X. (2004) Immobilized probe and glass surface chemistry as variables in microarray fabrication. *BMC Genomics* **5,** 53.
8. Thanks to Rahel Schaub and Damaris Bausch for preparation of the compounds.
9. Wrobel, G., Schlingemann, J., Hummerich, L., Kramer, H., Lichter, P., and Hahn, M. (2003) Optimization of high-density cDNA-microarray protocols by 'design of experiments'. *Nucleic Acids Res.* **31,** E67.
10. Rickman, D. S., Herbert, C. J., and Aggerbeck, L. P. (2003) Optimizing spotting solutions for increased reproducibility of cDNA microarrays. *Nucleic Acids Res.* **31,** E109.
11. Wang, H. -Y., Malek, R. L., Kwitek, A. E., et al. (2003) Assessing unmodified 70-mer oligonucleotide probe performance on glass-slide microarrays. *Genome Biol.* **4,** R5.
12. André Gerber, ETH Zurich, private communication.
13. http://cmgm.stanford.edu/pbrown/protocols/3_post_process.html.
14. Diehl, F., Grahlmann, S., Beier, M., and Hoheisel, J. D. (2001) Manufacturing DNA microarrays of high spot homogeneity and reduced background signal. *Nucleic Acids Res.* **29,** E38.
15. Many thanks to M. Schutkowski (JPT Peptide Technologies) for a gift of the peptide.
16. Angenendt, P., Gloekler, J., Sobek, J., Lehrach, H., and Cahill, D. J. (2003) Next generation of protein microarray support materials: Evaluation for protein and antibody microarray applications. *J. Chromatography A,* **1009,** 97–104.
17. Angenendt, P., Gloekler, J., Murphy, D., Lehrach, H., and Cahill, D. J. (2002) Toward optimized antibody microarrays: a comparison of current microarray support materials. *Anal. Biochem.* **309,** 253–260.
18. Kusnezow, W., Jacob, A., Walijew, A., Diehl, F., and Hoheisel, J. D. (2003) Antibody microarrays: an evaluation of production parameters. *Proteomics* **3,** 254–264.
19. Lee, C. -S. and Kim, B. -G. (2002) Improvement of protein stability in protein microarrays. *Biotechnol. Lett.* **24,** 839–844.

20. MacBeath, G. and Schreiber, S. L. (2000) Printing proteins as microarrays for high-throughput function determination. *Science* **289,** 1760–1763.
21. Cleland, J. L., Lam, X., Kendrick, B., et al. (2001) A specific molar ratio of stabilizer to protein is required for storage stability of a lyophilized monoclonal antibody. *J. Pharm. Sci.* **90,** 310–321, and literature cited therein.
22. Wang, D., Liu, S., Trummer, B. J., Deng, C., and Wang, A. (2002) Carbohydrate microarrays for the recognition of cross-reactive markers of microbes and host cells. *Nat. Biotechnol.* **20,** 275–281.
23. Thanks to Samu Melkko, ETH Zurich, for kindly providing a sample of the antibody.
24. Melkko, S., Scheuermann, J., Dumelin, C. E., and Neri, D. (2004) Encoded self-assembling chemical libraries. *Nat. Biotechnol.* **22,** 568–574.

5

Specific Detection of Bacterial Pathogens Using Oligonucleotide Microarrays Generated From Hydrolysis PCR Probe Sequences

Philip Day

Summary

A move away from high-density screening array formats and the implementation of probes specifically identifying targets for application in low-density hybridization capture analyses is growing in importance. Some of the highest specificity for bioassays that encompasses use of hybridization has enlisted hydrolysis probes in real-time PCR. It is demonstrated that by employing 5'-end-tethering to glass of hydrolysis PCR probe sequences that these can be employed as capture probe sequences. The probes retain hybridization binding specificity to their PCR amplicons and specificity of hybridization was achieved across all probes at equivalent stringency. This suggests that probes identified for use in hydrolysis PCR using one set of temperature cycling parameter can also be applied to a microarray, where again the probes all exhibit similar specificity of interact at analogous hybridization conditions. The procedure permits the exact same PCR amplicon sequences to be used in both quantitative PCR and qualitative microarray assay formats.

Key Words: Hybridization; microarray; oligonucleotides; real-time PCR; specificity; stringency.

1. Introduction

Gene microarrays most typically permit the simultaneous interaction of multiple immobilized characterized gene probes with labeled nucleic acid analytes. The immobilized probes can be comprised from the longer more complex nucleic acids (*1*), to shorter, less complex PCR products and oligonucleotides (*2*). The more complex probes are typically deposited onto a solid surface using one of a number of arraying devices while oligonucleotides can be arrayed either postsynthesis or by *in situ* synthesis (*3*). Oligonucleotide arrays provide

From: *Methods in Molecular Biology, vol. 382: Microarrays: Second Edition: Volume 2*
Edited by: J. B. Rampal © Humana Press Inc., Totowa, NJ

the highest density of nucleic acid microarrays and sequences are chosen to interact with high specificity following comparisons to sequence data bases, including single base polymorphic base alterations *(4)*. The same specificity cannot be easily achieved with more complex probes. Furthermore, when genomes are present within uncharacterized mixtures, computer-generated predictions of nonspecific hybridization interactions lose merit. Indeed, typically during the design of microarrays, nonspecific cross-hybridization interactions of probe sequences are not considered *(5)*. This situation contrasts with the extreme high specificity observed in real-time (q)PCR formats using hydrolysis (TaqMan™) probes *(6)*. In qPCR, recognition sequences are targeted with high specificity and sensitivity from within highly heterogeneous biological specimens *(7)*. A general scheme for microarray selection of oligonucleotide probes with high hybridization specificity for application among highly heterogeneous targets is not readily available. Characterized qPCR assays were chosen for several bacterial pathogens associated with infections of the upper respiratory tract to determine if TaqMan probe-selection software could be used to generate oligonucleotide probe sequences for application on a spotted microarray format and if this procedure also better safeguarded against nonspecific array hybridization interactions. These qPCR systems were already optimized in terms of specificity, sensitivity, and ability to perform in multiplexed reactions employing pertinent clinical specimens *(8–10)*.

2. Materials

2.1. Culturing of Bacterial Strains

1. Reference strains obtained from the following sources; *Streptococcus pneumoniae* strain ATCC 49619, *Moraxella catarrhalis*, *Haemophilus influenzae type b*, and *Mycoplasma pneumoniae* antigen strain ATCC 15531.
2. Culturing: sheep blood agar is supplied from Oxoid, Basel, Switzerland, cat. no. CM0854. Chocolate agar (Hardy Diagnostics, Santa Maria, CA., cat. no. E11) is used for growing fastidious bacteria including *H. influenzae.*

2.2. DNA Extraction

1. Digestion buffer (1X): 50 mM Tris-HCl, pH 8.5, 1 mM EDTA, 0.5% sodium dodecyl sulphate (SDS), 200 mg/mL proteinase K. Store at room temperature, add proteinase K just before use.
2. DNA is purified using the QIAamp tissue kit (QIAamp, Crawley, UK; cat. no. 51304).

2.3. Oligonucleotide Probe Design

1. All oligonucleotide probes are designed by primer express software (Applied Biosystems, Foster City, CA) and were already extensively tested in real-time PCR to display high sensitivity.

2. The probes are synthesized with a C6 aminolinker at the 5′-end (Cruachem, Glasgow, Scotland), and unlike the PCR probes from where they are derived, the oligonucleotide probes possessed no fluorochromes and are obtained following reverse-phase high-performance liquid chromatography purification from Microsynth GmBH, Balgach, Switzerland.

2.4. Spotting the Array

1. 2X NoAb printing buffer (NoAb Diagnostics Inc. Mississauga, Ontario, Canada; cat. no. UAS 0001PB).
2. Epoxy covalent binding slides (NoAb Diagnostics Inc. Mississauga, cat. no. UAS0005 E).
3. DNA samples are arrayed using a pin and ring GMS 417 Arrayer (Affymetrix Inc., Santa Clara, CA).
4. SYBR™ Green I, 10,000X concentrate (Molecular Probes, Leiden, Netherlands).

2.5. Labeling the PCR Amplicons

1. Design PCR primers using Primer Express© software for use with hydrolysis probes (*see* **Subheading 2.3.**) and synthesized by Microsynth GmBH (Balgach, Switzerland). Use primers at 1 mM during PCR.
2. 1X PCR Thermophilic buffer (Promega, Madison, WI): 500 mM KCl, 100 mM Tris-HCl (pH 9.0 at 25°C), 1% Triton X-100, 2.5 mM MgCl$_2$, 0.2 mM dATP, dGTP, and dTTP, 0.15 mM dCTP, 0.025 mM Cy3- or Cy5-labeled dCTP (Amersham Pharmacia Biotech, Dübendorf, Switzerland), 1 U of DNA Taq polymerase (Promega, Madison, WI).
3. Light mineral oil (Sigma–Aldrich, Poole, Dorset, UK; cat. no. cm 5904).
4. Perform PCR amplification in a GeneAmp PCR system 9600 (Perkin-Elmer, Norwalk, CT).
5. Purify labeled PCR products using the QIAquick™ nucleotide removal kit (Qiagen, Basel, Switzerland) according to the instructions of the suppliers.
6. Mix labeled PCR products with an equal volume of 50% glycerol.
7. Analysis of label incorporation: use 5 mL of a 1% (w/v) agarose gel caste on a microscope slide. Electrophoresis at 80 V for 1 h, dry at 70°C on a heating block for a further 1 h.
8. Image microscope slides in a ScanArray 5000 Scanner (Perkin Elmer/Packard Biosciences, Zurich, Switzerland), using a laser power of 80% and photon multiplier (PMT) gain set at 75%.
9. Universal 32 Hettich centrifuge (Hettich & Company, Dallas, TX).

2.6. Hybridization

1. All hybridization and stringency tests are performed in the automated slide processor (ASP) (Amersham Pharmacia Biotech, Little Chalfont, Buckinghamshire, UK).
2. Nonspecific hybridization blocking buffer is from NoAb (NoAb Diagnostics Inc. Mississauga).

3. SlideHyb no. 1 hybridization buffer (Ambion, Huntingdon, Cambs, UK).
4. Sample injection into the ASP hybridization chamber uses a Microliter™ Syringe (Hamilton, Bonaduz, Switzerland).
5. Stringency washing: 0.5X standard sodium citrate (SSC), 0.01% SDS, at temperatures ranging from 26 to 34°C.
6. Slide drying: automatic flushing in the ASP with isopropanol and drying in an airstream.

2.7. Specificity

1. Four sequential increments of stringency were analyzed by using the ASP (Amersham Pharmacia Biotech) and a temperature series: 26, 28, 30, and 34°C.

2.8. Image Capture and Data Analysis

1. Scanning: ScanArray 5000 Scanner (Perkin Elmer/Packard Biosciences) at a laser power of 100% and a photomultiplier gain of 95%.
2. Subsequent data analysis was performed using ImaGene™ software 3.0 (BioDiscovery Inc., Los Angeles, CA).

3. Methods

The detection of pathogenic microbes in nasopharyngeal aspirates is growing in popularity as a means of inferring infection of the lower respiratory tract. However the current routine procedure is arduous requiring culturing before strains can be identified and subsequently treated. Quantitative PCR methods have been developed that accurately identify and measure specific pathogens from within mixtures of other typical coinfecting strains of bacteria. The assays however are mainly useable qualitatively for the administration of antimicrobial treatment. Therefore given this scenario, once real-time PCR assays are fully characterized, their utility for pathogen detection of nasopharyngeal flora could be better employed by excluding the use of costly hydrolysis probes, performing endpoint PCR and then undertaking hybridization of these amplicons to a low density oligonucleotide microarray comprised from tethered hydrolysis probe sequences. The analysis of the array is purely qualitative, multiplexed, and not prone to subjective interpretation. Oligonucleotide microarrays have mainly lent themselves to high density and multiparallelized and simultaneous screening analysis of gene presence. Typically microarrays do not safeguard against nonspecific hybridization to other probe sequences. Because of the limited quantitative application of microarrays, oligonucleotide arrays might well find enhanced application for the qualitative screening of a lower number of disease-related targets, where the interaction of every target or group of targets is linked to a specific therapeutic regime. For the new procedure to work, the hydrolysis probe sequences need to retain binding specificity when immobilized on the surface of an array and furthermore all probes need to perform with analogous conditions of stringency (*see* **Fig. 1**).

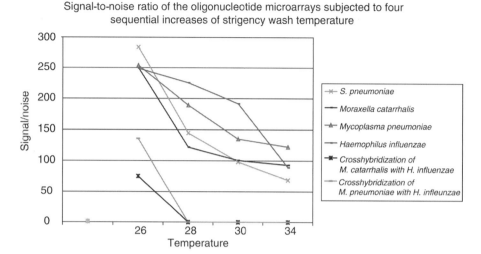

Fig. 1. Fluorescently labeled PCR products were hybridized to 5′-tethered oligonucleotide sequences derived from hydrolysis probes, and using a sequential increment of stringency washing temperature the signal-to-noise ratio was calculated using Quant Array® software. Average values (from triplicate data) are plotted and standard deviations did not exceed 12%. Nonspecific hybridization was never observed for *S. pneumoniae* and *H. influenzae* and rarely at ≤26°C for *M. catarrhalis* and *M. pnuemoniae*. Hybridization was totally specific at 28°C.

3.1. Culturing

S. pneumoniae, was grown aerobically on sheep blood agar plates at 37°C for 24 h. The bacteria were then washed from the plates using agitation in saline (0.9% NaCl) and further diluted with saline until a MacFarland standard of 0.5 representing 10^8 microorganisms/mL was reached. One mL of this dilution was used for DNA extraction. To determine the sensitivity for *M. catarrhalis* and *H. influenzea type b*, they were tested in the same manner as described earlier for *S. pneumoniae* apart from chocolate blood agar plates were used as growth media. To culture *M. pneumoniae*, one ampoule of lyophilized *M. pneumoniae* antigen (strain ATCC 15531) was resuspended in 1 mL of 0.85% NaCl solution, diluted to 10^{-2} and centrifuged at 14,000g for 10 min. Most of the supernatant was discarded and the remaining pellet was dried for 1 h in a vacuum centrifuge and stored at 4°C until DNA extraction was performed.

3.2. DNA Extraction

The extraction of the bacterial DNA was performed by subjecting 1 mL of liquid culture to centrifugation at 12,000g for 10 min. The pellet was resuspended in 200 µL of digestion buffer and incubated with shaking for 1 h at

55°C. The DNA was then purified with the QIAamp tissue kit according to the instructions of the supplier and without modification. Extracts were stored at –20°C until required for analysis.

3.3. Oligonucleotide Probe Design

For detection of these different bacteria by hybridization to microarrays, hydrolysis (TaqMan™) oligonucleotide probes targeting a specific gene were designed using the default settings of the Primer Express software v1.5 (*see* **Note 1**) The probe targeted the pneumolysin gene *(11)* of *S. pneumoniae* (GenBank accession no. M17717) with the sequence 5′-ACC CCA GCA ATT CAA GTG TTC GCG-3′ (positions 556–580) *(8)*, the P1 adhesion protein gene sequence of *M. pneumoniae* (GenBank accession no. X07191) with the sequence 5′-TCA ACT GAA TAA CGG TGA CTT CTT ACC ACT G-3′ (position 572–603) *(12,13)*, and the outer membrane protein gene of *M. catarrhalis* (GenBank accession no. U69982) with the sequence 5′-TGC TTT TGC AGC TGT TAG CCA GCC TAA-3′ *(14)* (positions 73–99). Also, the gene encoding for serotype b capsular *(15)* of *H. influenzae type b* (GenBank accession no. X78559) was used to provide the probe sequences. All of the oligonucleotide probes had a C6 aminolinker at the 5′-end (*see* **Note 2**).

3.4. Spotting the Array

The oligonucleotide probes were dissolved in water and were mixed with an equal volume of 2X NoAb printing buffer to a final concentration of 40 μ*M*. 5′ end-tethering of oligonucleotide probes was facilitated by spotted onto epoxy covalent binding slides using a pin and ring GMS 417 Arrayer (*see* **Note 3**). In addition, 1 μg/μL of an empty pUC 18 vector was spotted as an unspecific hybridization control. After arraying the quality of the spots on the first and last slides within a print series was controlled by staining for 5 min in 1X SYBR Green I. Surplus dye was washed off for 5 min in double-distilled water. The slides were then dried by placing into 50-mL Falcon tubes and centrifuged in a Hettich Universal 32 centrifuge at 2650 rpm (1000 rcf) for 2 min. Slides were image analyzed following scanning in a ScanArray 5000 Scanner (*see* **Note 4**).

3.5. Labeling the PCR Amplicons

A PCR assay was used to co-label targets for subsequent hybridization on a microarray. Amplimers matched to TaqMan probes were chosen using Primer Express software, v1.5. For *S. pneumoniae* the forward primer corresponds to sequences 531–552 (5′-AGC GAT AGC TTT CTC CAA GTG G-3′) *(8)* of the pneumolysin gene and the reverse primer to bases 605–583 (5′-CTT AGC CAA CAA ATC GTT TAC CG-3′). For the copB gene of *M. catarrhalis* *(14)* the forward primer was from base 50 to 70 (5′-GTG AGT GCC GCT TTA CAA CC-3′)

and the reverse primer was from base 121 to 102 (5′-TGT ATC GCC TGC CAA GAC AA-3′). For the P1 adhesion protein gene of *M. pneumoniae* (*9*) the forward primer from base 549 to 569 (5′-CCA ACC AAA CAA CAA CGT TCA-3′) and the reverse primer from base 624 to 605 (5′-ACC TTG ACT GGA GGC CGT TA-3′). Fluorochrome was incorporated during 50 μL volume PCRs employing a GeneAmp PCR system 9600 using a reaction mixture containing 5 μL of 10X Thermophilic Buffer supplemented with 0.025 m*M* Cy3- or Cy5-labeled dCTP, 1 m*M* primers, 1 U of DNA Taq polymerase, 2.5–100 ng/uL DNA and finally over-laid with approx 15 μL of light mineral oil. Thermal cycling conditions were performed using 10 min at 95°C, followed by 40 cycles of 15 s at 95°C and 1 min at 60°C. After PCR amplification, the labeled PCR products were purified using the QIAquick nucleotide removal kit according to the instructions of the suppliers. For checking the efficiency of the labeling reactions, 1–2 μL of the labeled PCR products were mixed with an equal volume of 50% glycerol before loading onto a 1% agarose gel caste on a microscope slide. The gel was then electrophoresed, dried at 70°C on a heating block and finally the microscope slide was imaged in a ScanArray 5000 scanner. Labeled PCR products were stored at −20°C until required for use.

3.6. Hybridization

The hybridization was carried out on the ASP. Before hybridization the spotted epoxy slides were incubated for 2 h with blocking buffer by gentle agitation at room temperature. The total volume (50 μL) of the labeled purified PCR products was concentrated to 20 μL by the use of vacuum centrifugation at 30°C, and subsequently denatured at 95°C for 5 min followed by chilling on ice. Two-hundred twenty micro-liters of SlideHyb no. 1 hybridization buffer was preheated to 65°C, mixed with 20 μL of the denatured PCR products and incubated at 65°C for 10 min. The mixture was injected with a microliter syringe into a hybridization chamber of the ASP and hybridized for 20 h at 30°C with continuous mixing. This was followed by a washing step with 0.5X SSC, 0.01% SDS for 2 min at temperatures ranging from 26°C to 34°C (*see* **Subheading 3.7.**). The slides were then automatically flushed with isopropanol and dried in an airstream.

3.7. Specificity

To determine the optimum treatment to achieve high hybridization specificity for each of the bacterial target gene sequences, DNA extraction, labeling, and hybridization were carried out as described earlier and the washing step was comprised from four different temperatures 26, 28, 30, and 34°C. Washing commenced at the lowest temperature, followed by drying and scanning before repeating the process at the next higher washing temperature.

3.8. Image Capture and Data Analysis

All of the hybridized slides were scanned using a ScanArray 5000 Scanner at a laser power of 100% and a photomultiplier gain of 95%. Subsequent data analysis was performed using ImaGene software 3.0. The signal-to-noise ratio was calculated using Quant Array and was shown to decrease for all hybrids in relation to increasing washing temperature (**Fig. 1**). Occasionally weak nonspecific hybridizations were seen for bacterial PCR products at 26°C with the exception of *S. pneumoniae* and *H. influenzae*, which always hybridized specifically and independent to the tested temperatures. When any nonspecific hybridization was observed, raising the stringency wash temperature to 28°C and greater removed all nonspecific hybridization signal to exclusively leave specific hybridization interactions (**Fig. 2**).

4. Notes

1. The default setting of Primer Express was used to primarily keep the T_m values of the different probes a constant. This was a necessary precaution to determine if all probes when applied to the surface of a gene chip are able to retain specific hybridization at constant stringency conditions.
2. Oligonucleotides representing TaqMan probes were synthesized to carrying a 5′ primary amine with a six carbon spacer (Cruachem, Glasgow, UK). This safeguards that only full-length synthesized oligonucleotides possess the means for reactive coupling to the epoxy surface of the glass.
3. Other coupling chemistries can be employed. Several specific hybridization were tried and discovered that it is plausible even without using a reactive end-coupling chemistry. However the benefit of the described protocol employing an epoxy-primary amine relates to enhanced signal-to-noise ratio, given the extremely low background signal. This has marked additional advantage for image analysis as the spot boundary definition is clear and pixel intensity is uniform, although analysis can be enhanced further by employing noncontact piezo spotting which increases the uniformity for pixel signal and spot dimension by the additional deposition of oligonucleotide probes.
4. Following washing the slides were not allowed to reside at room temperature and were swiftly placed into a 50-mL Falcon tube and spun-dried. This helped to greatly avoid the background smearing that can prevent assessment of spot morphology.

Acknowledgments

Thanks are extended to Dr. Felix Niggli and Dr. Oliver Greiner, Division of Oncology, Kinderspital, University of Zurich, the Swiss Research Foundation for Children and Cancer, Swiss Bridge, and the Swiss Cancer League (grant 961-09-1999) for financial support. Reference strains were obtained from D. Nadal, Division of Infectious Diseases, Kinderspital, University of Zurich, and M. Altwegg, Department of Medical Microbiology, University of Zurich.

Hybridization of PCR amplicons to a micro-array comprised
from hydrolysis (TaqMan™) oligonucleotide probes

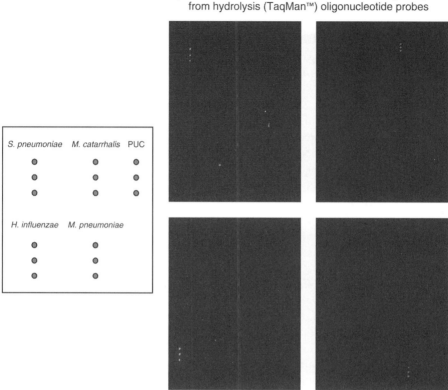

Fig. 2. The binding specificities of PCR products to 5′ tethered hydrolysis probe sequences that were selected using Primer Express software were shown to be devoid of any nonspecific cross-hybridization at 28°C in 0.5X SSC, 0.01% SDS. The pUC vector was used to demonstrate absence of nonspecific binding of PCR products to larger immobilized complex probes. Moreover, the temperature at which all probes demonstrate specificity is ≥28°C. The hydrolysis probes were designed by Primer Express to function at 60°C, therefore the procedure of 5′-end-tethering to a glass support appears to universally reduce the T_m of interaction to these oligonucleotide probe sequences and that lowering of T_m is independent of probe sequence.

References

1. Schena, M., Shalon, D., Davis, R. W., and Brown, P. O. (1995) Quantitative monitoring of gene expression patterns with a complementary DNA microarray. *Science* **270,** 467–470.
2. Yoshida, K. -I., Kobayashi, K., Miwa, Y., et al. (2001) Combined transcriptome and proteome analysis as a powerful approach to study genes under glucose repression in *Bacillus subtilis. Nucleic Acids Res.* **29,** 683–692.

3. Hughes, T. R., Mao, M., Jones, A. R., et al. (2001) Expression profiling using microarrays fabricated by an ink-jet oligonucleotide synthesizer. *Nat. Biotechnol.* **19,** 342–347.

4. Solinas, A., Brown, L. J., McKeen, C., et al. (2001) Duplex scorpion primers in SNP analysis and FRET applications. *Nucleic Acids Res.* **29,** E96.

5. Li, F. and Stormo, G. D. (2001) Selection of optimal oligos for gene expression arrays. *Bioinformatics* **17,** 1067–1076.

6. Livak, K. J., Flood, S. A. J., Marmaro, J., Giusti, W., and Deetz, K. (1995) Oligonucleotides with fluorescent dyes at opposite ends provide a quenched probe system for detecting PCR product and nucleic acid hybridisation. *PCR Methods and Applications* **4,** 357–362.

7. O'Donovan, M., Silva, I., Uhlmann, V., et al. (2001) Expression profile of human herpesvirus 8 (HHV-8) in pyothorax associated lymphoma and in effusion lymphoma. *Mol. Pathol.* **54,** 80–85.

8. Greiner, O., Day, P. J. R., Bosshard, P., Imeri, F., Altwegg, M., and Nadal, D. (2001) Quantitative detection of Streptococcus pneumoniae in nassopharyngeal secretions by real-time PCR. *J. Clin. Microbiol.* **39,** 3129–3134.

9. Marty, A., Greiner, O., Day, P. J. R., Gunziger, S., Mühlemann, K., and Nadal, D. (2004) Detection of *Haemophilus influenzae* type b by real-time PCR. *J. Clin. Microbiol.* **42,** 3813–3815.

10. Greiner, O., Day, P. J. R., Altwegg, M., and Nadal, D. (2003) Quantitative detection of *Moraxella catarrhalis* in nasopharyngeal secretions by real-time PCR. *J. Clin. Microbiol.* **41,** 1386–1390.

11. Walker, J. A., Allen, R. L., Falmagne, P., Johnson, M. K., and Boulnois, G. J. (1987) Molecular cloning, characterisation, and complete nucleotide sequence of the gene for pneumolysin, the sulfhydryl-activated toxin of Streptococcus pneumoniae. *Infect. Immun.* **68,** 1374–1382.

12. Dallo, S. F., Su, C. F., Horton, J. R., and Baseman, J. B. (1988) Identification of P1 gene domain containing epitope(s) mediating Mycoplasma pneumoniae cytoadherence, *J. Exp. Med.* **167,** 718–723.

13. Hardegger, D., Nadal, D., Bossart, W., Altwegg, M., and Dutly, F. (2000) Rapid detection of Mycoplasma pneumoniae in clinical samples by real-time PCR. *J. Microbiol. Methods* **41,** 45–51.

14. Aebi, C., Cope, L. D., Latimer, J. L., et al. (1998) Mapping of a protective epitope of the CopB outer membrane protein of Moraxella catarrhalis. *Infect. Immun.* **66,** 540–548.

15. Van Eldere, J., Brophy, L., Loynds, B., et al. (1995) Region II of the *Haemophilus influenzae* type b capsulation locus is involved in serotype-specific polysaccharide synthesis. *Mol. Microbiol.* **15,** 107–118.

6

Uses of Microarray Platforms in Cancer

A Correlative Study Between Genomic Copy Number Changes and Their Expression at mRNA and Protein Levels

Fahd Al-Mulla and Raba Al-Tamimi

Summary

With the completion of the Human Genome Project, the microarray technology has evolved into a sophisticated platform by which complex diseases such as cancer, can be studied at the genome, transcriptome, and proteome levels. Here, various microarray platforms, namely comparative genomic hybridization, cDNA, oligonucleotide, and protein-based microarrays are exploited to study genomic copy-number changes in a human cancer cell line and correlate these genomic aberrations with their expression at mRNA and protein levels. The protocols described therein can be assimilated for the study of other human tissues including cancerous ones.

Key Words: Cancer; CGH-microarrays; Colo320; expression; genomic amplification; genomic copy number; microarrays; protein microarrays.

1. Introduction

Genomic instability is a hallmark of cancer. During the last decade several techniques such as comparative genomic hybridization, fluorescent *in situ* hybridization, and multiplex ligation probe amplification have established the role of chromosomal copy-number changes, gene and exonal alterations in tumorigenesis, and cancer progression. However, these techniques have several disadvantages ranging from low resolution (5–20 Mb), technical limitation, such as the ability to study few genes in a single experiment to the requirement of tumor culture for some. Nevertheless, their widespread use has succeeded in localizing numerous amplifications that are thought to harbor oncogenes and deletions that pinpointed tumor suppressor genes. The issue whether these genetic alterations, particularly amplifications, lead to increased expression at mRNA and protein levels remains controversial. Traditionally, few genes lying

From: *Methods in Molecular Biology, vol. 382: Microarrays: Second Edition: Volume 2*
Edited by: J. B. Rampal © Humana Press Inc., Totowa, NJ

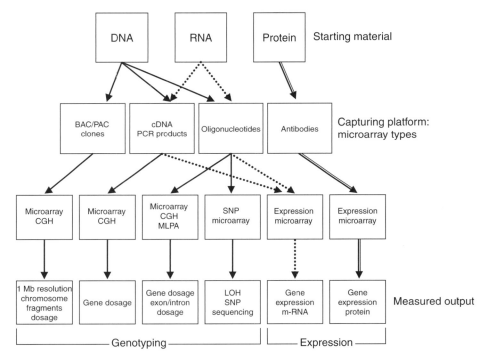

Fig. 1. Flow diagram depicting the type of starting biological materials currently used in various microarray platforms. The figure emphasizes the diverse biological processes that could be measured by various microarray paltforms.

within amplicons or deletions are selected and studied at transcriptional levels. This is understandable because techniques that are used to study genomic copy-number alterations could not be matched, until recently, with those that explore the transcriptome or proteome at a global level. We, at the cytogenetic level, and others have previously established that genomic amplifications are largely silenced and house only few highly expressed genes (*1,2*).

The emergence of microarray technology as a platform to study cancer represents a major step forward and offers a technology that has the potential to cope with the complexities of cancer genomes, transcriptomes, and proteomes and to correlate their relationships (**Fig. 1**). The use of BAC clones (*3*), cDNA (*4*), or oligonucleotides (*5*) as capturing microarray platforms for DNA offers a genomic resolution of 1 Mb (for comparative genomic hybridization [CGH]-microarrays) or up to identifying single nucleotide polymorphisms in SNP-chips (in oligonucleotides-based microarrays) (*6,7*). Moreover, recent probe capturing designs include oligonucleotides capable of distinguishing methylated from nonmethylated forms of DNA, which are relevant to the issue of amplicon silencing (*8*). Similarly, cDNA or oligonucleotides offer versatile

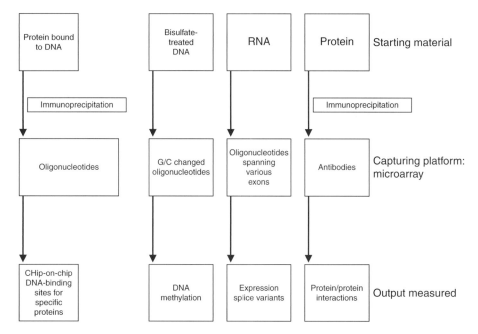

Fig. 2. The flow diagram depicts various modified microarray designs focused on measuring specific biological properties.

capturing platforms for assessing the transcriptome and mRNA splice variants *(9)*. Antibody-based microarrays offer an additional platform that interrogate the proteome directly thus bypassing biological processes such as nonsense-mediated decay, transcriptional, and posttranscriptional controls that could interfere with microarray-based experiments *(10)* (**Fig. 2**).

Utilizing microarray-platforms to study DNA, mRNA, and proteins from the same sample or cell line in parallel provide a straight-forward approach for assessing the relationship between amplicons and their expression at the level of single genes. Here, such an approach is decribed in Colo320 cell line using reliably tested protocols and evidence is provided to show that, although deleted genes are largely underexpressed, gene amplifications do not necessarily induce overexpression at mRNA or protein levels.

2. Materials

2.1. Cell Culture and Lysis

1. Human female colon adenocarcoma cell line, Colo320 *(11)*.
2. Roswell Park Memorial Institute-1640 Medium (Invitrogen, Carlsbad, CA), supplemented with 10% fetal bovine serum (FBS, Invitrogen), 1% (v/v) L-glutamine, and 1% (v/v) antibiotic–antimycotic solution (Invitrogen).

3. Trypsin 2.5% (10X) (Invitrogen, cat. no. 15090046). Working solution is prepared by dilution in sterile ethylenediamine tetra-acetic acid solution (EDTA) (1 mM) to 0.25% (1X, v/v), aliquots stored at −20°C.
4. Phosphate-buffered saline (PBS) powder (Sigma, Munich, Germany; cat. no. P3813) dissolved in 1 L to prepare a working solution of 0.01 M PBS (0.138 M NaCl, 0.0027 M KCl), pH 7.4, store at 25°C.
5. Angled lifter cell scrapers (Techno Plastic Products, Trasadingen, Switzerland).
6. Haemocytometer.
7. Trypan blue reagent: 0.2% (w/v) trypan blue in phosphate buffer saline (Sigma, cat. no. T6146).

2.2. DNA Extraction and Labeling

1. Puregene DNA isolation kit (Gentra Systems, Minneapolis, MN; cat. no. D-5500): cell lysis solution, RNase A solution, protein precipitation solution, and DNA hydration solution.
2. Puregene RBC lysis solution (Gentra Systems).
3. Proteinase K 20 mg/mL (Gibco/BRL, Scotland, UK) in 10 mM Tris-HCl, pH 7.5, 20 mM calcium chloride and 50% glycerol as stabilizers, store at 4°C.
4. Isopropanol.
5. Absolute ethanol. Aliquots of diluted 70% ethanol using molecular grade water and absolute ethanol are stored at −20°C.
6. Tris-(EDTA) (TE) buffer: 10 mM Tris-HCl, pH 7.0, 1 mM EDTA, pH 8.0.
7. *Dpn*II restriction endonuclease (10,000 U/mL) (BioLabs Inc., Cozad, NE) and 1X *Dpn*II buffer (50 mM Bis-Tris-HCl, 100 mM NaCl, 10 mM MgCl$_2$, 1 mM dithiothreitol, pH 6.0 at 25°C).
8. Buffered phenol:chloroform:isoamyl alcohol (25:24:1, v/v/v).
9. 3 M Sodium acetate solution. Dissolve 24.1 g of anhydrous sodium acetate in 100 mL of distilled water; adjust pH to 5.2 using glacial acetic acid. The solution is autoclaved and stored at 4°C.
10. BioPrime™ (Invitrogen) array CGH genomic labeling system (Invitrogen): Exo-Klenow fragment (40 U/μL), 2.5X random primer solution (octamers), 10X dUTP nucleotide mix (1.2 mM dATP, dGTP, and dCTP, and 0.6 mM dTTP in 10 mM Tris-HCl, pH 8.0, 1 mM EDTA), 10X dCTP nucleotide mix (1.2 mM dATP, dGTP, and dTTP, and 0.6 mM dCTP in 10 mM Tris, pH 8.0, 1 mM EDTA), and stop buffer (0.5 M EDTA, pH 8.0). All reagents are stored at −20°C.
11. Cy3- and Cy5-2′-deoxyuridine 5′-triphosphate or Cy3- and Cy5-2′-deoxycytidine 5′-triphosphate are aliquoted and stored at −20°C (Amersham Biosciences, Little Chalfont, UK).
12. BioPrime array CGH genomic labeling system purification module (Invitrogen): purification buffer A and purification buffer B. Working purification buffer B solution is prepared by adding 24 mL of 100% ethanol to stock buffer B. Purification buffers are stored at room temperature.
13. Refrigerated microcentrifuge 5417 R (Eppendorf, Germany).

2.3. RNA Extraction and Labeling

1. RNase zap (Ambion, Austin, TX).
2. TRIzol™ (Invitrogen).
3. Chloroform (Rathburn Chemical Ltd., Walkerburn, Scotland).
4. Isopropanol.
5. Absolute and 70% ethanol.
6. Nuclease-free water (Ambion).
7. DNA-free kit (Ambion): recombinant DNase I (2 U/µL), 10X DNase I buffer and DNase inactivation buffer.
8. 1% Denaturing agarose gel: 1 g of agarose (Sigma-Aldrich) dissolved in 85 mL diethyl pyrocarbonate (DEPC)-treated water and microwaved for 2 min. While stirring 5.4 mL of 12.3 M formaldehyde, 10 mL of 10X 4-morpholinepropane-sulfonic acid (MOPS) buffer (0.4 M MOPS, 0.1 M NaOAc, 10 mM EDTA, all at pH 7.0), and 10 µL of ethidium bromide are added. Gel is allowed to solidify for 1 h.
9. Running buffer: 1X MOPS.
10. Loading buffer: formamide, formaldehyde, 10X MOPS, and bromophenol blue in 50% glycerol.
11. Normal human colon total RNA (BD Biosciences Clontech, San Jose, CA).
12. Atlas™ PowerScript fluorescent labeling kit (BD Biosciences Clontech) Box 1: BD PowerScript reverse transcriptase, 10X fluorescent labeling dNTP mix (dATP, dCTP, dGTP, dTTP, and aminoallyl-dUTP), 5X first strand buffer, 100 mM DTT, 2X fluorescent labeling buffer, 10 U/µL RNase H, 0.1 µM coupling reaction control oligo, cDNA synthesis control (0.1 µg/µL RNA, 0.1 µM cDNA synthesis primer). Box 2: DMSO, QuickClean™ purification resin, 3 M NaOAc, FluorTrap™ matrix.
13. Cy3 and Cy5 mono reactive dye packs store at 4°C (Amersham).
14. 100-µL Quartz cuvets, 10 mm path-length.

2.4. Protein Extraction, Labeling, and Hybridization

1. Protein extraction and labeling kit (BD Biosciences Clontech): extraction/labeling buffer, blocking buffer, and 10X desalting buffer.
2. Normal colon tissue protein extract (US Biological, Swampscott MA).
3. BCA protein assay kit (Pierce Biotechnology, Inc., Rockford, IL): BCA reagent A (500 mL containing sodium carbonate, sodium bicarbonate, bicinchoninic acid, and sodium tartrate in 0.1 M NaOH), BCA reagent B (4% cupric sulfate), and 2 mg/mL albumin standard; bovine serum albumin (BSA) at 2.0 mg/mL in 0.9% saline and 0.05% sodium azide.
4. Cy3 and Cy5 mono reactive dye packs (Amersham).
5. Protein desalting spin columns (Pierce Biotechnology, Inc.).
6. BD Clontech Antibody glass microarrays (BD Biosciences Clontech).
7. BD Clontech Antibody microarray buffer kit (BD Biosciences Clontech): incubation buffer, background reducer, wash buffers 1–7.

2.5. DNA Hybridization and Washing

1. 16K cDNA array glass slides (Finnish DNA Microarray Centre, Turku, Finland).
2. 20X Standard saline citrate solution (SSC). 175.3 g of NaCl and 88.2 g of Na$_3$Citrate·2H$_2$O dissolved in 1 L of sterile water, adjust pH to 7.0 using 1 M HCl and stored at room temperature.
3. cDNA array pretreatment BSA blocking solution: 1% (w/v) BSA fraction V (Amersham Biosciences), 5X SSC, and 0.1% (w/v) sodium dodecyl sulfate (SDS) dissolved at 50°C for 30 min.
4. cDNA array pretreatment washing buffers: 2X SSC and 0.2X SSC.
5. Probe nonspecific hybridization blockers: yeast t-RNA (Invitrogen), polyadenylic acid-dA (poly[dA]) (ResGen, Invitrogen), and Human Cot-1 DNA (Invitrogen).
6. 10X DIG blocking buffer (Roche).
7. cDNA array hybridization buffer: 3X SSC, 0.3% SDS.
8. cDNA array wash buffer 1: 2X SSC, 0.03% SDS; wash buffer 2: 1X SSC; wash buffer 3: 0.2X SSC.
9. Tecan hybridization station and Tecan LS200 scanner (Tecan®, Grödig, Austria).
10. Array-Pro Analyzer v4.5 software (MediaCybernetics, Inc., Silver Spring, MD).

2.6. RNA Hybridization and Washing

1. 7.6 K 80-nt oligonucleotide glass arrays (BD Biosciences Clontech).
2. BD Atlas™ glass microarrays kit: random primer mix, GlassHyb™hybridization solution, GlassHyb™ wash solution.
3. BD Atlas™ glass wash chambers (cat. no. 7908-1).
4. Hybridizer HB-1 oven (Techne, Burlington, NJ).

3. Methods

3.1. Preparation of Samples for DNA

3.1.1. Extraction

1. Colorectal cancer cell-line (Colo320) is grown and propagated in 25-cm^2 tissue culture flasks and propagated into 75-cm^2 culture flasks when 70% confluence is reached to acquire sufficient cells for storage and experimental manipulation. Confluence cultures are terminated by replacing medium with 9 mL PBS and 1 mL trypsin/EDTA and incubated at 37°C for 5 min or until the cells are visibly detached. One milliliter of supplemented medium is added to inhibit the activity of trypsin and cells are pelleted by centrifugation at 550g for 10 min. Collected pellet is resuspended in 10 mL of PBS and centrifuged under the same previous settings. This washing step is repeated twice. Final cell pellet is used for seeding new flasks supplemented with full medium, or for storage by resuspending it in storage medium (FBS:DMSO, 9:1), or the pellet is kept on ice (4°C) for immediate use.
2. The final pellet is resuspended gently and 0.5–1 × 10^6 cells are used in a 1.5-mL centrifuge tube for DNA extraction. The cell suspension aliquot is centrifuged at

13,000–16,000*g* for 5 s to pellet cells. Remove the supernatant leaving a 5–10 μL of liquid overlay. Vortex the cell pellet vigorously and add 150 μL of cell lysis solution. Resuspend the cell pellet by pipetting till the solution is homogenous; if clumping appears incubate at 37°C till homogeneity is achieved.

3. RNA is removed by adding 0.75 μL of RNase A solution to the cell lysate. Mixing is achieved by inverting the tube 25 times followed by an incubation period of 15–60 min at 37°C. The sample is then cooled to room temperature. Add 50 μL of protein precipitation solution to the cell lysate and vortex vigorously for 20 s. Centrifuge at 15,000*g* for 3 min, a protein pellet is visible if not repeat centrifugation after an incubation period of 5 min on ice. The supernatant containing DNA is transferred into a new microcentrifuge tube containing 150 μL of 100% isopropanol, mix by gentle inversion 50 times. Centrifuge tube at 15,000*g* for 1 min, DNA will form a white pellet. Decant supernatant and drain tube on a clean filter paper. Wash the DNA pellet with 150 μL 70% ethanol and mix by inversion several times. Centrifuge tube at 15,000*g* for 1 min; decant 70% ethanol supernatant carefully to avoid loosing the pellet. Drain the tube on clean filter paper and leave to air-dry for 10–15 min (*see* **Notes 1** and **2**).

4. DNA pellet is rehydrated with 25 μL DNA hydration solution, followed by incubation at 65°C for 1 h with periodic tapping. The DNA concentration is quantified using spectrophotometric measurement at wavelength 260 nm, and stored at 4°C in O-ring tubes to minimize evaporation. The DNA prepared can be used in BAC/PAC, cDNA or oligonucleotide-based microarrays (*see* **Note 3**).

5. Genomic DNA is extracted from whole blood as a normal reference DNA. Add 600 μL of whole blood to a 2-mL microcentrifuge tube containing 900 μL RBC lysis solution. Invert the tube once and incubate at room temperature for 1 min, with interrupted gentle inversions every 20 s. Centrifuge for 20 s at 15,000*g*. Decant the supernatant leaving a small liquid overlay. Vortex the tube vigorously and add a fresh 900 μL of RBC lysis solution. Invert for mixing and incubate for 1 min at room temperature with less agitation. Centrifuge again for 20 s at 15,000*g*, decant supernatant leaving a visible white pellet and few microliters of liquid. The pellet is vigorously vortexed to facilitate better cell lysis for subsequent DNA extraction procedure. The DNA extraction steps for whole blood are the same as for cultured cells with one exception that the volume of all solutions and reagents are quadrupled.

3.2. DNA Labeling, Hybridization, and Washing

1. Aliquots (4 μg) of each DNA stock are used in separate probe synthesis and labeling reactions. Each DNA aliquot is fragmented by mixing it with 2 μL of *Dpn*II endonuclease and 4 μL of *Dpn*II buffer. The reaction mix is brought up to 20 μL with distilled water and incubated at 37°C for 2 h. The reaction is inactivated at 65°C for 20 min. One-hundred sixty microliters of TE (pH 8.0) buffer is added after cooling the mixture on ice.

2. Phenol:chloroform:isoamyl alcohol (25:24:1) purification is initiated by adding 200 μL of the reagent in each reaction tube and vortexed vigorously. The mixture is centrifuged for 5 min at 11,000*g*. The top layer is aspirated, transferred into a

new labeled tube, and 20 μL of 3 *M* sodium acetate is added to it, followed by 500 μL of cold absolute ethanol. The mixture is left at −80°C for 45 min to allow nucleic acid precipitation. Reaction tubes are centrifuged at 20,000g for 20 min at 4°C. Discard the supernatant and add 500 μL of 70% ethanol to wash the pellet and spin for 5 min; discard the supernatant. Vacuum-dry the pellet for 5 min and resuspend it in 21 μL of nuclease-free water.

3. Every labeling reaction and subsequent manipulation of fluorescently labeled probes must be performed in the dark or a dim-lighted room, if unfeasible always keep the reaction tubes securely foiled, amber tubes are another alternative. Start the labeling reaction by adding 20 μL of 2.5X random primers solution to the fragmented DNA and incubate at 95°C for 5 min, followed by an immediate cooling on ice for 5 min. On ice, add the following to each tube keeping in mind the differential labeling of samples; 5 μL of 10X dUTP nucleotide mix, 3 μL of Cy3-dUTP or Cy5-dUTP, and 1 μL of exo-Klenow fragment. Mix the reaction tubes by pipetting, spin briefly (10 s) and incubate at 37°C for 2 h. The reaction is terminated by adding 5 μL of stop buffer and placing the tubes on ice (if necessary reactions can be stored overnight at −20°C).

4. Purify the labeled probes in the dark (dim lights) by adding 45 μL of TE (pH 8.0) along with 400 μL of purification buffer A, to each tube and vortex for 30 s. Collection tubes provided by the kit are fitted into purification columns. Load entire samples into them and centrifuge at 10,000g for 1 min at room temperature. Discard the flowthrough and add 600 μL of purification buffer B to the column and centrifuge under the same settings. Perform an additional wash with 200 μL of purification buffer B under the same centrifugation settings. Discard used collection tube and fit a clean 1.5-mL collection tube in place, add 50 μL of sterile water to the center of the column to elute labeled probes, centrifuge under the same previous settings.

5. Assay the differentially labeled probes for quantity and quality using formulas (a, b, c). Optimal results for concentration must be ≥2.8 μg, optimal amount of incorporated labeled dUTPs should be ≥100 pmoles, and optimal base/dye ratios should be 40–80 for both cyanine dyes.
   a. DNA concentration = $(A_{260} - A_{320}) \times 50 \times 0.05$ (elution volume).
   b. Cy3 (pmole) = $(A_{550} - A_{650}/0.15) \times 50$ μL,
      Cy5 (pmole) = $(A_{650} - A_{750}/0.25) \times 50$ μL
   c. Base/dye for Cy3 = $(A_{260} \times 150{,}000)/(A_{550} \times 6600)$
      Base/dye for Cy5 = $(A_{260} \times 250{,}000)/(A_{650} \times 6600)$.

6. If probes are of sufficient quantity and quality proceed by pooling the two labeled probe sample/reference together (**Fig. 3**). Add to the reaction mix 10 μg of poly(dA), 10 μg of Cot-1 DNA, and 42 μg of yeast t-RNA to block nonspecific hybridization. Speed vacuum the reaction mixture to dryness, and resuspend the pellet in 100 μL of 10X DIG blocking buffer. Denature the reaction mixture for 90 s at 100°C and then incubate at 37°C for 30 min to anneal.

7. Pretreat and block the cDNA array slides to remove excess salts and residual DNA and to prevent nonspecific hybridization. Immerse the slides in a sterile BSA

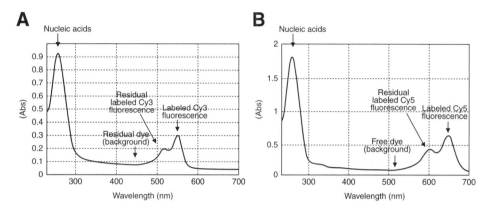

Fig. 3. An example of expected wavelength scans for Cy3 (**A**) and Cy5 (**B**) fluorescently labeled sample.

blocking solution for 30 min at 50°C inside a hybridizer. The slides are then washed with 2X SSC and 0.2X SSC for 3 min each at room temperature, respectively. Slides are dried by centrifugation at 1000*g* for 5 min at room temperature and used on the same day. One can use 50-mL Falcon tubes, if specialized tubes for the slides are not available.

8. Stock solutions of hybridization and posthybridization buffers are freshly prepared. Tecan hybridization station bottles are filled, tube lines are primed, and hybridization buffer warmed to 37°C. Dry slides are inserted into the slide adaptor and inserted into the hybridization chamber. Initiate predefined program, and wait until probe injection step. In the meantime, take your labeled probes mixture and keep on ice for 5 min. Inject probe when prompted by the machine and leave to hybridize for 20 h at 37°C. Slides are washed with wash buffer 1 at 65°C for 5 min, followed by washing with wash buffers 2 and 3 at room temperature for 5 min each. Slide drying is initiated for 2 min (*see* **Note 4**).

When the run ends, take out the slide adaptor out of the hybridization chamber, make sure the slides are dry (HS 4800 is equipped with nitrogen gas cylinder for drying slides). Insert the slide adaptor with slides into the Tecan LS200 scanner opened flap platform. Initiate prescan at wavelengths 543 and 633 nm, for Cy3 and Cy5, respectively. Select scanning area, adjust focus area, set desired resolution and start acquiring scanned images at different resolutions.

3.3. CGH-Microarrays Analysis

1. Capture multiple images of the array for each dye at its specific wavelength in a parallel (simultaneous) or sequential (one at a time) mode. Save the images in a designated experiment file to avoid mix up. Identify printed blocks on the slide and orient a fitting grid on it. Upload the file that contains label and information of each cDNA spotted on the slide in the correct orientation. Make sure the grid is well aligned on the spotted blocks and that the circumference of each

spot is within the circle inner margins to avoid loss of intensity. Define the background cells by choosing appropriately distributed spots that are not spotted with target cDNAs. Recommended background subtraction method is subtracting the raw intensity from global mean signals of selected background cells with a bilinear polynomial approximation. However, this can vary and depends on the operator. If any spots contain too much noise or hybridization artifacts such as residual material, designate to the ignored cell group. According to array noise level, clarity, cleanliness, and density set the background subtraction from raw signal intensities criteria.

2. Signal normalization was performed under the assumption that to some extent some genes preserve the same level of copy number in samples of similar tissue origin, and thus can be used as a normalization factor. Therefore, colo320 expression/normal colon expression ratios that are within the range of 0.95 to 1.15 are used as control cells for array platform normalization. This normalization approach is measured with a bicubic polynomial approximation.

3. Detected signal intensities <0.01 are omitted from the analysis after normalization. Cy3/Cy5 signal ratios are calculated by the software for all array platform replicas on the slide. Threshold log values for amplification and deletion ratios are user defined.

3.4. RNA Extraction

1. Cultured colo320 cells are harvested at 70% confluence, washed twice and final pellet is kept on ice with a residual fluid overlay, for RNA extraction. Laminar flow unit pipetes and glass ware used must be wiped carefully with RNase Zap, and all tips and tubes must be nuclease-free (or DEPC-treated before sterilization). The cell pellet is lysed with 500 µL of TRIzol and vigorous vortexing until the pellet is completely dissolved. Add 500 µL of cold chloroform and vortex patiently 10–15 min with interrupted cooling to avoid heat damage. centrifuge the cell lysate at 7000g for 30 min at 4°C. Transfer the upper layer with minimal disturbance of the lower layer, into a clean DEPC-treated 1.5-mL tube. Uniformly mix by inversion and keep tube on ice to add 500 µL of cold isopropanol to precipitate total RNA. The mixture is incubated on ice for 20 min and then centrifuged at 4°C, 21,000g for 30 min. The RNA pellet is washed with 600 µL of 70% alcohol, tap the tube four times and centrifuge at 4°C, 21,000g for 10 min. Decant supernatant; and keep the tube inverted on clean filter paper for 10 min. Vacuum dry the pellet for 5 min at no more than 55°C. Dissolve the pellet in 70 µL of nuclease-free water for 30 min. If necessary extracted RNA can be stored at −80°C for later.

2. Quantify extracted RNA at 260 nm, and according to resultant make a stock of 2 µg/µL concentration. Add 5 µg of total RNA in a new tube and mix with 10 µL of loading buffer, incubate sample at 75°C for 5 min followed by immediate cooling on ice. Prepare a 1% denaturing agarose gel to assay RNA quality, load samples in designated wells with marker and positive control. The gel is run for 160 min at 70 V. The gel is viewed under UV light; an expected result would be two clear bands corresponding to 28S and 18S ribosomal RNA (rRNA), and smear representing

messenger RNA. The reader is advised to watch for RNA degradation as it could bias microarray results *(12)*.

3. For microarray purposes RNA samples used must always be DNase treated to eliminate any DNA contamination that may interfere with expression results. According to sample concentration, aliquot 50 µg of RNA and add to it 2 µL of DNase I. Add 0.1 of volume of the DNase buffer to the sample and incubate for 25 min at 37°C. It is recommended to add 1 µL of RNase inhibitor (Invitrogen). Terminate the reaction by adding 5 µL or 0.1 of sample volume (whichever is more) of DNase deactivation buffer. Adequately mix the suspension by pipetting and incubate at room temperature for 2 min. Proceed to RNA purification by adding an equal volume of phenol:chloroform:isoamylalcohol, mix vigorously for 10 min with interrupted cooling. Cold centrifuge at 7000*g* for 10 min, transfer upper layer in a new tube and add to it an equal volume of chloroform and recentrifuge under similar settings. Transfer upper layer into a new tube; precipitate RNA with 2.5 of sample volume of cold 100% ethanol and 0.1 vol of sodium acetate. It is recommended if low RNA yield to add 1 µL of glycogen or acrylamide to aid in RNA precipitation; however, it does have a minimal effect on absorbance readings. Cold centrifuge at 21,000*g* for 30 min. Decant supernatant and concentrate pellet by adding 700 µL of cold 70% ethanol centrifuge at full speed for 5 min. Decant supernatant and air-dry pellet at room temperature. Dissolve pellet in adequate volume nuclease-free water to reach a concentration optimal for the labeling reaction (3–5 µg of pure total RNA, according to sample abundance in up to 7 µL).

3.5. RNA Labeling, Hybridization, and Washing

1. Prepare a master mix of the following Atlas PowerScript Kit components; 4 µL of 5X first-strand buffer, 2 µL of 2X dNTP mix (dCTP, dGTP, dATP, dTTP, aminoallyl-dUTP), 2 µL of DTT, 1 µL BD PowerScript reverse transcriptase, and 1 µL of deionized water for a final volume of 10 µL and store on ice (*see* **Note 5**). If multiple reactions are performed, increase according to number of reactions. Five micrograms aliquots of sample and reference total RNA that must not exceed 7 µL. Add 2 µL of random primer mix and 1 µL of cDNA synthesis control (*see* **Note 6**) to the RNA samples. If sample final volume is <10 µL then make it up with sterile distilled water. RNA samples are incubated for 5 min at 70°C in a thermal cycler. Cool the tubes at 37°C and add the entire 10 µL of master mix. Incubate reactions at 37°C for 60 min, and later terminate the reaction by heating it up to 70°C for 5 min. Cool down at 37°C for 1 min, add 0.2 µL of RNase H to degrade RNA templates, mix well by pipetting, vortex, and incubate at 37°C for 15 min. Spin the tube briefly to collect and inhibit the action of RNase H with 0.5 µL of EDTA. Purify the reaction mixture by adding 2 µL of QuickClean resin, vortex for 1 min and centrifuge to collect contents. Supernatant is transferred into a 0.22-µm spin filter tube and centrifuged at 12,000*g* for 1 min. The eluted synthesized cDNA is precipitated by adding 2.2 µL of 3 *M* sodium acetate and 55 µL of cold absolute ethanol. Vortex the mixture and leave to precipitate for 2 h at −80°C.

2. Centrifuge the tube at 12,000g for 20 min to pellet cDNA after incubation. Wash and concentrate the pellet with 200 µL of 70% ethanol, centrifuge at 12,000g for 10 min. Decant the supernatant and preferably air-dry the pellet, or dry in vacuum for 7 min at 55°C. Dissolve the pellet in 10 µL of 2X fluorescent labeling buffer and add 0.5 µL of coupling reaction oligonucleotides (*see* **Note 7**). Proceed with the remaining steps in the dark or a dim lighted room, if unfeasible always keeps the reaction tubes securely foiled, amber tubes are another alternative. Dissolve each stock of Cy3- and Cy5-free fluorescent dyes in 45 µL of DMSO to prepare a 5-mM stock. Ten microliters of each Cyanine dye are added to cDNA samples, bearing in mind that this is a differential labeling experiment. Incubate the labeling reaction mix at room temperature in the dark for 1 h to allow fluorescent dye coupling to aa-dUTPs. To purify the labeled probes, vortex the FluorTrap matrix vigorously for 1 min, and add 100 µL of it to the coupling reaction mix and vortex for a min. Insert the 0.22-µm spin filter into its accompanying collection tube and load the reaction mix into the spin filter. Centrifuge the tubes at 7000g for 1 min. The eluant will contain the purified labeled probes without the unlabeled cDNA fragments.

3. Assess probe quantity and quality by performing a 260–800 nm spectrum scan on the entire 100 µL undiluted labeled probe sample. Cy3 and Cy5 fluorescence intensities are measured at 550 nm and 650 nm, respectively (expected absorbance 0.1–0.4). Fluorescent dyes baseline is measured at 750 nm (expected absorbance <0.05), and cDNA quantity is measured at 260 nm (expected absorbance 0.5–1.5). Determine the volume of probe required for hybridization using the following formula;

$$V_{opt} \ (\mu L) = \frac{1000 \ OU_{opt}}{A_\lambda}$$

where V_{opt} is the optimal volume of labeled probe to be used in microliters, and OU_{opt} is the absolute optimal optical units of the probe that are achieved when; Cy3: $OU_{550 \ nm} = 0.010$ or Cy5: $OU_{650 \ nm} = 0.010$. A_λ is the absorbance reading of the probe sample at its corresponding fluorescent label's wavelength (i.e., 550 nm for Cy3 and 650 nm for Cy5).

4. Transfer 2.1 mL of BD GlassHyb hybridization solution into a 2-mL tube, and warm the solution to 50°C. Add the predetermined sample and reference probe volumes to the warmed hybridization solution and mix by pipetting. Unscrew the cap of the hybridization chamber with the enclosed array. Add the hybridization solution containing your labeled probe into the chamber by pipetting it into the round opening at the top of the insert without removing the plastic insert or the glass slide. While screwing the cap tightly on squeeze the wide sides of the chamber together using your thumb and fingers to apply pressure that will eliminate excess air and allow room for expansion during incubation. Place the hybridization chamber into a test tube rack or equivalent for support inside the hybridizer and incubate overnight (16 h or more) at 50°C in an upright position (with the cap on top). Do not agitate.

5. For washing a special green capped wash chambers are required, these can fit up to four slides; however, a wash is more effective one slide at a time. All washes are

performed at room temperature on an orbital shaker for 10 min each, and wash chambers are to be rinsed with distilled water after every wash. Label the wash chambers and fill them up with their designated wash buffers as follows; wash 1: 22 mL BD GlassHyb wash solution, wash 2a: 2 mL BD GlassHyb wash solution + 20 mL 1X SSC, wash 2b: 2 mL BD GlassHyb wash solution + 20 mL 1X SSC, and wash 3: 22 mL of 0.1X SSC. While holding the hybridization chamber in an upright position carefully unscrew the cap, remove the plastic insert using forceps as a lever by placing the end of the forceps in the round opening at the top of the insert. Gently pry and lift the insert, the slide might come out with the insert owing to surface tension or might not. If it does, be careful that the slide does not slip off the insert and fall outside the chamber, if it does not come out with the insert then carefully lift it out of the chamber and immerse it immediately in the first wash buffer. Always keep the printed face of the slide facing inward with the barcode or slide label facing upward. Maintain the same precautions throughout the washing series. When wash 3 is terminated, remove the slide out of the wash chamber and rinse it briefly with distilled water. Dry the slide rapidly to avoid passive water evaporation by placing the slide in a dry wash container and centrifuging it at 1500*g* for 5 min. Remember to balance the wash container with another one containing a mock glass slide to avoid slide breaking (*see* **Note 8**). The array is now ready for scanning.

3.6. Expression Microarray Analysis

1. Acquire images of the microarray slide on different wavelengths after setting the scanning area, resolution, pinhole, and whether a sequential or parallel capturing is preferred. Clontech glass microarray slide have orientation marker spots that outline the start of each spotted block in the array platform. Background cells are spotted at the last cell of each block. Therefore, grid the microarray platform using orientation markers for a guide and select predotted background cells for background subtraction. Background subtraction equals raw intensity subtracted from global mean signals of appropriately selected background cells with a bilinear polynomial approximation.

   BD Atlas glass microarray user manual specifies the block, row, and column coordinates of the spotted control oligonucleotides that will be used for normalization. These oligos are housekeeping genes that are universally expressed in all tissues. All housekeeping genes exhibit relatively constant expression levels in different tissues, cells, developmental stages, and diseases. The expression levels of one or more housekeeping genes as standards for measuring the expression levels of other genes is widely used approach for normalization. Normalization is measured for the entire platform with a bicubic polynomial approximation using signal intensities of selected housekeeping genes.

3.7. Antibody Array Probe Preparation and Hybridization

1. All the following protocol steps are performed rapidly with minimal exposure to heat. All items must be chilled and left on ice. Cultured colo320 cells are harvested

at 70% confluence, washed twice with PBS and final pellet is collected in a preweighed centrifuge tube. Weigh the cell pellet and approximate it for 15–25 mg of cells for small-scale protein extraction. Freeze the pellet at −80°C for 30 min to handle protein extraction procedures with minimal degradation. The control sample could be normal colonic tissue snap frozen in liquid nitrogen and powdered using pestle mortar, weighed and subsequently treated as with cell pellet protocol. Alternatively, normal colonic protein extract is commercially available.

2. Place the samples at room temperature and add 500 μL of extraction/labeling buffer (20 μL of extraction/labeling buffer for every milligram of cells). Vortex vigorously until homogeneity is achieved. Do not attempt to add any protease inhibitors as they could interfere with labeling. Incubate samples at room temperature for 10 min with constant rotation. Centrifuge the suspension at 10,000g for 30 min at 4°C. Transfer the supernatant into a new tube and place on ice. Determine the protein concentration in the supernatant using the BCA protein assay. Dilute each sample to 1.1 mg protein/mL by adding the appropriate volume of extraction/labeling buffer, the final volume must be ≥200 μL.

3. The labeling steps must be prepared rapidly without stopping, once the fluorescent dyes are dissolved in buffer they must be used immediately. A dye swap experiment is designed in which each sample and reference will be labeled with both fluorescent dyes and crossed against each other. Make sure that all tubes are labeled adequately so that each sample is dually labeled (A-Cy3, A-Cy5, B-Cy3, and B-Cy5). Dissolve each cyanine dye in 110 μL of extraction/labeling buffer by adding the buffer directly to the tube in which the dye is supplied. Vortex thoroughly for 20 s, and centrifuge to collect for 10 s. Add 10 μL of Cy3 fluorescent dye to tubes A-Cy3 and B-Cy3, and 10 μL of Cy5 fluorescent dye to tubes A-Cy5 and B-Cy5. Add 90 μL of extracted protein samples A and B into each tube, so that each sample is labeled separately with each dye. Mix tubes by pipetting and spin to collect for 10 s. Incubate tubes at 4°C for 90 min with interrupted gentle vortexing every 20 min. Terminate labeling reaction by adding 4 μL of blocking buffer to each tube, and incubate at 4°C for 30 min with gentle vortexing every 10 min. Proceed with removal of unbound dye directly.

4. Set and label a microspin desalting column and a collection tube for each labeled sample. Prepare 5 mL of 1X desalting buffer by diluting 10X stock aliquot in sterile distilled water. Adjust pH to 7.4 using dilute HCl or NaOH, and store on ice in a plastic tube. Spin the desalting columns at 1500g for 2 min to remove matrix storage buffer. Add 400 μL of 1X desalting buffer into each column and spin under the same previous settings. Repeat the previous step again and discard the flowthrough. Attach the prelabeled collection tubes to their corresponding desalting columns. Load the entire volume of each labeled protein sample into its corresponding column. Allow samples to pass and centrifuge columns at 1500g for 2 min. Detach collection tubes and keep them on ice to determine the labeled protein sample concentration. Because cyanine dyes absorb at 562 nm (for protein detection), a protein blank of labeled protein sample in 1X desalting buffer substituted for the BCA reagent is measured. Subtract the value of the protein blank

from the protein sample reading. Determine the protein/dye ratios according to Amersham's fluorescent dyes product specification sheet.

5. Prepare 45 mL of incubation buffer by mixing 4.5 mL of background reducer with 40.5 mL of stock incubation buffer. Store working incubation buffer in a clean plastic bottle or tube. Set up the array incubation tray, designate two chambers for slide 1 incubation and wash and the other two for slide 2 incubation and wash. Add 5 mL of incubation buffer in each incubation chamber. Label two clean tubes as slide 1 mix and slide 2 mix. Combine 100 µg of Cy3-reference with 100 µg of Cy5-sample as slide 1 mix. Combine the same amount of Cy5-reference and Cy3-sample as slide 2 mix. Transfer 20 µg of each mix and disperse into its corresponding incubation chamber. Leave the tray at room temperature for 30 min; meanwhile prepare the antibody array slides. Decant the antibody array storage buffer and add 30 mL of stock incubation buffer, cap the vial and slowly invert the vial for 10 times. Decant buffer and refill vial with 20 mL of working incubation buffer, repeat inversions for 10 times. Stand the vial in an upright position, and annotate slide lot numbers to identify slide 1 and slide 2. Transfer each slide into its pre-assigned incubation chamber array print side up. Incubate the slides at room temperature for 30 min with gentle rocking. Every 10 min lift the tip of the slide while keeping it submerged to allow liquid exchange. In each wash chamber add 5 mL of working incubation buffer and transfer slides into their respective wash chambers. Incubate the slides at room temperature for 5 min with gentle rocking. Remove the working incubation buffer from the wash chambers by aspiration and transfer 5 mL of wash buffer 1. Incubate the slides under the same conditions. Repeat all wash steps for all wash buffers 2–7 in the same manner described previously.

6. Dry the slides rapidly with minimal passive water evaporation. Carefully lift slides without touching their array surface and insert them into an empty, clean storage vial (a tissue culture 50-mL Falcon tube is appropriate as well). Cap the vial and balance it with a mock slide vial before centrifugation. Centrifuge at 1000*g* for 25 min at room temperature. Carefully, nudge the vial to protrude the tip of slides. Avoid slide falling out, as well as avoid touching the array surfaces, gently pull out each slide from its edge. Mount the slides on the adaptor rack in preparation for array scanning. Slides can be stored in a dry chamber away from light for up to 24 h before considerable signal intensities are lost.

3.8. Antibody Microarray Analysis

1. Adjust scanned array images to their correct orientation using the predotted Cy3/Cy5-labeled albumin as an orientation marker and a control spot. Create a grid for the array platform and save it. Negative control spots of unlabeled albumin are dotted in two different places on the microarray. These spots are selected for as background cells, their signal intensities are to be subtracted from the raw intensity of array platform. Before analysis BD Clontech Ab microarray analysis workbook must be downloaded (available at http://bioinfo.clontech.com). This workbook contains all the information about the array-specific platform, info-links

for each protein spotted, and formulas in an excel format of the arithmetic operations required for fluorescence data results.

2. Once slide grid is well aligned, and background subtraction is performed for both slides. Cy5/Cy3 ratios for both slides are calculated by the ArrayPro Analyzer software. Where ratios for slide 1 correspond to Cy5-reference/Cy3-sample, and ratios for slide 2 equal Cy5-sample/Cy3-reference. The ratios are copied into the import and analyses worksheet in the Ab microarray analysis work book. Once ratios are pasted the ratio 1/ratio 2 are calculated and the internally normalized ratios (INR) values are generated for each spot. Because each protein is spotted in duplicates in the array platform, average INR values are formulated and final results are obtained.

In theory, an INR >1 indicates that an antigen is more abundant in sample A (test) than in sample B (reference). Conversely, an INR <1 indicates that an antigen is less abundant in sample A (test) than in sample B (reference). The two slide approach addresses the majority of variability that could be introduced. Although internal normalization improves the quality of your data and addresses potential differences in dye labeling between samples, it is still not advisable to accept any INR value less than or greater than 1 as being a valid change. In our experience, INR values that are ≥1.3 or <0.77 indicate valid changes that signify differences in protein abundance.

3.9. Cross-Platform Comparison

It is not uncommon that microarray platforms from different vendors are used to measure different aspects of the same biological sample. Because different manufacturers could utilize different immobilized targets (cDNA, oligonucleotide) to capture genes, the use of different microarray platforms on the same samples could represent a significant problem for cross-platform data comparison. However, this can be minimized by matching either the dotted probe sequence on different platforms, if available, or matching their Unigene ID. This approach has been shown to improve data consistency across platforms *(13,14)*. Utilizing this approach for 714 altered genes; amplified genes are largely silenced while, as expected, deleted genes are underexpressed as shown here (**Fig. 4**).

4. Notes

1. It is recommended to use nuclease-free molecular biology grade water in all reactions.
2. It is paramount to wear disposable gloves and handle every step in a manner to avoid contamination with DNases and RNases. The procedure described here can also be applied for DNA extraction from frozen tissues after grinding or homogenizing by Pestle mortar.
3. The DNA was applied to 16 K cDNA-based microarrays, which are dotted in duplicate. The capturing platform represents PCR products ranging in size from 200 to 300 bp and verified by sequencing.

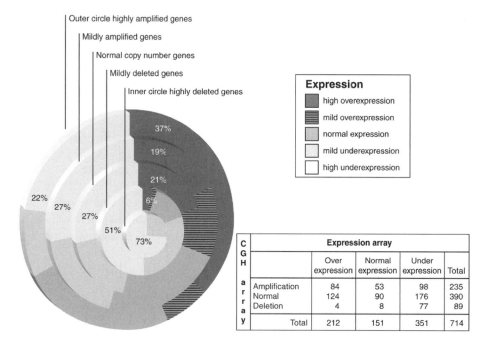

Fig. 4. Correlation between gene copy-number changes and expression status of 714 genes matched by Unigene ID. The majority of deleted genes are underexpressed while less than half of amplified genes are overexpressed, indicating the presence of silencing mechanisms. High copy number and overexpression indicate test to control ratio of >2, mild amplification or overexpression depict ratios between 1.35 and 2, normal copy number or expression between 0.75 and 1.35, mildly deleted genes or underexpression have ratios between 0.75 and 0.45 and highly deleted and underexpressed genes have ratios <0.45. The pie chart representing percentages of expression states within different gene copy number states in the colo320 cell line genome. The table shows actual gene numbers studied by microarray-CGH and expression microarrays and their correlation.

4. A number of microarray slide vendors have customized manual hybridization and washing chambers. It has been found that using an automated hybridizer and washer (Tecan HS 4800 hybridization station [Tecan, Austria]) minimizes slide handling and slide to slide quality variations. In addition to assuring consistency and reproducibility as well as the possibility for high throughput automation.

5. It is possible to add Cy5 or Cy3 dUTP directly to the cDNA synthesis step. However, it is found that aminoallyl-dUTP incorporates move efficiently into the cDNA and does not interfere with cDNA synthesis as much as directly labeled dUTP nucleotides do.

6. The cDNA synthesis control consists of a synthetic RNA, synthesized from phage λ DNA, mixed with a cDNA synthesis primer that is complementary to the λ RNA. The cDNA synthesis control mixture is spiked into the RNA sample to

supervise the cDNA synthesis reaction. When adequately labeled and hybridized to any BD Atlas glass microarray, spiked probes will produce a signal on a control spot corresponding to the synthetic RNA. This λ sequence has no homology to any described human or expressed sequence tags. The microarray printed cDNA synthesis control oligosequence (single strand) is: 5'-CCT TTA ACG GTG AAC TGT TCG TTC AGG CCA CCT GGG ATA CCA GTT CGT CGC GGC TTT TCC GGA - CAC AG-3'.

7. This is a short oligonucleotide from phage λ RNA already containing the amino groups necessary for coupling, it hybridizes to a this complementary sequence on the BD Atlas microarray: 5'-ACG GTA TGG TCA CCG GGA TGG CGG AA-G ATA TGC AGA GTC TGG TCG GCG GAA CGG TGG TCC GGC GTA AG-3'.

8. The manufacturer's procedure for microarray hybridization and washing using their supplied chambers to give the reader experience in utilizing them were maintained. It is also possible to use the Tecan HS hybridization instrument for these or any other glass slides.

Acknowledgments

This work is supported by Kuwait Foundation for the Advancement of Sciences (KFAS) grant no. 990707 and Shared Facility Grant Number GM/0101.

References

1. Platzer, P., Upender, M. B., Wilson, K., et al. (2002) Silence of chromosomal amplifications in colon cancer. *Cancer Res.* **62,** 1134–1138.
2. Al-Mulla, F., Al-Maghrebi, M., and Varadharaj, G. (2003) Expressive genomic hybridisation: gene expression profiling at the cytogenetic level. *Mol. Pathol.* **56,** 210–217.
3. Chung, Y. J., Jonkers, J., Kitson, H., et al. (2004) A whole-genome mouse BAC microarray with 1-Mb resolution for analysis of DNA copy number changes by array comparative genomic hybridization. *Genome Res.* **14,** 188–196.
4. Pollack, J. R., Perou, C. M., Alizadeh, A. A., et al. (1999) Genome-wide analysis of DNA copy-number changes using cDNA microarrays. *Nat. Genet.* **23,** 41–46.
5. Brennan, C., Zhang, Y., Leo, C., et al. (2004) High-resolution global profiling of genomic alterations with long oligonucleotide microarray. *Cancer Res.* **64,** 4744–4748.
6. Johnson, J. M., Edwards, S., Shoemaker, D., and Schadt, E. E. (2005) Dark matter in the genome: evidence of widespread transcription detected by microarray tiling experiments. *Trends Genet.* **21,** 93–102.
7. Karaman, M. W., Groshen, S., Lee, C. C., Pike, B. L., and Hacia, J. G. (2005) Comparisons of substitution, insertion and deletion probes for resequencing and mutational analysis using oligonucleotide microarrays. *Nucleic Acids Res.* **33,** E33.
8. Kimura, N., Nagasaka, T., Murakami, J., et al. (2005) Methylation profiles of genes utilizing newly developed CpG island methylation microarray on colorectal cancer patients. *Nucleic Acids Res.* **33,** E46.

9. Fehlbaum, P., Guihal, C., Bracco, L., and Cochet, O. (2005) A microarray configuration to quantify expression levels and relative abundance of splice variants. *Nucleic Acids Res.* **33,** E47.

10. Glokler, J. and Angenendt, P. (2003) Protein and antibody microarray technology. *J. Chromatogr. B Anal. Technol. Biomed. Life Sci.* **797,** 229–240.

11. Quinn, L. A., Moore, G. E., Morgan, R. T., and Woods, L. K. (1979) Cell lines from human colon carcinoma with unusual cell products, double min, and homogeneously staining regions. *Cancer Res.* **39,** 4914–4924.

12. Auer, H., Lyianarachchi, S., Newsom, D., Klisovic, M. I., Marcucci, G., and Kornacker, K. (2003) Chipping away at the chip bias: RNA degradation in microarray analysis. *Nat. Genet.* **35,** 292–293.

13. Carter, S. L., Eklund, A. C., Mecham, B. H., Kohane, I. S., and Szallasi, Z. (2005) Redefinition of Affymetrix probe sets by sequence overlap with cDNA microarray probes reduces cross-platform inconsistencies in cancer-associated gene expression measurements. *BMC Bioinformatics* **6,** 107.

14. Mecham, B. H., Klus, G. T., Strovel, J., et al. (2004) Sequence-matched probes produce increased cross-platform consistency and more reproducible biological results in microarray-based gene expression measurements. *Nucleic Acids Res.* **32,** E74.

7

Microarray Technology for Use in Molecular Epidemiology

Suzanne D. Vernon and Toni Whistler

Summary

Microarrays are a powerful laboratory tool for the simultaneous assessment of the activity of thousands genes. Remarkable advances in biological sample collection, preparation and automation of hybridization have enabled the application of microarray technology to large, population-based studies. Now, microarrays have the potential to serve as screening tools for the detection of altered gene expression activity that might contribute to diseases in human populations. Reproducible and reliable microarray results depend on multiple factors. In this chapter, biological sample parameters are introduced that should be considered for any microarray experiment. Then, the microarray technology that we have successfully applied to limited biological sample from all our molecular epidemiology studies is detailed. This reproducible and reliable approach for using microarrays should be applicable to any biological questions asked.

Key Words: Biological sample; epidemiology; microarray methods; microarray; molecular epidemiology; clinical specimen; gene expression; oligonucleotide array.

1. Introduction

Epidemiology is the distribution and determinants of diseases in human populations. Along with the development of powerful, high-throughput molecular technologies, the field of molecular epidemiology has evolved. As the name implies, molecular epidemiology focuses on the distribution and determinants of diseases at the molecular level. Now, studies of diseases in human populations can interrogate the contributions of genes, gene activity, proteins, and the environment.

Because population-based molecular epidemiology studies have the greatest potential to positively impact public health, the importance of the biological sample cannot be understated. The quality and standardization of the sample are as much a consequence of collection, handling, and storage as they are of isolation, extraction, and purification procedures (**Fig. 1**). The standardization of

From: *Methods in Molecular Biology, vol. 382: Microarrays: Second Edition: Volume 2*
Edited by: J. B. Rampal © Humana Press Inc., Totowa, NJ

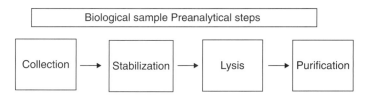

Fig. 1. The optimization and standardization of the preanalytic steps for the biological sample ensures reproducible and reliable results for any assay.

these first steps should be optimized to ensure representative and reproducible results. Each of these steps might vary depending on the sample type, what is analyzed (e.g., RNA, DNA, or protein) and the assay that is used and should be optimized prior to commencing any molecular epidemiology study.

The most logical and convenient biological sample for large, population-based molecular epidemiology studies is collected using noninvasive sampling methods and uniformly applied to a population. Examples of noninvasive samples commonly used in molecular epidemiology studies are peripheral blood, saliva, stool, urine, and exfoliated cells obtained by scraping or brushing. Peripheral blood is an attractive sample because it likely reflects ongoing systemic processes. Exfoliated cells, such as those collected for a Pap smear, might more closely reflect the disease but might have variable quality nucleic acids because of the dead and dying cells naturally present in this sample type *(1,2)*. Different collection medium with various storage and extraction procedures should be attempted to extract the best possible quality RNA (**Fig. 2**).

The yield of nucleic acids from biologic samples is another important consideration for microarray technology. For example, exfoliated cervical cytology samples typically yield a low concentration of RNA so extracting total nucleic acid (RNA and DNA) to improve RNA recovery might be necessary *(2)*. Our laboratory regularly uses microarray technology on peripheral blood samples and exfoliated cervical cytology. However, because exfoliated cervical cytology samples are more particular and are not as commonly used as peripheral blood for gene expression profiling, this chapter will focus on microarray technology as it applies to peripheral blood.

Peripheral blood serves as a representative sample of the systemic state allowing for evaluation and profiling of multiple pathological and physiological pathways *(3,4)*. It is often desirable to use peripheral blood cell samples in multiple platforms, for example, total RNA in microarrays and live cells in functional assays. Typical venous collection provides live peripheral blood cells that are an ideal source of material for different assays. The ideal collection system would be blood tube that collects and separates the cells while maximizing cell viability, such as the BD Vacutainer® CPT™ (BD, Franklin Lakes, NJ) cell preparation

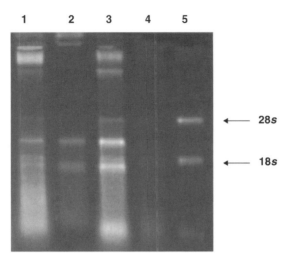

Fig. 2. RNA quality and quantity is variable depending on the sample source. **Lanes 1–3** have total RNA derived from exfoliated cervical cytology. **Lanes 4** and **5** have total RNA derived from peripheral blood cells.

tube. Because the cells are live following collection, steps should be taken to minimize the ex vivo cell stimulation to minimize alterations in gene expression *(5)*. If only RNA from peripheral blood sample is necessary, then integrated blood collection systems such as the Paxgene™ Blood RNA System (Qiagen Inc., Valencia, CA; PreAnalytix, http://www.preanalytix.com/RNA.asp) can be used. It should be noted that because this collection system is RNA specific, additional blood samples should be collected for other non-RNA assay types.

It is universally accepted that RNA quality and quantity are of paramount importance for assuring reliability of microarray results. Many different total RNA extraction methods are commercially available and each offers advantages. Given the different extraction chemistries involved, the single step phenol/guanidinium isothiocyanate methodology that is the basis to TRIzol® reagent (Invitrogen Corp., Calsbad, CA) and an RNA binding glass filter methodology for total RNA extraction was examined *(6)*. Both methods produced similar quality microarray results but the single step method resulted in slightly greater total RNA yields and less DNA contamination. Use of a single method total RNA extraction method is one way to reduce experimental variability. RNA quality and quantity can be simultaneously assessed by evaluating 28S:18S rRNA using denaturing agarose gel electrophoresis. With appropriate standards and imaging software, this method is quantitative (**Fig. 3**) and has the advantage of indicating the presence of contaminating DNA. Recently, microfluidic-based platforms for the analysis of RNA have been developed (Agilent 2100 Bioanalyzer, Agilent Technologies, Palo Alto, CA). These "lab-on-a-chip" technologies are excellent

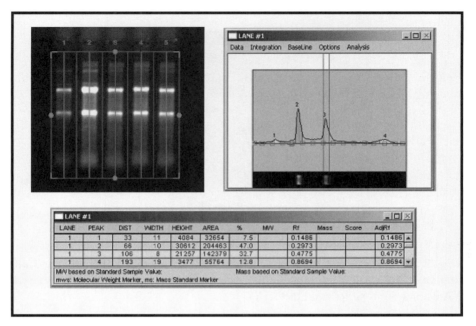

Fig. 3. Quantitation of ribosomal RNA (rRNA) bands by denaturing agarose gel electrophoresis using a control RNA of known concentration (**Lane 1**, 100 ng control RNA loaded). Using the Alpha Innotech gel imaging system (Alpha Innotech, San Leandro, CA), the area under the curve of the 28S and 18S rRNA bands are calculated, for the control RNA this area equates to 100 ng. The areas of each sample are then compared to the standard and the amount of RNA present in that sample determined. RNA concentration is determined by volume of RNA loaded onto the agarose gel. Intact total RNA run on a denaturing gel will have sharp, clear 28S and 18S rRNA bands. The 28S rRNA band should be approximately twice as intense as the 18S rRNA band, a good indication that the RNA is completely intact. From clinical samples we rarely get such high ratios and have found that rRNA with a 28S:18S = 1 gives high quality microarray data.

alternatives to gel electrophoresis because they are sensitive, reproducible, and fast. The disadvantage is that residual DNA is not detected, which is an important consideration if the total RNA is subjected to any enzymatic amplification steps.

Although population-based studies offer the luxury of collecting hundreds to thousands of biological samples, the limitation is the quantity (e.g., volume) and the yield of the total RNA necessary for microarray applications. It is found that the average total RNA from an 8-mL blood collection tube is 5 μg *(3,6)*. One of the more common microarray methods uses RNA from two sources (reference and sample), each labeled with a different fluorophore that is hybridized to a microarray. This dual intensity method calculates the ratio of fluorescence signals to determine the over- or underexpression of each gene. Although common, this method is not suited or feasible for population-based, high-throughput molecular epidemiology studies because 5–50 μg of total RNA is recommended for target synthesis (http://www.protocol-nline.org/prot/Genetics___Genomics/Microarray/).

Therefore, in addition to optimizing sample collection and RNA extraction procedures to maximize yield, an alternative target-labeling technique and hybridization protocol *(6)* has been evaluated, adopted, and optimized. A biotin-labeled cDNA was generated from 2 μg of total RNA from each sample that is used as the target. Each target is hybridized to one microarray, generating a single intensity signal. The hybridized biotinylated cDNA is detected using the highly sensitive GeniconRLS™ (Resonace Light Scattering) system (Invitrogen Corp.). This employs gold particles for detection and has been demonstrated to have greater sensitivity than fluorescent detection systems *(7)*. This higher sensitivity allows for the identification of low abundant mRNAs and is easily used with limited sample. Importantly, the signal is stable and does not photobleach, quench, or decay, thus hybridized microarray experiments can be archived indefinitely.

Automated hybridization of microarrays provides consistent hybridization conditions in between runs, low experimental variability and is high through-put. An automated platform such at the Ventana Discovery System™ (Ventana Medical Systems, Tucson, AZ) provides individual temperature-controlled slide pads, low hybridization volumes, and continuous mixing (**Fig. 4**). One disadvantage is the high cost; the hybridization stations are often expensive and not always feasible for a laboratory. In a comparison of manual with automated hybridization systems using the protocols described in this chapter, it is found that the reproducibility can be as good for manual hybridization as for auto-mated hybridization if five times higher probe concentration is used in the manual protocol compared with the automated hybridization (5 μg compared with 2 μg, respectively).

Our molecular epidemiological studies using microarrays show low experimental variability and have proven that microarrays are powerful tools for screening for perturbations in multiple biological pathways *(8,9)*. Great efforts have been given to optimize and standardize sample collection and microarray procedures (**Figs. 5** and **6**). All microarray experiments should adhere to the MIAME (Minimal Information About a Microarray Experiment) recommendations *(10)*. In this chapter, the microarray experimental design used in all our molecular epidemiology studies has been provided in detail.

2. Materials

For all protocols use RNAase-free reagents, water, and plasticwares to min-imize RNase contamination. Nonpowdered gloves are used for all manipula-tions and the best quality reagents are recommended for all manipulations.

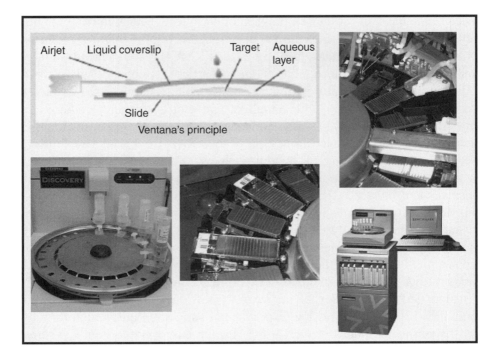

Fig. 4. Automated hybridization using the Ventana Discovery System.

2.1. RNA Extraction Using TRIzol

1. TRIzol Reagent (Invitrogen, cat. no. 15596-026).
2. Chloroform.
3. Isopropyl alcohol (2-propanol).
4. Ethanol (EtoH).

2.2. RNA Quantitation

1. An RNA sample of known concentration will work as the standard for the quantitation. Stratagene offers high quality, pure RNA from multiple sources.
2. UltraPure™ agarose (Invitrogen, cat. no. 15510-019).
3. Formaldehyde (Sigma, Aldrich, St. Louis, MO; cat. no. F8775).
4. Northern Max 10X MOPS (3-[N-morpholino] propane sulfonic acid) gel running buffer (Ambion Inc., Austin, TX cat. no. 8671).
5. RNA sample loading dye with ethidium bromide, (Sigma cat. no. R4268).

2.3. Direct Labeling of Target cDNA

1. Superscript II Reverse Transcriptase (Invitrogen, cat. no. 18064-014) supplied with 0.1 M DTT and 5X first strand buffer.
2. RNaseOut™ Recombinant Ribonuclease Inhibitor (Invitrogen, cat. no. 10777-019).
3. Random Primers, $pd(N)_6$ (Invitrogen, cat. no. 48190-011).
4. Oligo$(dT)_{12-18}$ (Invitrogen, cat. no. 18418-012).
5. Biotin-11-dUTP (Enzo Life Sciences, Farming dale, NY; cat. no. 42806).

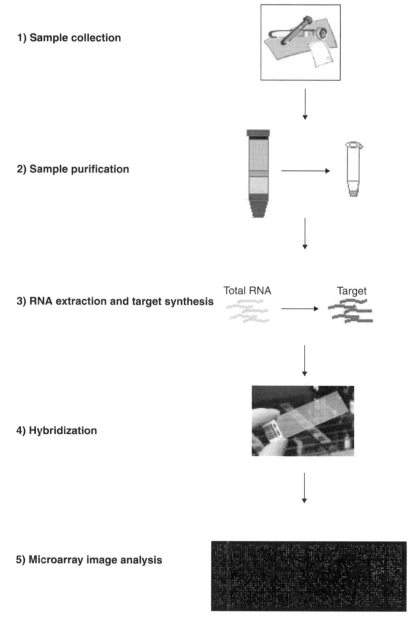

1) Sample collection

2) Sample purification

3) RNA extraction and target synthesis

Total RNA   Target

4) Hybridization

5) Microarray image analysis

Fig. 5. Optimized and standardized procedures for sample collection and microarray methods results in reproducible and reliable results from limited biological sample.

6. dNTP Mix: 10 mM each of dGTP, dATP, and dCTP, and 2 mM dTTP. Aliquot and store in single use volumes. (Invitrogen, cat. nos. 10218014, 10216018, 10217016, and 18255018, respectively).

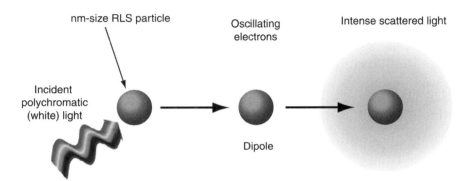

nm-size RLS particle

Oscillating electrons

Intense scattered light

Incident polychromatic (white) light

Dipole

Scattered light is a function of particle size, composition, and shape

Fig. 6. GeniconRLS technology overview. When illuminated with configured white light, spherical gold and silver RLS Particles of uniform dimension (between 40 and 120 nm diameter) generate intense, scattered light signals. The signal obtained from an individual RLS Particle is up to 10^4–10^6 times greater than that obtained with the most sensitive fluorescent molecules. This improved signal intensity translates at the assay level to an approximate 10-fold greater sensitivity with RLS as compared with fluorescence. (Reproduced with kind permission of Invitrogen Corp.)

7. QiaQuick PCR purification kit (Qiagen, cat. no. 28106).
8. 2 *M* NaOH.
9. 1 *M* Tris-HCl, pH 7.5.
10. Ethanol.

2.4. Hybridization

1. Blocking buffer (400 mL): 4X saline sodium citrate buffer (SSC), 0.5% sodium dodecyl sulfate (SDS), and 1% bovine serum albumin.
2. Lint-free paper towels.
3. Tissue-Tek® slide staining dishes (Sakura, Torrance, CA; cat. no. 4456).
4. A 24-slide capacity holder with a detachable handle for staining jars previously described (Sakura, cat. no. 4465A).
5. cDNA and oligonucleotide microarrays with compatible glass surface for hybridization and detection system (*see* **Note 1**).

2.4.1. Manual Hybridization

1. Water bath at 42°C.
2. Hybridization oven at 42°C.
3. Hybridization chamber (Rainin disposable pipet tip rack).
4. LifterSlips™ 25 × 60 mm² (Erie Scientific, Portsmouth, NH; cat. no. 25x60I-2-4789).
5. Hybridization cocktail: 50% formamide, 10X SSC, 0.2% SDS.
6. Wash buffer 1: 2X SSC, 0.1% SDS.

7. Wash buffer 2: 1X SSC.
8. Wash buffer 3: 0.1X SSC.
9. GeniconRLS Au particles and diluent (Invitrogen, *see* **Note 2**).

2.4.2. Automated Hybridization

1. Ventana Discovery System (Ventana Medical Systems Inc.).
2. ChipMap™ Kit: this contains optimized reagents for microarray applications on the Ventana Discovery System. It contains ChipPrep™ 1 reagent (Ventana Medical, ChipPrep™ 2 reagent Systems), ChipClean™ reagent, and ChipHybe™ reagent (Ventana Medical Systems, cat. no. 760-106).
3. Liquid cover slip (LCS) (Ventana Medical Systems, cat. no. 650-010).
4. RiboWash reagent (Ventana Medical Systems, cat. no. 760-105).
5. Reaction buffer (Ventana Medical Systems, cat. no. 950-300).
6. GeniconRLS Au particles and diluent (*see* **Note 2**).
7. Scanner: GSD-501™ (Invitrogen Corp.).
8. Pipet tips (Rainin Instrument, LCC, Woburn, MA).

2.5. Archiving

1. Archiving solution (Invitrogen, cat. no. 106225) (*see* **Note 2**).
2. 0.1% Tween-20.

3. Methods
3.1. RNA Extraction

1. Add 1 mL of TRIzol per 5×10^6 cells, gently resuspend the cells. Place the suspension in a microfuge tube and leave at room temperature for 5 min, insuring complete dissociation of protein complexes (*see* **Note 3**).
2. Following complete sample lysis add 200 µL chloroform.
3. Cap tightly and vortex for 15 s.
4. Centrifuge at 12,000g for 15 min at 4°C.
5. Transfer the upper aqueous phase into a microfuge tube. Be careful not to incorporate any of the interphase.
6. Discard the tube with remaining organic phenol–chloroform phase into a hazardous waste container.
7. Precipitate the RNA from the aqueous phase by adding 0.5 mL isopropanol and mixing.
8. Incubate samples at room temperature for 10 min.
9. Centrifuge at 12,000g for 10 min at 4°C. The RNA forms a gel-like pellet on the side and bottom of the tube.
10. Remove the supernatant and wash the pellet by vortexing with 1 mL of 75% ethanol.
11. Collect the pellet by centrifugation at 7500g for 5 min at 4°C.
12. After removing the supernatant, air-dry the pellet and resuspend in 20 µL RNase-free water.

3.2. Determining Quality and Quantity of Extracted RNA by Denaturing Agarose Gel Electrophoresis

These instructions assume the use of a 7×10-cm^2 agarose gel tray. They are easily adaptable to other formats.

3.2.1. Pouring the Agarose Gel (see **Note 4**)

1. Add 36 mL H_2O to 0.5 g agarose in 250-mL flask.
2. Microwave for 45 s until agarose is melted.
3. Place flask in a fume hood to cool down for 5 min.
4. Add 5 mL 10X MOPS buffer followed by 9 mL formaldehyde.
5. Mix and pour the gel into the tray (7×10 cm^2).
6. Apply comb and let the gel harden for 30 min.

3.2.2. Preparing Samples to Load Gel

1. Prepare samples by placing 1 µL specimen RNA (or 100 ng control RNA), 4 µL H_2O, and 5 µL loading dye into a tube.
2. Heat at 70°C for 10 min. This will denature the RNA allowing it to migrate to its true molecular weight.
3. Spin.

3.2.3. Load and Run

1. Pour 1X MOPS buffer into the gel apparatus to cover the solidified gel.
2. Place the gel tray in the apparatus so samples migrate to the positive cathode.
3. Remove the comb.
4. Add 10 µL of sample per well.
5. Run at a constant 90 V with a 400 mA setting, for 60 min. Two running dyes are evident (xylene cyanol and bromophenol blue), the lower should be two-thirds of the way down the gel.
6. Visualize and quantitate RNA.
7. Examine the gel with ultraviolet (UV) illumination (*see* **Note 5**).
8. Densitometric measurements of band intensities of the 18S and 28S ribosomal RNA bands (as measured by the area under the curve) are used to determine amount of specimen RNA loaded. Calculated from the area under the curve of the known standard RNA (*see* **Fig. 1** and **Note 6**).

3.3. Preparation of Target cDNA

1. Adjust 2 µg of specimen total RNA to a final volume of 8 µL with RNase-free water (*see* **Note 7**).
2. To each 8 µL of RNA mix, add the following (*see* **Note 8**):
   1 µL Stock random primers (3 µg).
   1 µL dNTP mix.
   1 µL Oligo(dT) primers (0.5 µg).
   1 µL Biotin-11-dUTP (1 mM).

3. Incubate tube at 65°C for 5 min, spin down to remove condensation.
4. Add the following to the tube (*see* **Note 8**):
   4 μL 5X First strand buffer.
   2 μL 0.1 *M* DTT.
   1 μL RNaseOUT (40 U/μL).
5. Mix gently and collect the contents of the tube by brief centrifugation.
6. Incubate at 42°C for 2 min.
7. Add 1 μL of SuperScriptII (200 U/μL) to each tube.
8. Incubate at 42°C for 60 min. After cDNA synthesis is complete, immediately perform the following hydrolysis reaction to degrade the original RNA.
9. Add 2 μL 2 *M* NaOH to each reaction tube.
10. Incubate the tube at 70°C for 15 min.
11. Add 5 μL 1 *M* Tris-HCl, pH 7.5 to neutralize the pH and mix gently.
12. Cleanup reaction products using the QiaQuick PCR Purification kit. Add 135 μL buffer PB (Binding buffer, supplied with kit) to each sample (27 μL), and mix well (*see* **Note 9**).
13. Apply the whole volume of sample (162 μL) to the QiaQuick column preinserted into a collection tube, and centrifuge for 1 min at 3500*g*.
14. Discard the flowthrough and place the column back in the same collection tube.
15. Wash the column with 750 μL buffer PE (Wash buffer supplied with kit) and centrifuge for 1 min at 3500*g*.
16. Discard flowthrough, and place the column back in the same tube, and centrifuge for 1 min at 16,000*g*.
17. Place the column in a clean 1.5-mL microfuge tube.
18. To elute the DNA, add 30 μL buffer EB (Elution buffer supplied with kit) to the column, let stand 1 min and centrifuge for 1 min at 16,000*g*.
19. Add a further 30 μL of buffer EB to the column, incubate at room temperature for 1 min, and then centrifuge at 16,000*g*.
20. The combined eluate contains the purified labeled cDNA. This can be stored at −20°C until used for hybridization. Avoid freeze/thawing.

3.4. Hybridization

3.4.1. Manual Hybridization

3.4.1.1. BLOCKING OF ARRAYS

1. Block the array slides in preheated blocking buffer at 42°C for 45 min (*see* **Note 10**).
2. Rinse the arrays by dipping them in water five times.
3. Dry the arrays with filtered compressed air.

3.4.1.2. PROBE PREPARATION

1. For manual hybridizations the probe volume is reduced by vacuum centrifugation to 40 μL and an equal volume of hybridization cocktail added for a final volume of 80 μL.
2. Vortex the probe and heat at 95°C for 5 min to denature the probe. Centrifuge to collect condensation.

3. Place the blocked, dry array on a flat surface.
4. Clean a LifterSlip™ with 70% EtOH and lint-free wipes. Place the LifterSlip on the array covering the spotted area, such that the Teflon strips of the LifterSlip are on the long axis of the array, touching the glass (*see* **Note 11**).
5. Add the probe by placing the pipet tip along the open edge of the slip and slowly pipet the probe out of the tip, allowing capillary action to draw the probe mix under the slip. Be careful not to let air bubbles form.

3.4.1.3. HYBRIDIZATION

1. Prepare a humid hybridization chamber. Disposable pipet tip boxes that have a removable "shelf" that carries the pipet tips are used. This is removed and several layers of Whatmann filter paper (cut to fit) are saturated in 2X SSC and placed at the bottom of the box. This ensures that saturated filter paper does not come in contact with the array in such a way that the target would be absorbed by the filter paper. The shelf is then put in place over the filters (*see* **Note 12**).
2. Place the in a horizontal and level position and then put the humid chamber into a hybridization oven at 42°C and incubate overnight (16 h). Humidity is maintained in the hybridization oven with a container of nuclease-free water.

3.4.1.4. POSTHYBRIDIZATION WASHES

1. All wash buffers are prewarmed to 42°C in the staining jars.
2. Remove the hybridization chamber and dry the outside to prevent water dripping on the array. Keep the chamber setup for use in the detection of target on the array.
3. Place the array with LifterSlip into a jar of prewarmed wash 1. Angle the array so that the LifterSlip can fall to the bottom of the jar under its own weight. Place the array into the wash tank of wash 1 buffer at 42°C. (*see* **Note 13**).
4. After all slides are in the tank wash the arrays for 5 min. Agitate the arrays by lifting the slide carrier up and down.
5. Repeat this wash step in a second container of prewarmed wash buffer 1 at 42°C for 5 min.
6. Transfer the slide carrier to the prewarmed tank of wash buffer 2 at 42°C. Wash for 5 min with agitation.
7. Repeat **step 6** once.
8. The final stringency wash is in prewarmed wash buffer 3 at 42°C for 5 min.

3.4.1.5. DETECTION OF BIOTINYLATED TARGET

1. After the stringency washes slides are again blocked in blocking buffer at room temperature for 10 min.
2. To the hybridization chamber add a lint-free paper towel that has been prewetted with nuclease-free water to the rack on which the slides are placed. This is now used as a hydration chamber for the detection step.
3. Immediately before use prepare the GeniconRLS particles by mixing the suspension

with gentle vortexing.

4. In a microfuge tube add 60 μL of GeniconRLS Au particles to 30 μL of GeniconRLS particle diluent for each array on which a detection reaction needs to be performed. Mix gently to ensure dispersion.
5. Work on one slide at a time: remove the slide from the block solution, and drain the slide on a lint-free paper towel by dabbing the bottom edge of the slide on the towel.
6. Place the slide on the wetted paper towel in the hybridization chamber—ensuring that the slide lies flat on the paper. The back of the slide needs to remain on the damp surface to prevent residue forming and drying.
7. Add 90 μL of the GeniconRLS particles to the bottom end of the slide.
8. Lower one edge of a LifterSlip onto the slide over the RLS particles and gently settle the other end of the LifterSlip onto the array, allowing the particles to wick over the full area.
9. Repeat **steps 6–8** until all arrays are set.
10. Seal the hybridization chambers and incubate at room temperature for 1 h.
11. Rinse the slide in wash buffer 3 and dry.

3.4.2. Automated Hybridization

The Ventana Discovery™ system is used for the automated hybridizations. This system offers the advantages of automated liquid-handling, individually controlled temperature pads for each microarray slide, and continuous mixing of hybridization solution under an oil-based LCS. This maximizes hybridization kinetics and allows for posthybridization stringency washes and target detection to be performed on the system.

3.4.2.1. METHODOLOGY FOLLOWS AS CLOSELY AS POSSIBLE THE CONDITIONS FOR MANUAL HYBRIDIZATION

1. The microarrays are blocked as for the manual hybridization (*see* **steps 1–3** in **Subheading 3.4.1.1.**).
2. A bar-coded label is added to each array and the slides loaded onto the Ventana Discovery.
3. The ChipPrep™1, ChipPrep™2, and ChipClean™ dispensers are loaded onto the reagent carousel.
4. The run is started.

3.4.2.2. AUTOMATED STEPS

1. Rinse slide with SSC solution
2. Apply ChipPrep™1 and LCS. Incubate at 70°C for 10 min.
3. Rinse slides.
4. Apply ChipPrep™ 2 and LCS and incubate for 30 min.
5. Rinse slides.
6. Apply cover slip.

3.4.2.3. MANUAL STEPS

1. Add 200 μL of ChipHybe™ to the 60 μL of biotinylated target, denature at 100°C for 2 min, spin to collect and keep at room temperature until applied to the array.
2. Apply the targets to the appropriately labeled arrays, by pipetting just under the LCS. Do not allow the pipet tip to touch the surface of the array.

3.4.2.4. AUTOMATED STEPS

1. Warm the slide to 70°C and incubate for 6 min (denaturation step).
2. Warm the slide to 42°C and incubate for 8 h.
3. Stringency wash 1: 2X SSC at 42°C for 6 min.
4. Stringency wash 2: 1X SSC at 42°C for 6 min.
5. Stringency wash 3: 0.5X SSC at 42°C for 10 min.
6. Apply LCS.
7. Apply GeniconRLS Au particles and incubate for 30 min.
8. Rinse slide in reaction buffer.
9. Apply ChipClean™ and incubate for 2 min at 37°C.
10. Rinse slide in reaction buffer.

3.4.2.5. MANUAL STEPS

1. Lift the microarray slide carefully from the platform and wipe off residual oil from the back of the slide with lint-free paper. Rinse the slide with a squash-bottle filled with reaction buffer.
2. Place slides into the slide holder, bar-code down and wash with vigorous plunging in three changes of reaction buffer, each of about 1 min.
3. Rinse slides in two changes of 1X SSC, about 20 dips per change.
4. Dip slides in 0.1X SSC for 20 plunges.
5. Dry arrays with compressed air.

3.5. Archiving (see Note 14)

1. Aliquot 45 mL archiving solution into a 50-mL tube and let stand so air bubbles will dissipate. Bubbles might cause unevenness in the optical coating.
2. Dip each array into archiving solution up to the bar code or label.
3. Withdraw the array slowly. Ensure no air bubbles are trapped on the surface. If bubbles are evident redip. A smooth even coating is desired.
4. Blot the bottom edge of the array on a lint free wipe. Repeat this blotting several times.
5. Lean the array against a clean surface barcode edge up, ensuring that the wet surface does not come in contact with anything. Use a dust-free location—this step is performed in a chemical cabinet. Allow to dry for 60 min.
6. Store in a well-sealed slide box.
7. To remove the archiving coating from the glass arrays. Immerse the array in a staining jar containing deionized water for 10 min. If the arrays are particularly dirty place them in 0.1% Tween-20 and agitate for 5 min. Briefly rinse in water and repeat the archiving procedure.

3.6. Scanning RLS™ Microarrays

1. The microarray images are captured using a GSD-501 scanner.
2. The slides are inserted face down with the slide label to the front.
3. In the image window mark the slide holders to be scanned and enter in file names.
4. In the preview window, select the particle type as gold and set exposure time in seconds (0.1–0.2 s is generally appropriate).
5. To determine the positioning of the array in the slide holder, set the area to be scanned by holding the mouse button and highlighting the space required.
6. Start the preview scan.
7. Open the "Brightness and Contrast" window select "mark saturated pixels." Count the number of features that are saturated and set a final scan time that will give 1–2% of features saturation. This effectively scales the arrays from 0 to 65,000 pixels.
8. In the layout window, align the blue box to surround the area of the array for the final scan.
9. Start imaging.
10. Remove arrays from the scanner and place in a dirt-free container.

4. Notes

1. Slide surfaces that are compatible with the Ventana Discovery System include: aldehyde, amine, GAPS2, MWG epoxy, and UltraGAPS. Those that were found to be incompatible are from Agilent and the Erie epoxy (the Erie aldehyde and aminosilane are compatible).
2. These reagents are available as part of the GeniconRLS One-color Kit from Invitrogen.
3. An insufficient amount of TRIzol will result in contamination of the isolated RNA with DNA.
4. **Caution:** Formaldehyde is a suspected carcinogen and must be used and disposed of in accordance with federal, state, and local regulations. Always use formaldehyde in a fume hood.
5. Wear UV-protective safety glasses or a full safety mask to prevent UV damage to the face and skin.
6. A drawback to using denaturing agarose gels to assess integrity RNA is the amount of RNA required for visualization. For this protocol, at least 50 ng of RNA must be loaded onto the gel in order to be visualized with ethidium bromide. Some RNA preparations, such as those from exfoliated cervical cells, needle biopsies or from laser capture microdissected samples result in very low yields. In these cases, it might be impossible to spare 200 ng of RNA to assess integrity before proceeding with the expression profiling application. Currently we use an alternative to this methodology, the Agilent Technologies 2100, a microfluidic system that provides detailed information about the condition of RNA samples using as little as 1 µL of 10 ng/µL per analysis.
7. RNA quantified by agarose gel electrophoresis or Agilent Bioanalyzer.
8. It is best to prepare master mixes when several reagents are to be added to each tube. The larger pipetting volumes increase accuracy and the reduced number of

steps saves time. When scaling calculations are made, generally an extra sample is added to the calculation to ensure adequate volumes are prepared.

9. Proceed with the purification of the labeled cDNA by removing any unincorporated nucleotides with the QiaQuick polymerase chain reaction purification kit. Before starting ensure ethanol has been added to the appropriate buffers in the kit.

10. It is found that blocking of glass arrays helped reduce the nonspecific background sometimes encountered with the RLS detection system. To ensure the blocking solution is clear and free from any precipitate filter it through a 0.40-µm membrane before use. Make sure that all slides are adequately covered by the blocking solution.

11. Use of the LifterSlip greatly improves data quality by preventing nonuniform hybridization. These cover slips have a little edge of Teflon along two opposing outer edges (available from Erie Scientific). This creates a platform, which allows even distribution of the hybridization solution across the array.

12. If more than one array is being hybridized, use several chambers and do not overcrowd arrays. It is important to make sure nothing touches each array, so that the target does not wick out from the LifterSlip. Process one chamber at a time before and after hybridization.

13. Do not include more than eight arrays per wash station (the plastic holder can carry 24 slides). This insures proper washing.

14. The archiving process places an optical coating on the slides reducing the signal to background ratio during scanning. It also allows for prolonged storage of the arrays. The optical coating can be replaced if slides become scratched or dirty.

References

1. Dimulescu, I., Unger, E. R., Lee, D. R., Reeves, W. C., and Vernon, S. D. (1998) Characterization of RNA in cytologic samples preserved in a methanol-based collection solution. *Mol. Diagn.* **3,** 67–71.

2. Habis, A. H., Vernon, S. D., Lee, D. R., Verma, M., and Unger, E. R. (2004) Molecular quality of exfoliated cervical cells: implications for molecular epidemiology and biomarker discovery. *Cancer Epidemiol. Biomarkers Prev.* **13,** 492–496.

3. Vernon, S. D., Unger, E. R., Dimulescu, I., Rajeevan, M., and Reeves, W. C. (2002) Utility of the blood for gene expression profiling and biomarker discovery in chronic fatigue syndrome. *Dis. Markers* **18,** 193–199.

4. Whitney, A. R., Diehn, M., Popper, S. J., et al. (2003) Individuality and variation in gene expression patterns in human blood. *Proc. Natl. Acad. Sci. USA* **100,** 1896–1901.

5. Baechler, E. C., Batliwalla, F. M., Karypis, G., et al. (2004) Expression levels for many genes in human peripheral blood cells are highly sensitive to ex vivo incubation. *Genes Immun.* **5,** 347–353.

6. Ojaniemi, H., Evengard, B., Lee, D., Unger, E., and Vernon, S. (2003) Impact of RNA extraction from limited samples on microarray results. *Biotechniques* **35,** 968–973.

7. Bao, P., Frutos, A. G., Greef, C., et al. (2002) High-sensitivity detection of DNA hybridization on microarrays using resonance light scattering. *Anal. Chem.* **74,** 1792–1797.

8. Whistler, T., Jones, J. F., Unger, E. R., and Vernon, S. D. (2005) Exercise responsive genes measured in peripheral blood of women with chronic fatigue syndrome and matched control subjects. *BMC Physiol.* **5,** 5.

9. Whistler, T., Unger, E. R., Nisenbaum, R., and Vernon, S. D. (2003) Integration of gene expression, clinical, and epidemiologic data to characterize chronic fatigue syndrome. *J. Transl. Med.* **1,** 10.

10. Brazma, A., Hingamp, P., Quackenbush, J., et al. (2001) Minimum information about a microarray experiment (MIAME)-toward standards for microarray data. *Nat. Genet.* **29,** 365–371.

8

Utilization of Microarray Platforms in Clinical Practice

An Insight on the Preparation and Amplification of Nucleic Acids From Frozen and Fixed Tissues

Fahd Al-Mulla

Summary

The last decade has witnessed an impressive upsurge in the utilization of microarray platforms for biomedical research. However, the application of this emerging technology in medical practice lagged behind. This lag is understandable because there are specific issues pertaining to the utilization of clinical samples, which has to be rigorously addressed and overcome before microarrays enter mainstream medical practice. Such issues include cost, ethics, the complexity and heterogeneity of human tissue architecture, and their corresponding diseases, the type of tissues to be used, nucleic acids amplification, and experimental variability. As microarrays enter, albeit cautiously, the frontline of clinical practice, investigators and clinicians require to set up protocols that address these issues. This chapter decribes the methods used for nucleic acids preparation from frozen and formalin-fixed paraffin-embedded human tissues using macro-and microdissection and show their suitability for use in microarray experiments.

Key Words: Formalin-fixed paraffin-embedded tissue; microarrays; microdissection; nucleic acids extraction; amplification; frozen tissue.

1. Introduction

Microarrays are miniaturized platforms made up of nucleic acid fragments that represent few thousand genes to whole genomes immobilized methodically on a solid surface *(1)*. The immobilized nucleic acid fragments can capture complementary sequences in the test or control samples by specific hybridization and allow simultaneous analysis of thousands of genes or their products in parallel. Many human diseases are complex entities and traditionally they were, and still are, studied by the analysis of a single or few "genes" or their products. Although such an approach has improved our understanding of the role specific-genes play

From: *Methods in Molecular Biology, vol. 382: Microarrays: Second Edition: Volume 2*
Edited by: J. B. Rampal © Humana Press Inc., Totowa, NJ

in human diseases, their complexity and compounding factors have hampered it. Therefore, microarrays represent a novel technology that probably can match the complexities of human diseases and offers a medium to view the genome, transcriptome and proteome, and their interactions in a single experiment.

Generally speaking, microarrays have been extensively used in biological research but their clinical applications have lagged behind. However, recently significant achievements in clinical research have been witnessed at the diagnostic, prognostic, and therapeutic levels brought about by the use of microarrays *(2–4)*.

With the ever increasing microarray platforms types and the availability of different nucleic acid amplification and labeling procedures, performing microarray-based analysis on clinical samples is becoming a daunting task for clinicians including pathologists *(5)*. The modern clinicians, who are usually part of a multidisciplinary team involved in the utilization and application of microarray platforms for clinical use, should be aware of the pitfalls and limitations of microarrays *(6–9)*. Questions that should be addressed by the team before embarking on microarray experiments include: what does the microarray intent to measure? Which microarray platform would serve the experimental objectives better? Could the widely available formalin-fixed paraffin-embedded tissue (FFPE) be used instead of frozen tissues? Was tissue heterogeneity accounted for? Is tissue microdissection necessary? Could the clinical samples be pooled? Are clinical samples limited and could whole genome amplification be used to circumvent the shortage? This chapter describes the use of microarray platforms in clinical settings with special emphasis on extraction of nucleic acids from processed tissues.

2. Materials

2.1. DNA Extraction and Purification From Frozen and FFPE Clinical Samples

1. Pestle mortar.
2. Fresh xylene.
3. Puregene® DNA Isolation Kit (Gentra Systems, Inc., Minneapolis, MN; cat. no. D-5500): cell lysis solution, RNase A solution, protein precipitation solution, and DNA hydration solution. Proteinase K (20 mg/mL, Gibco/BRL, Scotland, UK) in 10 mM Tris-HCl, pH 7.5; 20 mM calcium chloride, and 50% glycerol as stabilizers, store at 4°C.
4. Isopropanol.
5. Absolute ethanol.
6. Tris-ethylenediamine tetraacetic acid (TE) buffer: 10 mM Tris-HCl, pH 7.0, 1 mM ethylenediamineteraacetic acid (EDTA), pH 8.0.
7. PicoPure DNA extraction kit (Arcturus, Mountainview, CA): lyophilized Proteinase K and reconstitution buffer.

8. Nuclease-free water (Ambion, Austin, TX).
9. Microcentrifuge 5415 C (Eppendorf, Germany).
10. GeneAmp® PCR System 9700 (Applied Biosystems, Framingham, MA).

2.2. DNA Amplification

1. Restriction enzyme N1aIII (10 U/µL, New England Biolabs, Beverly, MA).
2. T4 DNA ligase (40 U/µL), 1X T4 DNA ligase reaction buffer (50 mM Tris-HCl, pH 7.5, 10 mM MgCl$_2$, 10 mM dithiothreitol [DTT], 1 mM ATP, 25 µg/mL bovine serum albumin [BSA] [New England Biolabs].)
3. Composite linkers LN1 and LN2 (2.8 µg/µL stock of each) in 5′–3′ orientation:
   LN1: AAC TGT GCT ATC CGA GGG AAA GGA CAT G
   LN2: AAC TGT GCT ATC CGA GGG AAA GAG CAT G
4. Common oligonucleotide P1, and specific primers P2a, and P2b (5′–3′):
   P1: AGG CAA CTG TGC TAT CCG AGG GAA
   P2a: AAC TGT GCT ATC CGA GGG AAA GGA
   P2b: AAC TGT GCT ATC CGA GGG AAA GAG
5. BD advantage 2 polymerase mix (BD Bioscience, San Jose, CA): 50X advantage 2 polymerase mix (50% glycerol, 15 mM Tris-HCl, pH 8.0, 75 mM KCl, 0.05 mM EDTA), 10X advantage 2 PCR buffer (400 mM Tricine-KOH, pH 8.7, 150 mM KOAc, 35 mM Mg[OAc]$_2$, 37.5 µg/mL BSA, 0.05% Tween-20, 0.05% Nonidet-P40), and 50X dNTP mix (10 mM dATP, dCTP, dGTP, and dTTP).
6. QIAquick PCR purification kit (Qiagen, Boston, MA): PB buffer, PE buffer, EB buffer (10 mM Tris-HCl, pH 8.5), and QIAquick columns with collection tubes.
7. PicoGreen assay (Molecular Probes, Eugene, OR): PicoGreen reagent, 20X TE buffer DNase-free, and Lambda dsDNA standard.
8. Titanium *Taq* PCR kit (BD Biosciences): 50X Titanium *Taq* DNA polymerase (50% glycerol, 20 mM Tris-HCl, pH 8.0, 100 mM KCl, 0.1 mM EDTA, pH 8.0, 0.25% Tween-20, 0.25% Nonidet-P40), 10X Titanium *Taq* PCR buffer (400 mM Tricine-KOH, pH 8.7, 160 mM KCl, 35 mM MgCl$_2$, 37.5 µg/mL BSA), and 50X dNTP mix (10 mM dATP, dCTP, dGTP, and dTTP).

2.3. DNA Labeling and Hybridization

1. BioPrime™ array CGH genomic labeling system (Invitrogen, Carlsbad, CA): Exo-Klenow fragment (40 U/µL), 2.5X random primer solution (octamers), 10X dUTP nucleotide mix (1.2 mM dATP, dGTP, and dCTP and 0.6 mM dTTP in 10 mM Tris-HCl, pH 8.0, 1 mM EDTA), 10X dCTP nucleotide mix (1.2 mM dATP, dGTP, and dTTP and 0.6 mM dCTP in 10 mM Tris-HCl, pH 8.0, 1 mM EDTA), and stop buffer (0.5 M EDTA, pH 8.0). All regents stored at −20°C.
2. Cy3- and Cy5-2′-deoxyuridine 5′-triphosphate or Cy3- and Cy5-2′-deoxycytidine 5′-triphosphate are aliquoted and stored at −20°C (Amersham Biosciences, Little Chalfont, UK).
3. BioPrime array CGH genomic labeling system purification module (Invitrogen): purification buffer A, and purification buffer B. Working purification buffer B solution is prepared by adding 24 mL of 100% ethanol to stock buffer B. Purification buffers are stored at room temperature.

2.4. RNA Extraction and Purification

1. RNase zap (Ambion).
2. Nuclease-free water (Ambion).
3. Pestle mortar.
4. Paradise reagent system (Arcturus, Carlsbad, CA) Staining components: 100, 95, and 75% ethanol, xylene, staining solution, and staining jars. RNA extraction/isolation components: conditioning buffer, absolute ethanol, wash buffers 1 and 2, elution buffer, Pro K mix, binding buffer, MiraCol (Arcturus) RNA purification columns with collection tubes, reconstitution buffer, DNase mix, and DNase buffer. Only the last three items are stored at −20°C the rest of the kit is kept at room temperature.
5. PicoPure RNA isolation kit (Arcturus): conditioning buffer, extraction buffer, wash buffer 1, wash buffer 2, elution buffer, and RNA purification columns with collection tubes.
6. Microcentrifuge 5415 C (Eppendorf).
7. GeneAmp PCR System 9700 (Applied Biosystems).

2.5. RNA Amplification

1. SenseAmp Plus™ RNA amplification kit (Genisphere, Hatfield, PA) Box 1: SenseAmp dT24 RT primer (50 ng/µL), SenseAmp random 9-mer RT primer (250 ng/µL), dNTP mix (10 mM each dATP, dCTP, dGTP, and dTTP), Superase-In™ (Ambion) RNase inhibitor, 10 mM dTTP, 10X reaction buffer, terminal deoxynucleotidyl transferase, SenseAmp T7 template oligo, Klenow enzyme, T7 nucleotide mix (ATP, GTP, CTP, and UTP), 10X T7 reaction buffer, and T7 enzyme mix. Box 2: ATP mix, 5X PAP buffer, PAP enzyme, and 25 mM MnCl$_2$. All kit components are stored at −20°C.
2. 0.1 M DTT.
3. MinElute® PCR Purification Kit (Qiagen, cat. no. 28006): PB buffer, PE buffer, EB buffer, MinElute buffer, and MinElute columns in collection tubes.
4. RNeasy® MinElute Kit (Qiagen, cat. no. 74204): RLT buffer, RPE buffer, RNase-free water, RNeasy MinElute spin columns in collection tubes, and 1.5- and 2-mL collection tubes.
5. SuperScript II (Invitrogen) or other reverse transcriptase enzyme.
6. 0.5 M EDTA.
7. 1 mM Tris-HCl, pH 8.0.
8. Nuclease-free water (Ambion, cat. no. 9934)
9. Absolute ethanol (Riedel-deHaën, Seelze, Germany). Aliquots of diluted 80% ethanol using molecular grade water and absolute ethanol are stored at −20°C.
10. TE buffer: 10 mM Tris-HCl, pH 7.0, 1 mM EDTA, pH 8.0.

2.6. Microarray Hybridization and Labeling

1. 3DNA Expression Array 900 Detection kit (Genisphere): Cy3/Alexa Fluor 546 or Cy5/Alexa Fluor 6473DNA array 900 capture reagent, RT primer Cy3/Alexa Fluor 546 or Cy5/Alexa Fluor 647 (1 pmole/µL), dNTP mix (10 mM of each), Superase-In RNase inhibitor, 2X enhanced cDNA hybridization buffer, 2X sodium dodecyl sulfate (SDS)-based hybridization buffer, 2X formamide-based hybridization buffer,

antifade reagent, LNA dT blocker, and RT primer Cy3/Alexa Fluor 546 or Cy5/Alexa Fluor 647 (5 pmole/μL).
2. Microarrays: commercial or in-house synthesized oligonucleotide, cDNA or BAC clones based microarrays.
3. cDNA synthesis stop solution: 1 M NaOH and 100 mM EDTA.
4. 1X TE buffer.
5. 2X SSC, 0.2% SDS buffer (wash buffer 1); 2X SSC buffer (wash buffer 2); and 0.2X SSC buffer (wash buffer 3).
6. SuperScript II (Invitrogen): 5X SuperScript II first strand buffer, 0.1 mM DTT, and SuperScript II enzyme (200 U/μL).
7. Glass cover slips or lifter slips.

3. Methods

3.1. Preparation of Frozen and FFPE Clinical Samples for Nucleic Acids Extraction

1. Fresh clinical samples must be immediately frozen by immersion in liquid nitrogen and then stored in liquid nitrogen or in a −80°C freezer (*see* **Note 1**). An alternative method is to embed the tissue in an optimal cutting temperature compound (OCT) in a cryomold cassette at −80°C. Before sectioning, the samples are taken out on dry ice and a precooled sectioning cryostat is prepared by covering it with OCT to maintain low temperature throughout sectioning. A sample is then sectioned and stained with hematoxylin and eosin to visualize the cellular composition and decide on whether the tissue is highly representative of the disease and is suitable for bulk extraction, macrodissection, or cellular microdissection. For bulk frozen tissue extraction the tissue is ground using a pestle mortar with constant addition of liquid nitrogen to minimize heat-caused degradation. Homogenized tissue is transferred into two DNase/RNase-free tubes placed on ice and containing solutions of tissue degradation properties in preparation for nucleic acid isolation (depending whether DNA or RNA is required).
2. Paraffin-embedded tissue slides required for nucleic acid extraction are selected and grouped in a separate slide box. It is generally recommended to select block not older than 5 yr and to ensure that the formalin used in fixation was buffered and samples were not incubated in it for prolonged periods *(10)*. Up to four slides are processed at each round of deparaffinization, staining, and dehydration. It is important to ensure that all consumables, glassware, and surfaces are nuclease-free. At first the slides are heated in a 60°C oven for 2 min. Immerse the slides in xylene for 2 min and transfer them into a second xylene jar for another 2 min. Remove the slides and place them in a jar containing 100% ethanol for 2 min while gently inverting the jar. Transfer the slides into 95% ethanol for 1 min, followed by 75% ethanol for 1 min. Finally immerse the slides in nuclease-free water for 30 s. Using a nuclease-free pipet tip dispense 100 μL of Paradise staining solution covering the entire section area (*see* **Note 2**). Stain for 35 s at room temperature, and tap off excess stain. Reimmerse the slides in 75 and 95% ethanol, respectively; for 30 s each. Transfer the slides to

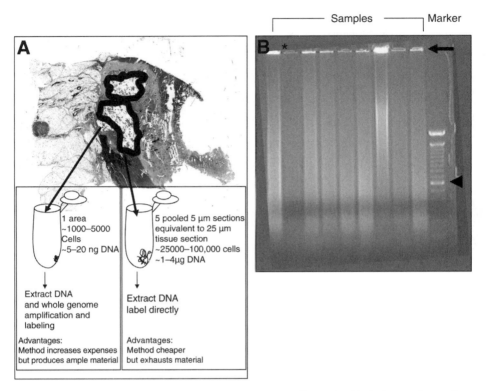

Fig. 1. DNA extraction from formalin-fixed paraffin-embedded tissue shown here stained with toluidine blue (**A**). The colorectal cancer cells microdissected from a 5-µm section are marked by thick black lines. To increase the number of cells to 100,000 or more, five similar sections were stained, microdissected, and tissues pooled in one tube. This method of collection ensures enough DNA is present to circumvent the need for amplification. Alternatively, smaller areas are microdissected, DNA extracted and amplified for use on microarrays. (**B**) One percent agarose gel electrophoresis of DNA extracted after microdissection from several sections without amplification. Note the smears, and the high-molecular weight DNA at the well (white arrow) indicating successful extraction. The star marks a sample that failed the extraction. The most important reasons for failure are no adequate removal of paraffin and limited or failed protein digestion. The arrow head on the size marker well indicates 600-bp size band.

100% ethanol for 1 min and then to xylene for 5 min, or up to 2 h (try to minimize this as much as possible. Do not exceed 2 h). Remove the slides from xylene and place them on Kimwipe in a laminar flow hood to dry. Make sure that macro- or microdissection is carried out within 2 h. Nucleic acid extraction is performed directly after acquisition of cells either by tissue scaring with a sterile, nuclease-free 12-gauge needle or laser capture microdissection for optimum nucleic acid recovery (*see* **Note 3** and **Fig. 1A**).

3.2. DNA Extraction and Purification

1. Frozen and FFPE scrapes undergo the same DNA extraction procedure, as well as bulk fresh tissue homogenate. It is recommended that Puregene kit is used for these tissues. Ground tissues or tissue scrapes (0.5–2 mg) are transferred into an O-ring tubes containing 100 μL of Puregene cell lysis solution (volume is scaled up according to mass of tissue). Homogenize tissue lysate further using a small pestle and place on ice. Incubate the tissue lysate at 65°C for 60 min. Adding 0.5 μL of Proteinase K solution followed by 25 tube inversions and incubation at 55°C for 3 h or up to 3 d in a rotating oven to ensure complete tissue digestion (incomplete digestion of proteins is one of the most important reasons for DNA extraction failures). It is recommended to add fresh Proteinase K (5 μL of 20 mg/mL) after 8 h of incubation elapse. This step can be repeated several times. It is important to ensure complete lysis of the tissues. Once this is achieved deactivate Proteinase K by heating at 95°C for 10 min. RNA is then digested by mixing 0.5 μL of RNase A solution followed by inverting the tube 25 times and then incubated at 37°C for 45 min.

2. DNA isolation is achieved by adding 33 μL of protein precipitation solution, vortex tube vigorously for 20 s and incubate on ice for 5 min. Centrifuge lysate at 16,000g for 3 min and make sure that a protein pellet is visible. Aspirate supernatant containing DNA to a tube containing 100 μL of isopropanol. When dealing with small tissue mass it is advisable to add 0.5 μL of glycogen to aid in DNA precipitation. Invert the tube 50 times and centrifuge at 16,000g for 5 min. Decant supernatant and wash the pellet with 100 μL of 70% ethanol, tap the pellet several times and centrifuge at 16,000g for 1 min. Gently pour off supernatant and air-dry the pellet over a clean filter paper or by vacuum centrifugation. Rehydrate pellet with adequate volume of DNA hydration solution according to expected DNA yield. Incubate DNA at 65°C for 1 h or overnight at room temperature for complete rehydration. Store DNA at 4°C for subsequent use or at −20°C for long-term storage. Quantitate DNA using standard spectrometer assuming that DNA optical density of 1 at 260 nm is equivalent to 50 μg/μL. The purity of DNA is calculated by 260/280 nm ratio. Ratio close to 2 is advisable. However, for FFPE tissues ratio of 1.7 or above can still be used for labeling. Run 3 μL of the DNA on 1% agarose gel stained with ethidium bromide (*see* **Note 4**).

3. Laser capture microdissectuion of cells from FFPE and frozen sections require alternative DNA extraction procedures that employ PicoPure DNA extraction. Microdissected cells are transferred into 1.5-mL centrifuge tube. Reconstitute a fresh Proteinase K solution by adding 155 μL of reconstitution buffer to a vial of lyophilized Proteinase K. Gently vortex to completely dissolve the reagent; do not mix excessively to avoid enzyme denaturation. Immediately store the reagent on ice.

4. Initiate tissue-scrape digestion by adding 50 μL of Proteinase K solution to the tube. Vortex the mixture gently to mix and incubate at 65°C for 18–24 h. Centrifuge tube at 1000g for 1 min and heat the tube at 95°C for 10 min to inactivate Proteinase K. Store extracted DNA at −20°C or proceed directly to DNA amplification.

3.3. DNA Amplification and Labeling

1. Extracted DNA from microdissected cells is quantified spectrophotometrically (if possible, using Picogreen; *see* **Subheading 3.2.**, and **Note 5**). Accordingly, 10 ng of extracted DNA is used in the amplification reaction. Genomic DNA is digested with 1 µL of N1aIII (10 U/µL) restriction enzyme in 1X T4 DNA ligase reaction buffer for a final 5 µL reaction volume. The reaction is incubated at 37°C for 2 h, and subsequently inactivated by incubation at 70°C for 1 h.

2. Composite linkers LN1 and LN2 (*see* **Subheading 2.2.**) are ligated to the digested tissue DNA and reference DNA, respectively. Add 0.3 µL of each linker to its specified sample with 0.5 µL of T4 DNA ligase and make up the final volume up to 10 µL with 1X T4 DNA ligase reaction buffer. Incubate the reaction at room temperature for 1 h; later inactivate the reaction at 65°C for 40 min.

3. Linker modified target and reference DNA are mixed together for PCR amplification using common primer P1 (*see* **Subheading 2.2.**). Amplification reaction mix is prepared by adding 5 µL of 10X advantage 2 PCR buffer, 2 µL of P1 primer, 1 µL of 50X dNTP mix, 21 µL of PCR grade water, and 1 µL of 50X advantage 2 polymerase mix. Mix gently by pipetting and centrifuge briefly to collect. Set thermo-cycling conditions as follows; 8 min at 72°C, 1 min at 95°C; 20 cycles (30 s at 95°C and 60 s at 72°C), and final extension at 72°C for 5 min.

4. Amplified DNA is purified from unincorporated primers by adding 250 µL of PB buffer to the reaction tube and mixing gently by pipetting. Place a QIAquick column into a collection tube and load the entire sample into the column. Centrifuge the column at 12,000g for 60 s, discard flowthrough and reuse the collection tube. Wash the column twice with 750 µL of PE buffer and centrifuge at 12,000g for 60 s. Discard flowthough and refit the same collection tube for a column drying centrifugation at maximum speed for 1 min. Fit the column in a clean, new 1.5-mL microcentrifuge tube and pipet 50 µL of nuclease-free water to the center of column filter membrane to elute the DNA. Let the column stand for 1 min and then centrifuge the column at 12,000g for 1 min.

5. PicoGreen reagent assay is used to quantify the PCR product. Prepare 10 mL of 1X TE buffer from the kit supplied stock, and dilute standard lambda dsDNA in a descending series (0, 20, 50, 100, 200, 300, 400, and 500 ng/mL). Unknown test sample is prepared by adding 1 µL of PCR product to 99 µL of 1X TE buffer. Prepare a 1:200 dilution of the PicoGreen reagent in 1X TE. The PicoGreen reagent is light sensitive and should be kept wrapped in foil while thawing and in the diluted state. Vortex the diluted PicoGreen reagent well and add 100 µL of it to each standard and unknown sample. Mix samples well with dye and incubate at room temperature for 4 min. Read the fluorescence measurements using a suitable spectrofluorometer after excitation at wavelength/bandwidth of about 485 nm/20 nm, and record emission at wavelength/bandwidth of about 520 nm/25 nm. Plot a standard curve of fluorescence (*y*-axis) against standards concentration (*x*-axis) and determine unknown sample concentration.

6. Reseparation of reference and target amplified DNA is performed using the specific primers P2a and P2b, which contain 2 nt at their ends to distinguish the two

samples. The genome specific amplification reaction requires 10 ng from the first PCR reaction for each target and reference genome reactions. Add to each reaction 5 μL of 10X Titanium *Taq* PCR buffer, 1 μL of primer mix, 1 μL of 50X dNTP mix, 1 μL of 50X Titanium *Taq* DNA polymerase, and make up the reactions up to 50 μL using PCR grade water. Mix the reactions by gentle pipetting and centrifuge briefly. Initiate the balanced PCR amplification reaction using the following settings; 1 min at 95°C, 10 cycles of 30 s at 95°C and 60 s at 72°C each, and a final extension at 72°C for 5 min.

7. Purify the PCR product thoroughly using QIAquick columns as previously described. Run the products on 1% agarose gel to ensure amplification in a form of a smear ranging between 500 bytes and 1 kb in size. The PCR products are then quantitated using PicoGreen as described in **Note 5.**

8. For labeling reactions use 2–4 μg of amplified products in a random priming reaction that employs BioPrime labeling kit as described in Chapter 6 (Al-Mulla), **Subheading 3.2.** Similarly, the Bioprime labeling procedure is used to label DNA extracted from frozen section or scraps macrodissected from FFPE tissues. However, it was noted that in labeling the latter, DNA digestion should be avoided or at least minimized. The labeled and purified DNA can be hybridized to cDNA/PCR fragments, oligonucleotides, and BAC-based microarrays successfully (*see* **Note 6**).

3.4. RNA Extraction and Purification

3.4.1. Frozen Tissue

Frozen tissue blocks homogenized by grinding are transferred into a tube containing 500 μL Trizol for cell digestion (*see* **Note 7**). Successive steps are the same as those carried out for cell-line RNA extraction and purification described in Chapter 6 (Al-Mulla), **Subheading 3.4.** Resultant RNA is generally of sufficient quantity (5–10 μg) and quality to undergo minimal amplification and labeling procedures. Microarray probe synthesis might involve direct incorporation of labeled nucleotides during a cDNA synthesis reaction, or incorporation of aminoallyl modified nucleotides for subsequent fluorescent dye coupling (*see* Chapter 6, **Subheading 3.5.**).

3.4.2. FFPE Scrapes

1. Stock Pro K mix is prepared by completely dissolving dried Paradise Pro K mix in 75 μL of Paradise reconstitution buffer. Pipet 25 μL of Pro K mix in a 0.5-mL microcenrifuge tube and store the remainder at −20°C. Tissue scrapes (not exceeding 0.5 × 0.5 cm²) are transferred into the Pro K aliquot tube. Vortex the tube slightly ensuring that the scrapes are in the Pro K mix. Incubate the tube in a 50°C oven for 16 h and proceed to RNA extraction or store at −70°C.

2. Precondition a MiraCol purification column by adding 200 μL conditioning buffer onto the column membrane and incubate for 5 min at room temperature. Centrifuge the column at 16,000*g* for 1 min. Pipet 87.5 μL of binding buffer into

the cell extract and mix well by pipetting. Add 162.5 μL of ethanol to the tube and mix well. Centrifuge the tube at 16,000g for 1 min. Load 200 μL of the supernatant into the preconditioned MiraCol column without agitating the debris pellet. Centrifuge the column at 100g for 2 min, then centrifuge at 16,000g for 1 min. Repeat the same step until the entire supernatant is eluted through the column. Discard flowthrough when it reaches the bottom of the column.

3. The column is loaded with 100 μL of wash 1 buffer and centrifuged at 16,000g for 1 min. Mix 2 μL of DNase with 18 μL of DNase buffer and load into the column. Incubate at room temperature for 20 min. Load 40 μL of wash buffer 1 into the column and centrifuge at 8000g for 1 min. Pipet 100 μL of wash buffer 2 and centrifuge at 16,000g for 1 min. Repeat the wash with a fresh 100 μL of wash buffer 2 and centrifuge at 16,000g for 2 min. Transfer the column to a new 0.5-mL tube and elute the RNA with 70 μL of elution buffer applied directly onto the filter membrane. Incubate at room temperature for 1 min and centrifuge at 1000g for 1 min first, then at 16,000g for 1 min. Extracted RNA is quantified spectrophotometrically and can be directly used for the amplification reaction or stored at −70°C.

3.4.3. FFPE Sections and Frozen Tissue Sections Microdissection

1. A protocol similar to that used for tissue scrapes is followed for microdissected cells from tissue sections. Minor differences involve using 10 μL of reconstituted Pro K solution for each microdissected extract. The reaction tube is incubated at 50°C for 16 h, and centrifuged at 800g for 2 min. The extract can be stored at −70°C or prepared for RNA isolation.

2. Purify microdissected cells RNA by adding 200 μL of conditioning buffer to the purification column filter membrane. Incubate the column for 5 min at room temperature and centrifuge with attached collection tube at 16g for 1 min. Load 35 μL of Paradise-binding buffer into the cell extract tube, mix thoroughly by pipetting. Add 65 μL of absolute ethanol to the extract and mix well. Transfer the entire extract volume into the preconditioned purification column and centrifuge for 2 min at 100g, followed by a second centrifugation at 16,000g for 1 min. Wash the column with 100 μL of wash buffer 1, and centrifuge at 8000g for 1 min.

3. Remove DNA by mixing 2 μL of DNase mix with 18 μL of DNase buffer and apply it to the column. Incubate the column at room temperature for 20 min. Pipet 40 μL of wash buffer 1 and spin the column at 8000g for 1 min. Add 100 μL of wash buffer 2 twice to the column in two separate steps followed by centrifugation, first at 8000g for 1 min and discard the flowthrough. The second wash is done by centrifuging at 16,000g for 2 min and if residual wash buffer still remains then recentrifuge at 16,000g for 1 min (*see* **Note 8**).

4. To elute membrane bound RNA transfer purification column into a new 0.5-mL microcentrifuge tube and load 12 μL of elution buffer directly on the filter membrane. Incubate column at room temperature for 1 min and centrifuge at 1000g for 1 min to disperse elution buffer through the membrane. Spin the column again at 16,000g for 1 min to elute RNA. RNA isolated can be stored at −70°C or directly processed into the amplification protocol.

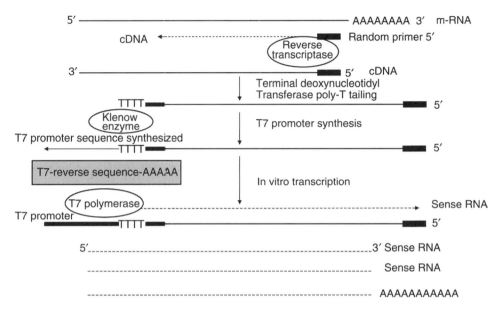

Fig. 2. An overview of the SenseAmp Plus mRNA amplification procedure. Notice that the end products are sense RNA similar to the original start material but in larger quantities and ready for further amplifications and labeling for the production of antisense capable of hybridization to sense complimentary sequences on various microarrays. X indicates a DNA polymerase blocker to prevent second strand synthesis in that direction.

3.5. RNA Amplification

3.5.1. From FFPE Tissue Scrapes and Microdissected Cells

3.5.1.1. FIRST STRAND CDNA SYNTHESIS WITH RANDOM AND DT PRIMERS

1. Several methods have been described for RNA amplification. For example, the Eberwine T7 methodology is widely used and utilized on several microarrays including long oligonucleotide-based platforms. However, labeling or linear amplification of RNA using T7-Eberwine methodology has been shown to be unreliable on long oligonucleotide-based microarrays *(11)*. On the other hand, the SenseAmp Plus from Genisphere has proven reliability with the added advantage of versatility *(12)*. The sense strand RNA synthesized is similar to the original RNA and the method ensures representation of both the 5'- and 3'- ends (**Figs. 2** and **3**). In addition, the sense strand synthesized RNA can be further amplified using other methods and can be utilized by most available microarray platforms (*see* **Note 9**).

2. Before starting the reaction, all the SenseAmp Plus kits are thawed at room temperature, vortexed, briefly centrifuged, and kept at room temperature except for dNTP mix, Superase-In RNase inhibitor, dTTP, terminal deoxynucleotidyl transferase, and Klenow, T7, and PAP enzymes. The latter components are thawed on ice, briefly centrifuged if necessary and kept on ice at all the times, never vortexed.

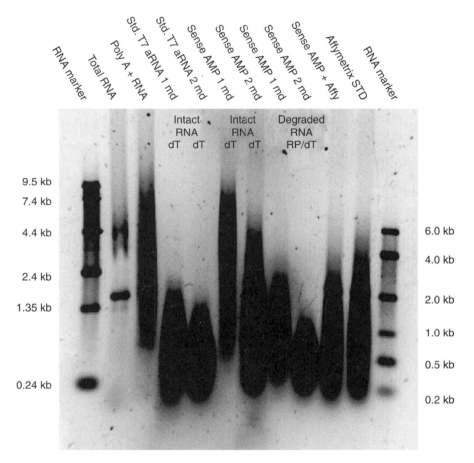

Fig. 3. Denaturing 1% Agarose Gel of RNA produced by SenseAmp Plus (Genisphere) vs Eberwine T7 amplification.

3. For degraded RNA extracted from FFPE tissues it is recommended to enhance the amplification process by adding random 9-mer RT primer. The random primer is diluted to a concentration of 1X by mass of input total RNA per microliter. For example, if using 5 ng of total RNA, dilute the random primer to 5 ng/μL.

4. On ice an RNA/primer mix is prepared for each RNA sample in a 0.5-mL micro-centrifuge tube; containing 1–7 μL (5–10 ng) RNA, 2 μL of SenseAmp dT24 RT primer (50 ng/μL), and 2 μL of diluted random 9-mer RT primer. The reaction is mixed and heated to 80°C for 10 min in a heated-lid thermal cycler. The tube is placed on ice immediately for 2 min, briefly centrifuged and returned on ice.

5. A master mix is prepared with the following components on ice; 4 μL of 5X first strand buffer, 2 μL of 0.1 *M* DTT, 1 μL Superase-In, 1 μL of dNTP mix, and finally 1 μL Superscript II for a final volume of 9 μL. For more than a single reaction multiply the volumes of the master mix components by the number of reactions carried out.

6. Combine 9 μL of master mix with 11 μL of the RNA/primer mix for a final volume of 20 μL, mix gently and briefly centrifuge to collect. Incubate the reaction at 42°C for 2 h in a heated-lid thermal cycler. Spin the tube briefly to collect, and add 80 μL of 1X TE for a final volume of 100 μL.

7. Purify the cDNA product by adding 500 μL of PB buffer to the 100-μL reaction volume. Transfer the entire volume to a MinElute column and centrifuge for 1 min at 12,000*g*. Discard the flowthrough, and place the column in the same collection tube. Add 750 μL of PE buffer into the column and centrifuge again under the previous settings. Discard the flowthrough and reuse the same collection tube. Add 500 μL of 80% ethanol and centrifuge for 2 min at 12,000*g*. Discard the flowthrough and drain the column using the same collection tube and keeping the column cap open, while centrifuging for 5 min at 12,000*g*. Replace the collection tube with a new 1.5-mL microcentrifuge tube. Elute cDNA by adding 10 μL of EB buffer to the center of the column membrane; incubate at room temperature for 2 min. Centrifuge the column for 2 min at 12,000*g*, discard the column and save the 10 μL of eluted cDNA. If the eluted volume is <10 μL, then make it up to 10 μL with nuclease-free water (*see* **Note 10**).

3.5.1.2. TAILING OF FIRST STRAND cDNA AND T7 PROMOTER SYNTHESIS

1. Heat the purified cDNA at 80°C for 10 min and place on ice immediately for 2 min. Spin briefly to pellet and return on ice. For each reaction a master mix is prepared containing; 2 μL of 10X reaction buffer, 2 μL of nuclease-free water, 4 μL 10 m*M* dTTP and 2 μL TdT enzyme. Mix the cDNA with the master mix for a volume of 20 μL and spin briefly to collect. Incubate the reaction at 37°C for 2 min only. Terminate the reaction by heating it to 80°C for 10 min. Centrifuge briefly to collect and cool at room temperature for 2 min.

2. Prepare a dilution of the T7 template oligo to be used in the first round of amplification by adding 2 μL of T7 template oligo to 6 μL of nuclease-free water, briefly vortex and centrifuge to collect. Take 2 μL of diluted T7 template oligo to the tailed cDNA for a reaction volume of 22 μL. Briefly vortex and centrifuge, then incubate at 37°C for 10 min to anneal the strands.

3. For each reaction add 1 μL of 10X reaction buffer, 1 μL of dNTP mix and 1 μL of Klenow enzyme for a final volume of 25 μL. Mix gently and spin briefly then incubate at room temperature for 30 min. Stop the reaction by heating it to 65°C for 10 min and place the tube on ice.

3.5.1.3. IN VITRO TRANSCRIPTION

1. Initiate the reaction using 12.5 μL of the promoter modified cDNA reaction and store the remaining half of the reaction at −20°C for future use. Incubate the 12.5 μL of cDNA at 37°C for 10 min to reanneal the strands. Thaw the T7 nucleotide mix and 10X T7 reaction buffer, vortex the reaction buffer, and leave both vials at room temperature until use.

2. For each reaction add the following: 8 μL of T7 nucleotide mix, 2.5 μL of 10X T7 reaction buffer, and 2 μL of T7 enzyme mix. The final reaction volume of 25 μL

is gently mixed and centrifuged briefly. The reaction is incubated in a thermal cycler at 37°C for 5 min, then transferred to a 37°C hybridizer oven and incubated for 16 h. It is essential to seal the reaction thoroughly to avoid evaporation and condensation.

3. Adjust the reaction volume to 100 µL by adding 75 µL of RNase-free water in preparation for sense RNA purification. Add 350 µL of RLT buffer and mix thoroughly. Add 250 µL of absolute ethanol to the reaction and mix thoroughly by pipetting. Load the entire volume to an RNeasy MinElute column fitted in a 2-mL Collection Tube. Gently cap the tube and centrifuge for 15 s at 8000*g*, and discard the flowthrough. Prepare a working solution of RPE buffer by adding four volumes of absolute ethanol to it. Change the collection tube with a new 2-mL tube and pipet 500 µL of RPE buffer working solution onto the spin column. Close the tube gently and spin for 15 s at 8000*g* to wash the column. Discard the flowthrough and reuse the collection tube. Add 500 µL of 80% ethanol to the column and centrifuge at 8000*g* for 2 min. Discard collection tube and the flowthrough.

4. Transfer the column carefully into a new 2-mL collection tube. Open the cap of the spin column gently and centrifuge at full speed for 5 min. Discard the flowthrough and the collection tube. To elute the sense RNA transfer the spin column into a 1.5-mL collection tube. Pipet 14 µL of RNase-free water onto the center of the column membrane. Close the tube and centrifuge at maximum speed to elute sense RNA. The recovered volume should be approx 12 µL. It is crucial to analyze sense RNA with gel electrophoresis and/or spectrophotometry to ensure sample amplification is substantial.

3.5.1.4. SECOND ROUND OF AMPLIFICATION

The first round of amplification is followed by a second round of amplification to procure sufficient amount of stock RNA. For the second round of second strand cDNA synthesis 8 µL of sense RNA are added to 2 µL of random 9-mer RT primer and 1 µL of SenseAmp dT24 primer (50 ng/µL). Proceed with the reaction following the same steps mentioned previously for the second strand synthesis reaction, cDNA purification, and tailing of first strand cDNA. However, for the T7 promoter synthesis add 2 µL of undiluted stock T7 template oligo. Proceed with the entire reaction as before ending with purification of stock sense RNA. Determine sense RNA quantity and quality using absorbance readings and 1% agarose denaturing gel electrophoresis (**Fig. 3**).

3.5.1.5. OPTIONAL: FOR FROZEN TISSUE RNA EXTRACT

An amplification protocol is performed if more RNA is required for further analyses or the actual extract yield is lower than expected. The first round of amplification in the SenseAmp procedure is followed with minor changes. The starting quantity of RNA should not be <25 ng, and not >2 µg in a volume not more than 9 µL. The reaction only includes the SenseAmp dT24 RT primer (50 ng/μL), the random 9-mer primer is not added. The rest of the reaction proceeds as previously stated except for the T7 promoter synthesis step. The T7 template oligo added into the first and only amplification round involves 2 μL of the undiluted stock. At the in vitro transcription step only 12.5 μL of the reaction can be used or the entire volume of 25 μL can be used but it needs to be purified first using the MinElute PCR purification kit as previously mentioned. The final volume is adjusted to 12.5 μL with 1 μL of T7 template oligo and nuclease-free water. The amplification ends with the final purification step without any further amplification rounds of sense RNA.

3.6. Expression Microarray

1. Sense RNA is quantified and according to resultant concentration a sample is aliquoted that is within 1–5 μL (100–1000 ng) range. Sense RNA aliquots from each sample are transferred to a 1.5-mL microcentrifuge tube. Each sample is assigned with a different dye and accordingly the choice of RT primer (5 pmole/μL) is designated. Sample mass governs RT primer volume, for 100–499 ng of sense RNA 1 μL of 5 pmole/μL RT primer is needed and for 500–1000 ng of sense RNA 2 μL is required per reaction. If needed nuclease-free water is added for a final volume of 6 μL. Mix components thoroughly and centrifuge to collect. Incubate the reaction at 80°C for 5 min and place on ice immediately for 3 min then recentrifuge briefly and return on ice.

2. Reverse transcription reaction master mix is prepared by adding 4 μL of 5X SuperScript II first strand buffer, 2 μL of 0.1 m*M* DTT, 1 μL of Superase-In RNase inhibitor, 1 μL of dNTP mix and 1 μL of SuperScript II enzyme; for one reaction. Gently mix and centrifuge and keep on ice. Combine 4 μL of the reaction master mix to the RNA-RT primer mix. Mix slowly and incubate at 42°C for 2 h. Terminate the reaction by adding 1 μL of cDNA synthesis stop solution. Heat the reaction to 65°C for 10 min to denature dsDNA/RNA and degrade the sample RNA. Adding 1.2 μL of 2 *M* Tris-HCl, pH 7.5 is sufficient to neutralize the reaction pH resulting in a final volume of 12.7 μL.

3. cDNA hybridization is governed by type of microarrays used, whether it is cDNA or oligo and if it requires SDS or formamide-based or enhanced hybridization buffer. Accordingly, thaw and resuspend the 2X hybridization buffer of choice by heating at 65°C for 10 min and vortex to full resuspension and centrifuge for 1 min. Prepare a cDNA hybridization mix per array by mixing 12.7 μL of sample cDNA, and 12.7 μL of reference cDNA (if single probe experiments then substitute with nuclease-free water), 29 μL of 2X enhanced hybridization buffer (other buffers 25 μL is sufficient), 1.6 μL of nuclease-free water, and 2 μL of LNA (augmented dT blocker) (might be excluded when using oligoarrays). For cDNA arrays it is preferred to add 1 μL of Cot-1 DNA denatured at 95°C for 10 min. Gently vortex to uniformly mix the reaction components and centrifuge briefly. Heat the reaction to 80°C for 10 min and then keep at the required array hybridization temperature until probe injection.

4. Microarray prehybridization handling might involve washes with blockers (e.g., BSA) or dilutions of detergent buffers, or it might not involve any depending on the type of microarray. Hybridization temperatures slightly vary according to length of the spotted target. However, prior to hybridization the microarray platform must be warmed up to the hybridization temperature of the reaction. However, lower hybridization temperatures are used when enhanced hybridization buffer is added. Apply the cDNA hybridization mix to the array and mount an appropriate cover slip (24×60 mm^2). Incubate the slide in a humidified chamber overnight at the optimal hybridization temperature of the array platform.

5. Before posthybridization washing, wash buffer 1 is prewarmed to 60°C for cDNA arrays or 42°C for oligoarrays. Wash array slide with prewarmed wash buffer 1 until the coverslip falls out. Wash array with prewarmed wash buffer 1 for 15 min, 15 min with wash buffer 2, and 15 min with wash buffer 3 at room temperature. Dry the slide by transferring the array label down into an empty 50-mL centrifuge tube and centrifuge with cap off for 2 min at 500g.

6. Prepare the 3DNA capture reagent by thawing at room temperature in the dark for 20 min. Vortex reagent at maximum for 3 s to ensure full resuspension and centrifuge briefly. Activate the reagent by heating at 55°C for 10 min, vortex at maximum for 5 s and centrifuge to collect. It is important to completely resuspend the reagent, repeat previous steps if necessary. Thaw and heat 2X SDS- or formamide-based hybridization buffer at 70°C for at least 10 min until complete resuspention is achieved. Vortex buffer vial thoroughly to eliminate sediments completely and centrifuge for 1 min. It is advisable to use antifade reagent by mixing 1 µL of the reagent with 100 µL of 2X hybridization buffer used for the 3DNA hybridization. The antifade/hybridization buffer stock can be stored at −20°C to be used within 2 wk.

7. For each 3DNA reaction mix 2.5 µL of 3DNA capture reagent 1 (Cy3) and the same amount of 3DNA capture reagent 2 (Cy5), 20 µL of nuclease-free water, and 25 µL of antifade/hybridization buffer mix. Gently vortex and centrifuge the mixture then incubate mix at 80°C for 10 min. Maintain the mixture at the required hybridization temperature and prewarm the array to that temperature. Hybridization temperature vary according to hybridization buffer used; SDS-based buffer requires 55–65°C, whereas formamide-based buffer requires 43–53°C depending on the microarray platform used. Apply the 3DNA hybridization mix to the array, cover with cover slip and incubate in a humidified dark chamber for 4 h at the desired temperature.

8. Post 3DNA hybridization wash steps must be performed in the dark to minimize fluorescence loss. Prewarm wash buffer 1 to 65°C (cDNA and >50 nt oligoarrays), or to 42°C (<50 nt oligoarray). Start with incubating the array in wash buffer 1 to remove the cover slip (2–5 min). Once the cover slip is removed incubate the array in prewarmed wash buffer 1 for 15 min. Proceed with array washings and drying as in the postarray hybridization step. Store the dry array slide in the dark until time for scanning.

4. Notes

1. The standard protocols for tissue collection and preservation in surgical pathology departments are geared primarily toward patients care. For this reason surgical biopsies or tissues are placed in buffered formalin immediately after excision. Tissues are fixed usually overnight or for longer if required. The fixative ensures that the architecture of the tissues are preserved for accurate diagnosis and staging. Any variation on this protocol should ensure that patients care is not compromised. With the introduction of molecular techniques in the diagnostic process, there is an overwhelming demand for frozen tissues because they are an excellent source of high-molecular weight nucleic acids and proteins. However, variations introduced to the current practices in pathology departments could potentially stretch resources to unmanageable levels Moreover, frozen tissues are hard to prepare and the cellular architecture is usually poorly preserved. It is of paramount importance that the pathologist and the team involved in the study to be aware of tissue heterogeneity and account for it during tissue selection. Some surgical pathology departments cut the specimen in two halves (or more sections), freeze one half and fix the other half for guidance and diagnosis. The team should also minimize the time from resection to freezing the tissue because it has been shown that gene expression could be altered with time *(13)*. Alternative fixation with 70% ethanol instead of formalin, which degrades nucleic acids and causes protein–nucleic acids crosslinking has been proposed as a method, which preserves tissue architecture as well as yields good quality nucleic acids for use in microarrays and other techniques *(14)*. However, this procedure is not widely used.

2. It is important to note that clean but noncoated glass slides be used to place the tissues on. The tissue sections vary in thickness from 4 to 7 μm but 7 μm is preferred. For manual micro- and macrodissection the tissues can also be stained with 0.1% (w/v) toluidine blue O for 3–10 s or 0.5% (w/v) methyl green for 1–5 min *(15)*. It is recommended to stain mock tissues to gain experience and optimize staining intensities. Hematoxylin and eosin is not recommended because it interferes with PCR. Slides are then immersed in nuclease-free water for washing. You can dry the slides in a laminar flow hood and then hydrate them before manual microdissection. Hydration with nuclease-free water or TE allows easier micro- or macrodissection. If frozen tissue sections are used as a source of nucleic acids Arcturus HistoGene frozen section staining kit is recommended.

3. The number of cells or tissue size microdissected is optional and should be discussed at the organizational level of the project. Obviously, microdissection of several hundreds to thousands of representational cells require whole genome amplification (assume that 1000 cells can yield 5 ng of DNA). Alternatively, if whole genome amplification is to be avoided for economical reasons or worry about amplification bias then microdissection of several sections can yield sufficient number of cells (50,000, 100,000) that can be used directly on microarrays after purification and labeling of DNA **(Fig. 1A)**.

4. Electrophoresis of high-molecular weight DNA on 1% agarose gel should produce a blob of DNA at or below the well. DNA extracted from FFPE tissue

would expectedly form a smear from 100 bp up to the well, which should also contain some high-molecular weight DNA **(Fig. 1B)**. Avoid using highly degraded DNA (samples with smears below 700 bp and devoid of high-molecular weight molecules).

5. There are several methods described for whole genome amplification *(16)*. These techniques have important clinical applications because of the limited tissues available from patients. It should be realized that all genome amplification procedures introduce some sort of bias to the original material. However, the bias could be tolerated if it is introduced to the test- and control samples equally. We have tested several whole amplification techniques and found "balanced-PCR amplification" method to be reliable and applicable to FFPE tissues *(17–19)*. Alternatively, GenomePlex technology from Sigma appears to be gaining popularity. Degenerate oligonucleotides priming and multiple displacement amplification should be avoided because the former method introduces significant bias and the latter is not suitable for low-molecular weight or short fragments that are found in FFPE tissues. Because of the added expense and possible bias, it is recommended that DNA extracted from whole representative sections *(see* **Note 3)** or frozen tissue to be used initially and compared with the results obtained from whole genome amplification. Alternatively, the microarray dosage results should always be confirmed by quantitative PCR, at least for some genes.

6. The BioPrime labeling described here is applicable to most commercially available or in-house manufactured microarrays. However, Affymetrix-based GeneChips require different processing and labeling. The reader is advised to consult the manufacturer for details. It is always advisable to do reverse labeling of the test and control samples to account for preferential incorporation of labeled nucleotides especially in BAC-based CGH-microarrays **(Fig. 4)**. Also, it is paramount that the microarrays used contain replicated spots for the assessment of reproducibility **(Fig. 5)**.

7. When working with RNA ensure that all glass and plastic ware are RNase-free. RNase is abundantly found on our fingertips. Therefore, it is paramount to wear disposable gloves and handle every step in a manner to avoid contamination with RNases. It is recommended to designate a laminar flow hood as an RNase-free zone by regular wiping with ethanol and RNase Zap. Also, it is advisable to keep pipets and nucleases-free tips dedicated for RNA work in the designated hood.

8. It is essential to use DNase as the SenseAmp methodology can also amplify DNA. It is equally important to inactivate the DNase not only by heating but also by using Paradise columns as directed or alternatively by use of RNeasy columns from Qiagen. Heating alone might not fully deactivate the DNase.

9. The microarray results obtained from FFPE tissues using the SenseAmp Plus method should be verified by qRT-PCR for at least some genes and preferably using the originally extracted RNA.

10. This is a modification on the Qiagen's standard protocol.

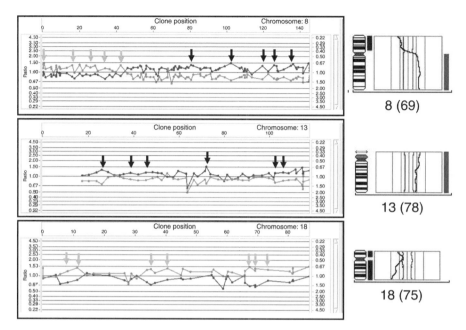

Fig. 4. Microarray CGH profiles of three chromosomes from DNA extracted by microdissection from formalin-fixed paraffin-embedded tissues of patients with colorectal cancer. Black lines show the ratio of test DNA to reference DNA from one microarray and gray lines show the ratios obtained by reverse labeling the test and the reference DNA. Notice the reciprocality of the two lines. The microarray CGH is represented by a ratio plot such that gains in DNA copy number at a particular locus are observed as the simultaneous deviation of the ratio plots from a modal value of 1, with the black ratio plot showing a positive deviation upward while the gray ratio plot shows a negative deviation at the same locus downward (black arrows). Conversely, DNA copy number losses show the opposite pattern (gray arrows). The linear order of the clones is reconstituted in the ratio plots consistent with an ideogram, such that the p terminus is to the left and the q terminus is towards the right of the plot. The experiment was performed twice and in reverse labelling. Inset shows corresponding chromosomal imbalances detected by metaphase comparative genomic hybridization for selected chromosomes that show similar patterns of chromosomal imbalances to microarray-CGH. Mean ratios are shown as thick lines. Bars on the right denote increased chromosome copy numbers or amplifications, while bars to the left represent reduced chromosome copy numbers or deletions. Numbers indicate the chromosome and those in brackets indicate chromosome counts. Notice how the two different techniques gave similar results. Genomic DNA microarrays (2–4 MB Spectral Genomics™, Houston, TX) containing 1500 nonoverlapping BAC and PAC clones printed in duplicate were used here. These microarrays provide an average of 3 Mb resolution for detection of chromosomal imbalances throughout the genome.

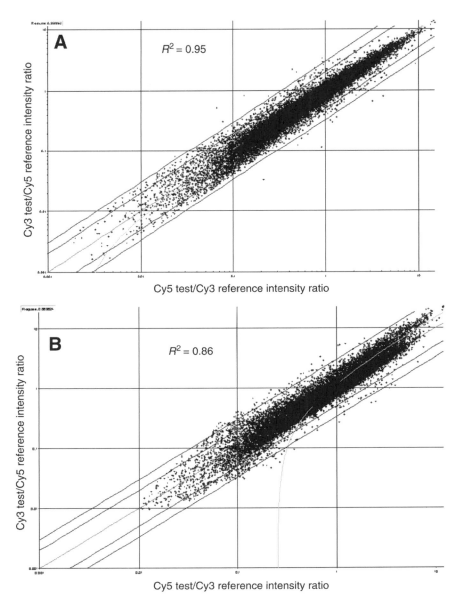

Fig. 5. Correlation scatter plots of two 16-K cDNA microarrays signal intensity ratios. (**A**) test DNA labeled Cy3 and reference DNA (normal blood) labeled Cy5 (*y*-axis) plotted against the same DNA labeled Cy5 and reference DNA labeled Cy3 (*x*-axis). (**B**) another test sample from different patient, test DNA labeled Cy3 and reference DNA (normal blood) labeled Cy5 (*y*-axis) plotted against the same DNA labeled Cy5 and reference DNA labeled Cy3 (*x*-axis). **A** and **B** plots show efficient reproducibility $R^2 = 0.95$ and 0.86, respectively.

Acknowledgments

I wish to thank Raba Al-Tamimi for the CGH-Microarray protocol optimization and Sindhu Jacob for her help with comparative genomic hybridization. My appreciation also goes to Dr. Bob Getts and Joan Ferola for their helpful information regarding the SenseAmp plus kit. This work is supported by Kuwait Foundation for the Advancement of Sciences grant number 990707 and Shared Facility grant number GM/0101.

References

1. Schena, M., Shalon, D., Davis, R. W., and Brown, P. O. (1995) Quantitative monitoring of gene expression patterns with a complementary DNA microarray. *Science* **270,** 467–470.
2. Golub, T. R., Slonim, D. K., Tamayo, P., et al. (1999) Molecular classification of cancer: class discovery and class prediction by gene expression monitoring. *Science* **286,** 531–537.
3. Alizadeh, A. A., Eisen, M. B., Davis, R. E., et al. (2000) Distinct types of diffuse large B-cell lymphoma identified by gene expression profiling. *Nature* **403,** 503–511.
4. van 't Veer, L. J., Dai, H., van de Vijver, M. S., et al. (2002) Gene expression profiling predicts clinical outcome of breast cancer. *Nature* **415,** 530–536.
5. Snijders, A. M., Meijer, G. A., Brakenhoff, R. H., van den Brule, A. J., and van Diest, P. J. (2000) Microarray techniques in pathology: tool or toy? *Mol. Pathol.* **53,** 289–294.
6. Forster, T., Roy, D., and Ghazal, P. (2003) Experiments using microarray technology: limitations and standard operating procedures. *J. Endocrinol.* **178,** 195–204.
7. Russo, G., Zegar, C., and Giordano, A. (2003) Advantages and limitations of microarray technology in human cancer. *Oncogene* **22,** 6497–6507.
8. Ein-Dor, L., Kela, I., Getz, G., Givol, D., and Domany, E. (2005) Outcome signature genes in breast cancer: is there a unique set? *Bioinformatics* **21,** 171–178.
9. Yuen, T., Wurmbach, E., Pfeffer, R. L., Ebersole, B. J., and Sealfon, S. C. (2002) Accuracy and calibration of commercial oligonucleotide and custom cDNA microarrays. *Nucleic Acids Res.* **30,** E48.
10. Srinivasan, M., Sedmak, D., and Jewell, S. (2002) Effect of fixatives and tissue processing on the content and integrity of nucleic acids. *Am. J. Pathol.* **161,** 1961–1971.
11. Al-Mulla, F., Al-Tamimi, R., and Bitar, M. S. (2004) Comparison of two probe preparation methods using long oligonucleotide microarrays. *Biotechniques* **37,** 827–833.
12. Goff, L. A., Bowers, J., Schwalm, J., Howerton, K., Getts, R. C., and Hart, R. P. (2004) Evaluation of sense-strand mRNA amplification by comparative quantitative PCR. *BMC Genomics* **5,** 76.
13. Auer, H., Lyianarachchi, S., Newsom, D., Klisovic, M. I., Marcucci, G., and Kornacker, K. (2003) Chipping away at the chip bias: RNA degradation in microarray analysis. *Nat. Genet.* **35,** 292–293.

14. Cole, K. A., Krizman, D. B., and Emmert-Buck, M. R. (1999) The genetics of cancer–a 3D model. *Nat. Genet.* **21,** 38–41.

15. Ehrig, T., Abdulkadir, S. A., Dintzis, S. M., Milbrandt, J., and Watson, M. A. (2001) Quantitative amplification of genomic DNA from histological tissue sections after staining with nuclear dyes and laser capture microdissection. *J. Mol. Diagn.* **3,** 22–25.

16. Lasken, R. S. and Egholm, M. (2003) Whole genome amplification: abundant supplies of DNA from precious samples or clinical specimens. *Trends Biotechnol.* **21,** 531–535.

17. Wang, G., Brennan, C., Rook, M., et al. (2004) Balanced-PCR amplification allows unbiased identification of genomic copy changes in min cell and tissue samples. *Nucleic Acids Res.* **32,** E76.

18. Makrigiorgos, G. M., Chakrabarti, S., Zhang, Y., Kaur, M., and Price, B. D. (2002) A PCR-based amplification method retaining the quantitative difference between two complex genomes. *Nat. Biotechnol.* **20,** 936–939.

19. Stoecklein, N. H., Erbersdobler, A., Schmidt-Kittler, O., et al. (2002) SCOMP is superior to degenerated oligonucleotide primed-polymerase chain reaction for global amplification of min amounts of DNA from microdissected archival tissue samples. *Am. J. Pathol.* **161,** 43–51.

9

A Microarray-Based Method to Profile Global microRNA Expression in Human and Mouse

Ranjan J. Perera

Summary

The microRNAs (or miRNAs) are small noncoding RNAs (21–25 nt) that are processed from large hairpin RNA precursors and are believed to be involved in a wide range of developmental and cellular processes, by either repressing translation or triggering mRNA interference (RNA interference). Over 200 of distinct genes encoding miRNAs have been identified through either computer-assisted approaches or complementary deoxyribonucleic acid cloning strategies in many organisms including worm, plants, flies, mouse, and human. Recently, a microarray based robust method to profile miRNAs expression in organs cell lines and tissues in mammalians were developed *(1)*. Using this method, we have identified a group of miRNAs preferentially expressed in human primary adipocytes and knocking down one such miRNA (miRNA 143) reverses the differentiation process *(2)*. Groups of kidney specific miRNAs share evolutionary conserved phylogenetic foot print Ets1 in the upstream of the miRNA, possibly important for kidney physiological maintenance were reported. A detail protocol of the method are discussed to develop miRNA profile for global gene expression in tissues, organs, and cell lines in eukaryotes.

Key Words: Expression profiling; gene regulation; microarray; microRNA.

1. Introduction

MicroRNAs (miRNAs) are approx 22- nt noncoding RNAs that can play important roles in cell-function and development by targeting the mRNA sequences of protein-coding transcripts, resulting in either mRNA cleavage or repression of productive translation *(3–6)*. Originally, miRNAs were discovered in the nematode *Caenorhabditis elegans* through genetic screens for mutants that lacked the ability to control the timing of specific cell fate switches during development *(7,8)*. Several hundred miRNAs from *C. elegans*, plants, *Drosophila melanogaster* and mammals have since been identified through computational and cloning approaches *(9–22)*.

From: *Methods in Molecular Biology, vol. 381: Microarrays: Second Edition: Volume 2*
Edited by: J. B. Rampal © Humana Press Inc., Totowa, NJ

There are currently estimated to be approx 283 miRNAs *(10)* present in the human genome. In *C. elegans*, there are more than 1000 molecules per cell, with some exceeding 50,000 copies per cell *(11)*. The experimental validation of miRNA target mRNA remain a challenge. Many miRNAs are limited in their expression to certain stages in development or to certain tissues and cell types *(6)*. Recently, it has been reported that human miRNA genes are frequently located in fragile sites and genomics regions involved in cancer *(23)*. Therefore, the ability to determine global miRNA expression in mammalian cell will prove valuable in helping to understand the putative roles played by miRNAs in cell function. The first use of an array-based technology to profile microRNA expression was documented *(24)*. Since then, number of miRNA array articles have been published *(25–31)*. In a recent review article different array-based detection approaches are compared *(31)*. Our method has some unique features compared to those already published articles *(24,25,28)*, and these differences in the results section were discussed.

In this study, robust array-based technique was described, which allows to identify expression of 254 miRNAs in mammalian cells. To do this, oligonucleotides (5′-Amino-Modifier-C6) corresponding to human and mouse mature sense miRNA sequences, were designed to hybridize to biotin end-labeled antisense miRNA targets. We have used this approach to compare miRNA expression in six different human organs, and our array results indicate the presence of differentially expressed groups of miRNA in different human tissues. Where possible, to compare the array data with these published results, and these were found to be in high concordance.

2. Materials

2.1. Arrays

1. 5′-amine-modified-C6-oligonucleotides.
2. 1X Micro Spotting Plus buffer (ArrayIt, Sunnyvale, CA).
3. CodeLink-activated slides (GE Health/Amersham Biosciences, Piscataway, NJ).
4. Pixsys7000 pin-based dispensing system (Genomics Solutions, Irvine, CA).
5. Random oligonucleotides.
6. Microarray hybridization and washing solutions.
7. Axon 4000B scanner (Axon Instruments Union city, CA).
8. GenePix Pro 4.0 software (Axon Instruments).
9. Data analysis software (S-plus and Matlab).

2.2. Labeling

1. 10 m*M* dNTPs mix.
2. Random primers, 3 µg/µL.
3. SuperScript II Reverse transcriptase, 200 U/µL.
4. SUPERase In™, 20 U/µL.
5. Nuclease-free water.

6. NaOH, 1 *N* solution.
7. HCl, 1 *N* solution.
8. QiaQuick Nucleotide removal kit (Qiagen, Valencia, CA; P/N 28304).
9. 10X One-Phor-All Buffer.
10. DNase I.
11. BioArray™ Terminal Labeling Kit with Biotin ddUTP (Enzo Bio, NewYork, NY).
12. 0.5 *M* Ethylenediaminetetraacetic acid (EDTA).
13. Absolute ethanol.
14. 80% Ethanol.

2.3. Target Hybridization

1. Water, molecular biology grade.
2. Acetylated bovine serum albumin (BSA) solution, 50 mg/mL.
3. Herring sperm DNA (Promega, Madison, WI; P/N D1811).
4. Micropure Separator (Millipore, P/N 42512) (optional).
5. NaCl (5 *M*) RNase-free, DNase.
6. MES-free acid monohydrate.
7. MES sodium salt.
8. EDTA disodium salt, 0.5 *M* solution.
9. Secure-Seal™ hybridization chamber.
10. Adhesive seal tabs.
11. Surfact-Amps X-100 (Tween-20), 10%.

2.4. Washing, Staining, and Scanning

1. Water, molecular biology grade.
2. Distilled water.
3. Acetylated BSA solution, 50 mg/mL.
4. R-Phycoerythrin streptavidin (Molecular Probes, Eugene, OR; P/N S-866).
5. NaCl (5 *M*), RNase-free, DNase-free.
6. PBS (0.02 *M*), pH 7.2.
7. 20X SSPE.
8. Goat IgG, Reagent Grade (Sigma-Aldrich, St. Louis, MO).
9. Antistreptavidin antibody (goat), biotinylated (Sigma-Aldrich).
10. 10% Surfact-Amps20 (Tween-20).
11. Bleach (5.25% sodium hypochlorite).
12. Immunopure streptavidin.

3. Methods

3.1. Micro-RNA Array

The array described here offers more comprehensive coverage and higher throughput than Northern blot approaches, and represents a valuable tool to better understand the recently identified class of gene regulating RNA molecules.

3.2. miRNA Array Design

5′-amine-modified-C6 oligonucleotides were resuspended in 1X Micro Spotting Plus buffer at 20 μ*M* concentration. Each oligonucleotide probe is printed four times on a CodeLink-activated slides by a Pixsys7000 pin-based dispensing system in 2×2 pin and 40×8 spot configuration of each sub-array, with a spot diameter of 120 μm (*see* **Note 1**). The array also contains several 23 bp U6 and *Drosophila* tRNA oligonucleotides specifically designed as labeling and hybridization controls (positive) whereas 23 bp random oligonucleotides are designed as negative controls. To a limited extent we have investigated the properties of the miRNA probes we printed on the chip. We studied the secondary structure of the probes as well as the possibility of binding to other closely related probes on the array. Computational searches were also done for global cross hybridization. Even after testing standard parameters for microarray probe designing, it is not possible to design an alternate probe to a given processed miRNAs owing to the limited sequence length. Therefore, all the available miRNA sequences on the chip were decided to be inculded.

3.3. Sensitivity and the Specificity of the miRNA Array

264 oligonucleotides (5′-Amino-Modifier C6) corresponding to human and mouse mature sense miRNA sequences designed to hybridize on to biotin end-labeled antisense miRNA targets, at a final concentration of 20 μM were spotted (four chip replicates) on 3-D CodeLink slides (*see* **Note 1**). The specificity and sensitivity of the array based technology was determined using an array designed for Let-7a miRNA detection (**Fig. 1**).

This array contains Let-7a probes and closely related probes sequences including Let-7a probes containing single, double and multiple mismatches to Let-7a. In detail, Let-7e and Let-7c contain one mismatch each and Let-7b and Let-7d contain two mismatches, whereas Let-7f and Let-7i contain multiple mismatches to Let-7a. As seen in **Fig. 1**, Let-7f and Let-7i transcripts are hardly detected in 1 pM biotin end label spike concentrations. However, Let-7b and Let-7d, which carry two mismatches each to Let-7a can detect in 1 pM levels but started to disappear producing signal intensities in 0.1 pM spike concentrations. Let-7a is clearly visible throughout all spike controls even up to 0.1 fM concentrations. Though we did not carry on further dilutions, we are certain that it is possible to detect Let-7a signals in lower dilutions than 0.1 fM. Hence, results of studies using the Let-7a array and a synthetic Let-7a target RNA for sensitivity studies can demonstrate that the array could undoubtedly differentiate perfect matches and mismatches at 10 fM target concentrations. U6 oligonucleotides were spotted on the array as negative controls.

3.4. Statistical Analysis

Axon 4000B scanner and the GenePix Pro 4.0 software were used to scan images. The median intensities of each feature and of the corresponding

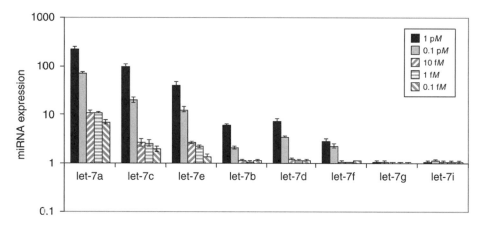

Fig. 1. miRNA array specificity. The array was probed with a *Let-7a* transcript and detection was performed as described in **Subheadings 2.** and **3.** Signal intensity values of the different *Let 7* mirs are plotted in **Fig 1**. In higher concentrations (1 p*M*) and above closely related Let 7 members (Let-7e and Let-7c) with single mismatch does not clearly separate their signal intensity values from their perfect match counter part: Let-7a. But however, during the spike dilution, remaining Let-7 members (let-7b, let-7d, let-7f, let-7g, and let-7i) clearly distinguish the expression patterns from let-7a.

background were measured. The median intensity of the background was subtracted from the median intensity of the feature. Outliers detected by the ESD procedure *(32)* were also removed at this stage. Resulting signal intensity values were normalized to per-chip median values. These signal intensity values were then used to obtain geometric means and standard errors for each miRNA. Each miRNA signal was transformed to log base 2 and 1-sample *t*-test was conducted. If the signal was significantly ($p \geq 0.05$) high or low compared to the chip median then a present (P) or absent (A) call is assigned. A marginal (M) call is given otherwise. Our method of miRNA array data analysis has some unique features compared to those published earlier. For example, our analysis method has built-in automatic outlier detection algorithms, called the ESD procedure *(32)*, and stringent expression call algorithms based on statistics.

3.5. Labeling and Hybridization Procedure

Sections after **Subheading 3.5.1.** are mainly adapted from Affymetrix labeling protocols (*see* **Note 2**).

For 1000 μL:

1. 25 mL 3 mg/mL Random primers.
2. 975 μL Nuclease-free H_2O.
3. Store at −20°C in a nonfrost-free freezer.

3.5.1. cDNA Synthesis

1. The following protocol starts with 20 µg of total RNA per sample. Incubation is performed in a heat block and water bath. Prepare the following mixture for primer annealing:

Primer Hybridization Mix

Reagent	Volume (µL)	Final concentration
Total RNA 20 µg, 1 µg/µL	20	0.33 µg/µL
Random primers (75 ng/µL)	20	25 ng/µL
Nuclease-free H$_2$O	20	
Total volume	60	

2. Incubate the RNA/primer mix at the following temperatures:
   a. Incubate at 70°C for 10 min.
   b. Incubate at 25°C for 10 min.
   c. Chill to 4°C (*see* **Note 3**).
3. Prepare the reaction mix for cDNA synthesis. Briefly centrifuge the reaction tube to collect sample to the bottom and start the cDNA synthesis.

Components	Volume/amount (µL)	Final concentration
RNA/primer hybridization mix		
(from previous step)	60	
5X First Strand Buffer	24	1X
100 m*M* DTT	12	10 m*M*
10 m*M* dNTPs	6	0.5 m*M*
SUPERase In (20 U/µL)	3	0.5 U/µL
SuperScript II (200 U/µL)	15	25 U/µL
Total volume	120	

4. Incubate the reaction at the following temperatures:
   a. Incubate at 25°C for 10 min.
   b. Incubate at 37°C for 60 min.
   c. Incubate at 42°C for 60 min.
   d. Inactivate SuperScript II at 70°C for 10 min.
   e. Chill to 4°C.
   f. Add 40 µL of 1 *N* NaOH and incubate at 65°C for 30 min.
   g. Add 40 µL of 1 *N* HCl for the neutralization.

3.5.1.1. cDNA CLEAN UP

Add 2 mL (10 vol) Buffer PN to the 200 µL (1 vol) single-stranded cDNA synthesis). Mix by vortexing for 3 s. Each sample (−20 µg starting total RNA) using two columns. Apply 700 µL of the sample to each QiaQuick Column sitting in a 2-mL collection tube, and centrifuge for 1 min at 6000 rpm, discard flow-through.

Reload 400 μL of mixture to each spin column and centrifuge as previously described. Discard flow-through and collection tube. Transfer spin column into a new 2-mL collection tube (supplied). Pipet 750 μL wash buffer PE onto the spin column. Centrifuge for 1 min at 6000 rpm. Discard the flow-through and place the Qiaquick column back in the same tube which should be empty. Centrifuge for additional 1 min at 13,000 rpm (17,900g).

Transfer spin column into a clean 1.5-mL collection tube, and 42 μL of dH$_2$O directly onto the spin column membrane. Incubate for 1 min at room temperature and centrifuge 1 min at 13,000 rpm (17,900g) to elute. Ensure that the dH$_2$O is dispensed directly onto the membrane. The average volume of elute is 40 μL from 42 μL elution buffer. Combine the two elutes from same sample together. The total volume should be 80 μL.

Quantify the purified cDNA product by 260 nm absorbance. 1 A260 unit = 33 μg/mL of single strand DNA. Typical yields of cDNA are 3–7 μg.

3.5.1.2. TERMINAL LABELING

Prepare the following reaction mix using the BioArrayTerminal Labeling kit with Biotin-ddUTP (*see* **Note 2**):

Components	Volume/amount (μL)	Final concentration
5X Reaction buffer	40	1X
10X CoCl$_2$	20	1X
100X Biotin-ddUTP	2	1X
50X Terminal deoxynucleotide transferase	4	1X
Fragmentation product (1.5–6 μg) up to	100	1.5–6 μg
Total volume	200	

When the amount of fragmentation product exceeds 3 μg, extend the terminal labeling reaction time up to 60 min.

1. Incubate the reaction at 37°C for 60 min.
2. Stop the reaction by adding 4 μL of 0.5 M EDTA.
3. Speed Vac to concentrated to 144 μL volume.

The target is now ready to be hybridized onto probe arrays, or it may be stored at −20°C for later use.

3.5.2. Target Hybridization

After determining that the fragmented cDNA is labeled with biotin, prepare the hybridization solution mix. The solution is stable for approx 6–8 h at 4°C. The following protocol can be used for freshly prepared or frozen hybridization cocktail (*see* **Note 4**).

3.5.2.1. Hybridization Solution Mix

Components	Volume/amount (µL)	Final concentration
2X MES hybridization buffer	150	1X
10 mg/mL Herring sperm DNA	3	0.1 mg/mL
50 mg/mL BSA	3	0.5 mg/mL
Fragmented labeled cDNA up to	144	1.3–12 µg
Water (molecular biology grade)	4	
Final volume	300	

Equilibrate probe array to room temperature immediately before use. After incubation at 99°C, transfer the hybridization cocktail to a 45°C heat block for 5 min. Spin the hybridization cocktail at maximum speed in a microcentrifuge for 5 min to remove any insoluble material from the hybridization mix. Add the hybridization solution mix (300 µL) to miRNA Array (mir sequence on the slide) array using Seal Secure Hybridization Chamber. Place probe array in the hybridization oven set at 45°C. After 12–16 h hybridization, proceed to the next **Subheading 3.5.2.2.**

3.5.2.2. Washing, Staining, and Scanning

1. Stringent wash buffer:100 mM MES, 0.1 M [Na+], 0.01% Tween-20:
   a. Preparation for 1000 mL wash buffer.
   b. 83.3 mL of 12X MES stock buffer.
   c. 5.2 mL of 5 M NaCl.
   d. 1.0 mL of 10% Tween-20.
   e. 910.5 mL of water.
   f. Filter through a 0.2-µm filter.
   g. Store at 2–8°C and shield from light.
2. Nonstringent wash buffer:6X SSPE, 0.01% Tween-20:
   a. Preparation for 1000 mL Nonstringent wash buffer.
   b. 300 mL of 20X SSPE.
   c. 1 mL of 10% Tween-20.
   d. 698 mL Water.
   e. Filter through a 0.2-µm filter.
   f. Store at room temperature.

3.5.2.3. 2X Stain Buffer

1. Preparation for 250 mL stain buffer (final 1X concentration: 100 mM MES, 1 M [Na+], 0.05% Tween-20):
   a. 41.7 mL 12X MES stock buffer.
   b. 92.5 mL 5 M NaCl.
   c. 2.5 mL 10% Tween-20.
   d. 112.8 mL Water.
   e. Filter through a 0.2-µm filter.
   f. Store at 2–8°C and shield from light.

g. 10 mg/mL Goat IgG stock.
h. Resuspend 50 mg in 5 mL of PBS.
i. Store at 4°C.
j. 1 mg/mL Streptavidin stock.
k. Resuspend 5 mg in 5 mL of PBS.
l. Store at 4°C.

3.5.2.4. STREPTAVIDIN SOLUTION MIX (FOR TWO SLIDES)

Components	Volume (µL)	Final concentration
2X MES stain buffer	9600	1X
50 mg/mL BSA	768	2 mg/mL
1 mg/mL Streptavidin	192	10 µg/mL
DI H$_2$O	8640	

3.5.2.5. ANTIBODY SOLUTION MIX (FOR TWO SLIDES)

Components	Volume (µL)	Final concentration
2X MES stain buffer	4800	1X
50 mg/mL BSA	384	2 mg/mL
10 mg/mL Normal goat IgG	92	0.1 mg/mL
0.5 mg/mL Biotin antistreptavidin	57.6	5 µg/mL
DI H$_2$O	4224	

1. Post Hyb Wash no. 1. For 1000 mL:
    a. 83.3 mL of 12X MES stock buffer.
    b. 5.2 mL of 5 *M* NaCl.
    c. 1 mL of 10% Tween-20.
    d. 910.5 mL of water.
    e. Wash four times with Wash no. 1 for 6 min at 30°C. Each time one slide use 120 mL buffer (240 mL buffer heated at Micro-oven about 20–25 s).
2. Post Hyb Wash no. 2.
    For 1000 mL.
    a. 300 mL of 20X SSPE.
    b. 1.0 mL of 10% Tween-20.
    c. 698 mL Water.
    d. Wash two times with 120 mL of Wash buffer no. 2 for 8 min at 45°C per slide at (240 mL buffer heated at Micro-oven about 45 s).

3.5.2.6. STAIN

Stain array for 10 min in 4.8 mL Streptavidin solution mix at 25°C.
Poststain wash: wash four times with Wash no. 1 for 6 min at 25°C. For each time use 120 mL of buffer per slide.

Second stain: stain the array for 10 min in 4.8 mL Antibody Solution Mix at 25°C.

Wash two times with Wash no. 1 for 5 min of at 25°C. For each time use 120 mL of buffer per slide.

Third stain: stain the array for 10 min in 4.8 mL SAPE Solution at 25°C.

Final wash: four times with Wash buffer no. 1 for 6 min at 25°C. For each time use 120 mL of buffer per slide.

Wash one time with 0.2X SSC for 5 min at room temperature.

Wash one time with 0.1X SSC for 5 min at room temperature.

Dry the slide by spinning at 750 rpm for 5 min. Scan the slide at 532 nm (*see* **Note 5**).

3.6. Conclusion

The microarray described here is a novel, selective and sensitive approach to monitoring miRNA expression in mammalian cells and tissues. We have identified miRNAs that are selectively expressed in kidney, heart, and skeletal muscle. By performing comparative sequence analysis we have proposed the presence of a putative transcriptional regulatory mediator for miRNAs in kidney. It is anticipated that future use of this array-based approach will lead to a greater understanding of recently identified class of gene regulating RNA molecules.

4. Notes

1. For additional information of printed slides and processing, follow manufacturer's recommendations.
2. The biotin end labeling protocol is mainly adapted from the Affymetrix protocols, and therefore, acknowledgments to Affymetrix for sharing the information publicly in their website.
3. The integrity of total RNA is essential for the success of the assay. Because of the sensitivity of the assay, removal of DNA contamination from the RNA preparation is crucial to prevent false-positive results.
4. Reuse of MiRNA hybridization sample has not been thoroughly tested and, therefore, is not recommended.
5. Unlike mRNAs, miRNAs (~22-nucleotide) are hard to manipulate, and therefore, to detect their expression levels in human and mouse organs have been difficult through Reverse transcriptase polymerase chain reaction and Northern. Current protocol was tested many times and the array data was confirmed by Northern analysis was described in **ref. 1**.

References

1. Yingqing, S., Seongjoon, K., Neill, W., et al. (2004) Development of a microarray to detect human and mouse microRNAs and characterization of expression in human organs. *Nucleic Acid Res.* **32,** E188.

2. Esau, C., Kang, X., Peralta, E., et al. (2004) MicroRNA-143 regulates adipocyte differentiation. MicroRNA-143 regulates adipocyte differentiation. *J. Biol. Chem.* **10**, 52,361–52,365.

3. Ambros, V. (2003) MicroRNA pathways in flies and worms: growth, death, fat, stress, and timing. *Cell* **113,** 673–676.

4. Bartel, B. and Bartel, D. P. (2003) MicroRNAs: at the root of plant development? *Plant Physiol.* **132,** 709–717.

5. Palatnik, J. F., Allen, E., Wu, X., et al. (2003) Control of leaf morphogenesis by microRNAs. *Nature* **425,** 257–263.

6. Bartel, D. P. (2004) MicroRNAs: genomics, biogenesis, mechanism, and function. *Cell* **116,** 281–297.

7. Pasquinelli, A. E., Reinhart, B. J., Slack, F., et al. (2000) Conservation of the sequence and temporal expression of let-7 heterochronic regulatory RNA. *Nature* **408,** 86–89.

8. Lee, R. C., Feinbaum, R. L., and Ambros, V. (1993) The C. elegans heterochronic gene lin-4 encodes small RNAs with antisense complementarity to lin-14. *Cell* **75,** 843–854.

9. Reinhart, B. J., Weinstein, E. G., Rhoades, M. W., Bartel, B., and Bartel, D. P. (2002) MicroRNAs in plants. *Genes Dev.* **16,** 1616–1626.

10. Lim, L. P., Glasner, M. E., Yekta, S., Burge, C. B., and Bartel, D. P. (2003) Vertebrate microRNA genes. *Science* **299,** 1540.

11. Lim, L. P., Lau, N. C., Weinstein, E. G., et al. (2003) The microRNAs of *Caenorhabditis elegans. Genes Dev.* **17,** 991–1008.

12. Ambros, V., Lee, R. C., Lavanway, A., Williams, P. T., and Jewell, D. (2003) MicroRNAs and other tiny endogenous RNAs in *C. elegans. Curr. Biol.* **13,** 807–818.

13. Grad, Y., Aach, J., Hayes, G. D., et al. (2003) Computational and experimental identification of *C. elegans* microRNAs. *Mol. Cell* **11,** 1253–1263.

14. Kim, J., Krichevsky, A., Grad, Y., et al. (2004) Identification of many microRNAs that copurify with polyribosomes in mammalian neurons. *Proc. Natl. Acad. Sci. USA* **101,** 360–365.

15. Lee, R. C. and Ambros, V. (2001) An extensive class of small RNAs in *Caenorhabditis elegans. Science* **294,** 862–864.

16. Lagos-Quintana, M., Rauhut, R., Yalcin, A., Meyer, J., Lendeckel, W., and Tuschl, T. (2002) Identification of tissue-specific microRNAs from mouse. *Curr. Biol.* **12,** 735–739.

17. Lagos-Quintana, M., Rauhut, R., Meyer, J., Borkhardt, A., and Tuschl, T. (2003) New microRNAs from mouse and human. *RNA* **9,** 175–179.

18. Mourelatos, Z., Dostie, J., Paushkin, S., et al. (2002) miRNPs: a novel class of ribonucleoproteins containing numerous microRNAs. *Genes Dev.* **16,** 720–728.

19. Dostie, J., Mourelatos, Z., Yang, M., Sharma, A., and Dreyfuss, G. (2003) Numerous microRNPs in neuronal cells containing novel microRNAs. *RNA* **9,** 180–186.

20. Lau, N. C., Lim, L. P., Weinstein, E. G., and Bartel, D. P. (2001) An abundant class of tiny RNAs with probable regulatory roles in *Caenorhabditis elegans*. *Science* **294,** 858–862.
21. Lagos-Quintana, M., Rauhut, R., Lendeckel, W., and Tuschl, T. (2001) Identification of novel genes coding for small expressed RNAs. *Science* **294,** 853–858.
22. Lai, E. C., Tomancak, P., Williams, R. W., and Rubin, G. M. (2003) Computational identification of Drosophila microRNA genes. *Genome Biol.* **4,** R42.
23. Calin, G. A., Sevignani, C., Dumitru, C. D., et al. (2004) Human microRNA genes are frequently located at fragile sites and genomic regions involved in cancers. *Proc. Natl. Acad. Sci. USA* **101,** 2999–3004.
24. Krichevsky, A. M., King, K. S., Donahue, C. P., Khrapko, K., and Kosik, K. S. (2003) A microRNA array reveals extensive regulation of microRNAs during brain development. *RNA* **9,** 1274–1281.
25. Liu, C. G., Calin, G. A., Meloon, B., et al. (2004) An oligonucleotide microchip for genome-wide microRNA profiling in human and mouse tissues. *Proc. Natl. Acad. Sci. USA* **101,** 9740–9744.
26. Calin, G. A., Liu, C. G., Sevignani, C., et al. (2004) MicroRNA profiling reveals distinct signatures in B cell chronic lymphocytic leukemias. *Proc. Natl. Acad. Sci. USA* **101,** 11,755–11,760.
27. Babak, T., Zhang, W., Morris, Q., Blencowe, B. J., and Hughes, T. R. (2004) Probing microRNAs with microarrays: tissue specificity and functional inference. *RNA* **10,** 1813–1819.
28. Miska, E. A., Alvarez-Saavedra, E., Townsend, M., et al. (2004) Microarray analysis of microRNA expression in the developing mammalian brain. *Genome Biol.* **5,** R68.
29. Thomson, J. M., Parker, J., Perou, C. M., and Hammond, S. M. (2004) A custom microarray platform for analysis of microRNA gene expression. *Nat. Methods* **1,** 47–53.
30. Nelson, P. T., Baldwin, D. A., Scearce, L. M., Oberholtzer, J. C., Tobias, J. W., and Mourelatos, Z. (2004) Microarray-based, high-throughput gene expression profiling of microRNAs. *Nat. Methods* **2,** 155–161.
31. Esquela-Kerscher, A. and Slack, F. J. (2004) The age of high-throughput microRNA profiling. *Nat. Methods* **2,** 106–107.
32. Rosner, B. (2000) *Fundamentals of Biostatistics*. Duxbury, New York.

10

Genotyping of Single Nucleotide Polymorphisms by Arrayed Primer Extension

Scott J. Tebbutt

Summary

Although the majority of microarray studies have been directed toward RNA expression profiling (functional genomics) and increasingly toward proteomics, a steady increase in the use of microarrays as platforms for DNA genotyping has occurred over the past 5 yr. Multiple array-based chemistries have been developed in order to genotype single nucleotide polymorphisms. Conceptually, the simplest of these microarray genotyping technologies is based on the dideoxynucleotide chemistry of mini-sequencing by arrayed primer extension, whereby oligonucleotide probes (preprinted on the array) are extended by a single nucleotide base. This enzyme-catalyzed single base extension reaction is dependent on the sequence (genotype) of the template nucleic acid (sample) that is temporarily hybridized to the probes. Utilization of all four dideoxynucleotides, each conjugated to a different fluorophore, allows genotyping by spectral differentiation of the single base extension reaction products.

Key Words: Arrayed primer extension (APEX); fluorescent dideoxynucleotide; genotyping; microarray; mini-sequencing; single nucleotide polymorphism (SNP); single base extension (SBE).

1. Introduction

The completion of the human genome project was an important step in the exploration of human genetics *(1,2)*. However, it is the present worldwide search for individual variation in the human genome that promises to elucidate how genetic variation interacts with the environment to confer individual resistance or susceptibility to disease, responsiveness to medical interventions, and drug toxicity *(3,4)*. The most common form of genetic variation between individuals is single nucleotide polymorphisms (SNPs), which are single base changes at specific DNA sites in the genome, occurring at a frequency of approx 1 SNP every 200–1000 bases *(5)*. Different combinations of SNPs in single or multiple

From: *Methods in Molecular Biology, vol. 382: Microarrays: Second Edition: Volume 2*
Edited by: J. B. Rampal © Humana Press Inc., Totowa, NJ

genes are hypothesized to interact with environmental factors to determine risk for disease as well as variability in how individuals respond to illness and medical therapy and whether they might develop adverse drug responses. Research directed toward discovering gene–gene and gene–environment interactions in disease causation and clinical outcome is increasing at an exponential rate and pharmacogenomics is often quoted as being poised for application to health care as "personalized medicine" *(3,6,7)*.

Of the many methods that have been developed for genotyping, those based on the use of microarrays offer the greatest potential for economic, patient-specific application, owing to their ability to simultaneously interrogate multiple genetic markers (SNPs) using genetic material (template) amplified from an individual using the PCR *(8)*. Genotyping microarrays are devices displaying specific oligonucleotide probes, precisely located on a small-format solid support such as a glass slide. Although, a number of different microarray genotyping chemistries exist *(9–11)*, we are concerned here with arrayed primer extension (APEX) *(12,13)*.

APEX is a mini-sequencing *(14)* method based on a two-dimensional array of oligonucleotides, immobilized through their 5′-ends on a microarray glass surface. The oligonucleotides (from 20- to 25-mers) are designed so that they are complementary to the gene up to, but not including, the base where the SNP exists, although allele-specific oligonucleotide probes (where the 3′ base is the complement of the allelic site) can also be used *(15–17)*. The Sanger-based "terminator" sequencing chemistry of APEX allows simultaneous genotyping of hundreds to a few thousand SNPs, with the array chemistry taking just minutes. APEX achieves this clinically relevant speed because it uses the catalytic ability of a DNA polymerase enzyme to carry out a single nucleotide base extension at the 3′-end of the arrayed oligonucleotide probes, specific to the SNP sites of interest in template DNA that is temporarily hybridized to these probes. Dideoxynucleotide "terminator" bases are prelabeled with fluorophores specific to each one of the four bases of DNA (A, C, G, and T). Hence the fluorescent "color" (wavelength of emitted light) at each of the probe sites (array spots) will give SNP-specific genotypic information (*see* **Fig. 1**). As a discovery research tool, APEX has been used to detect β-thalassemia *(16)*, p53 *(18)*, and BRCA1 mutations *(19)*. Importantly, APEX has also been shown to be efficient at analyzing genome-wide SNP markers *(17,20)*.

The scope of this chapter is to describe the procedures for APEX, including: (1) APEX probe and PCR primer design; (2) microarray printing; (3) PCR amplification; (4) PCR product purification and fragmentation; (5) APEX reaction; (6) microarray imaging and spot channel intensity extraction; and (7) data management and genotyping (*see* **Fig. 2**).

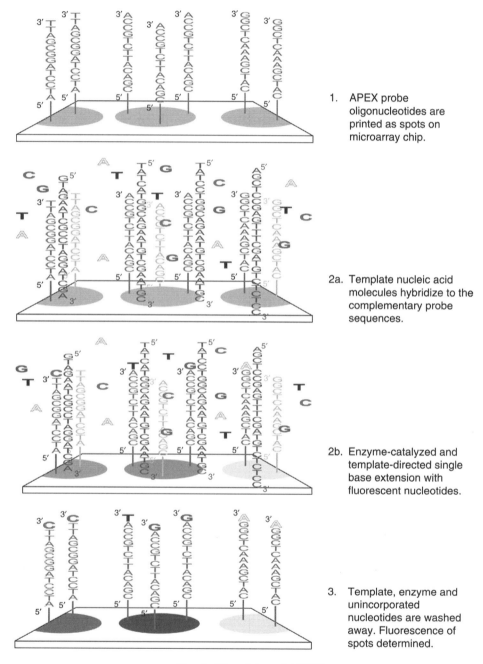

1. APEX probe oligonucleotides are printed as spots on microarray chip.

2a. Template nucleic acid molecules hybridize to the complementary probe sequences.

2b. Enzyme-catalyzed and template-directed single base extension with fluorescent nucleotides.

3. Template, enzyme and unincorporated nucleotides are washed away. Fluorescence of spots determined.

Fig. 1. Chemistry of APEX.

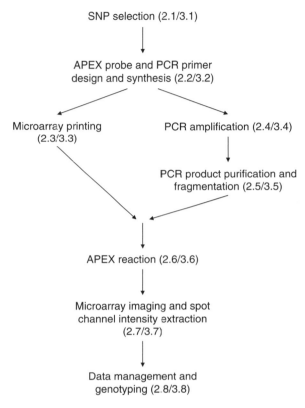

Fig. 2. Flow chart of methodology.

2. Materials

Safety Precautions: Read the material safety data sheet for each reagent before beginning experiments. Wear appropriate personal protective equipment required according to the material safety data sheet, such as lab coats, protective gloves, and eyewear. In particular, use double wash chambers to avoid injury by hot water during the APEX experiment (*see* **Note 1**), and handle glass slides and racks with care.

2.1. SNP Selection

1. NCBI SNP website (http://www.ncbi.nih.gov/SNP/).
2. The SNP Consortium website (http://snp.cshl.org/).
3. The Coriell Institute Cell and DNA Repository website (http://coriell.umdnj.edu/).

2.2. APEX Probe and PCR Primer Design and Synthesis

1. Assistance with APEX probes, and PCR primer design, for higher throughput use can be obtained from Biodata Ltd., Estonia (http://www.biodata.ee).

2. A positive control/four channel normalization APEX probe has been designed ("Npg1 Positive control") *(17)* that is based on the sequence of a plant gene *(21)*. The sequence of this probe is 5′-TCGTGTGATACCTAGTCAAGATAAT-3′. It is designed to give an equal probability incorporation of any of the four nucleotides, driven by hybridization to the following template oligonucleotide: 5′-TAATCT-NATTATCTTGACTAGGTATCACACGAAGTTCTAAA-3′.

3. APEX probes require 5′ amino-linker modification and can be ordered at the lowest synthesis level available because so little of the material is used on each array (*see* **Note 2**). Store all oligonucleotides at −20°C.

2.3. Microarray Printing

1. Microarray spotting robot—e.g., BioRobotics MicroGrid™ spotters (Genomic Solutions, Ann Arbor, MI) with split pins.
2. CodeLink™-Activated microarray slides (Amersham Biosciences, Piscataway, NJ).
3. Printing buffers and blocking reagents recommended by slide manufacturer (*see* **Note 3**).
4. Slide blocking buffer (50 m*M* ethanolamine, 0.1 *M* Tris-HCl, pH 9.0, 0.1% sodium dodecyl sulfate).
5. General laboratory equipment required: Wheaton slide dishes and racks, plastic sandwich boxes, water bath, shaking platform, centrifuge capable of handling glass slide racks (*see* **Note 4**).

2.4. PCR Amplification

1. PCR thermocycler.
2. HotStarTaq® DNA polymerase and buffer: 10X PCR buffer (Tris-HCl, ammonium sulfate—$(NH_4)_2SO_4$, 15 m*M* $MgCl_2$, pH 8.7) (Qiagen, Valencia, CA). Store at −20°C.
3. dATP, dCTP, dGTP, dTTP, dUTP—molecular biology grade. Store at −20°C.

2.5. PCR Product Purification and Fragmentation

1. Ethanol (70 and 99%). Store at −20°C in small quantities.
2. 10 *M* ammonium acetate. Store at room temperature.
3. 10X Uracil N-Glycosylase (UNG) digestion buffer: 0.5 *M* Tris-HCl, pH 9.0, 0.2 *M* $(HN_4)_2SO_4$. Store at −20°C.
4. Shrimp alkaline phosphatase, SAP (Amersham Biosciences). Store at −20°C.
5. UNG (EPICENTRE, Madison, WI). Store at −20°C.

2.6. APEX Reaction

1. HyPro100 incubation plate (Thermo Electron, Milford, MA).
2. Parafilm®.
3. Stainless steel hot water kettle.
4. Thermo Sequenase™ DNA polymerase (Amersham Biosciences). Store at −20°C.
5. Texas Red®-ddATP, Cy™3-ddCTP, Cy5-ddGTP, R110-ddUTP (Perkin Elmer Life Sciences, Boston, MA). Store at −20°C.

6. 10X Thermo Sequenase reaction buffer (260 mM Tris-HCl, pH 9.5, 65 mM MgCl$_2$). Store aliquots at −20°C.
7. 0.3% Alcanox solution. Store at room temperature.

2.7. Microarray Imaging and Spot Channel Intensity Extraction

1. Four channel microarray scanner—*arrayWoRx®e Auto* Biochip Reader (Applied Precision, LLC, Issaquah, WA).
2. Filter sets: 1. A488—Ex. 480/15x—Em. 530/40 (R110 dye); 2. Cy3 (narrowband)— Ex. 546/11—Em. HQ570/10m (Cy3); 3. Texas Red—Ex. 602/13—Em. 631/23 (Texas Red); 4. Cy5—Ex. 635/20—Em. 685/40 (Cy5) (Chroma Technology Corporation, Rockingham, VT).
3. Microarray TIFF image gridding, segmentation, and intensity extraction software. E.g., *softWoRx* Tracker software V.2.23.02 (MolecularWare Inc., Cambridge, MA), and Genorama™ software V.4.2 (Asper Biotech, Tartu, Estonia).

2.8. Data Management and Genotyping

1. Microsoft Excel®.
2. SNP Chart© genotyping tool *(22)* (http://www.snpchart.ca).

3. Methods

The procedures described next have been tested and validated through our laboratory's experience with developing an APEX microarray resource known as the "Genomic Control SNP Chip" *(17)*, which allows genotyping of more than one hundred and twenty random ("null") SNPs throughout the human genome. This resource can be used to control for spurious associations (false-positives) that may arise owing to hidden population stratification in gene association studies of complex diseases *(23–25)*. An innovative microarray genotyping analysis and quality control tool was also developed (the SNP Chart application *[22]*). This software utility is a data management and visualization tool for array-based genotyping by single base primer extension from multiple oligonucleotide probes.

3.1. SNP Selection

1. A great deal of useful genetic polymorphism information can be obtained from NCBI, The SNP Consortium, and related websites, in order to assist in the selection of SNPs from the human genome. Researchers studying other organisms will know the websites that curate genetic information appropriate to those organisms.
2. For validation purposes, Coriell DNA samples from multiple ethnic groups can be obtained for which a substantial number of SNPs have already been genotyped by other laboratories. It is highly recommended that researchers developing new APEX genotyping chips test and validate those using Coriell samples.

3.2. APEX Probe and PCR Primer Design

1. Six oligonucleotide probes (25-mers) for each SNP are designed: two classical APEX probes (one probe per DNA strand), plus four additional allele-specific (AS-APEX) probes per SNP marker (two probes per strand) that include the actual SNP site at the 3′-end of the probe (*see* **Note 5**).

2. PCR primers are designed based on a T_m of 62°C plus/minus 3°C (at 20 mM monovalent salt concentration in PCR buffer) with a length between 20 and 26 bases. PCR amplicon product length range should be between 150 and 1000 bp. All primer pairs should be computationally tested against the human genome in order to check that they amplify single product. Any SNP primers found to be located in repeated regions of the genome should be removed from further analysis and hence from the final chip design, unless alternate primers that amplify a unique PCR product for that SNP can be found.

3. Oligonucleotides are synthesized at 10–50 nmol scale and aliquotted into 96-well plates at 200 pmol/μL stock concentration in water. Each of the APEX and AS-APEX probes should be synthesized with a 5′ amino-linker.

4. To keep PCR costs to a minimum, algorithms can be used to suggest the grouping together of multiple PCR primer pairs that would multiplex the entire amplicon set in a minimum number of PCR reactions (e.g., "MultiPLX" [http://www.biodata.ee]).

3.3. Microarray Printing

1. The APEX and AS-APEX probe oligonucleotides (50 pmol/μL in 150 mM sodium phosphate print buffer, pH 8.5) are printed to specific grid positions on the microarray slides, following the manufacturer's recommended protocols. The 5′-end of each probe oligonucleotide is amino-modified, allowing its covalent attachment to the slide's preapplied surface chemistry. Each grid consists of duplicate spots of each of the six probes per SNP, as well as multiple buffer-only spots and positive control spots. The latter can comprise two types of positive controls: multiple combinations of self-extending control oligonucleotides, designed to extend to one or more of the four DNA bases, A, C, G, and T (SeqN probes—*[13]*); and the positive control/four channel normalization APEX probe ("Npg1 Positive control").

2. Each Npg1 Positive control probe can be spotted at least 80 times onto the grids, at regular physical intervals. Each one of the six probes for each SNP should be printed at a reasonably wide distance apart from any other probe for the same SNP within the grid (as are their duplicate spots). This enables a useful degree of robustness in the system, especially helpful in cases of high local background and hybridization problems.

3. Following printing of the arrays, the slides are placed in a glass rack and are incubated overnight at room temperature at 75% relative humidity to drive the covalent coupling reaction between the probe's 5′ amino group and the CodeLink-Activated microarray slide chemistry to completion.

4. Blocking of the residual reactive groups on the slides is facilitated by soaking in 250 mL prewarmed blocking buffer at 50°C for 15 min.

5. Discard the blocking buffer and rinse the slides twice with 250 mL water in a plastic container.
6. Wash slides with 250 mL 4X SSC buffer with 0.1% sodium dodecyl sulfate (prewarmed to 50°C) for 40 mL on a shaking platform.
7. Discard SSC buffer and rinse the slides twice with 250 mL water (room temperature) in a plastic container.
8. Place the glass rack containing the slides into a microplate carrier of a centrifuge, putting Kimwipes® tissues between the carrier and rack. Cover the rack with foil, then centrifuge at 100*g* for 3 min to dry the slides. The printed and blocked slides can be stored dry in slide boxes for up to 6 mo.

3.4. PCR Amplification

1. Multiplex PCR amplification is performed on genomic DNA samples (plus a negative PCR control sample that contains no genomic DNA).
2. Each PCR reaction is performed in a total volume of 25 µL containing 2.5 µL 10X PCR buffer, 3 m*M* MgCl$_2$, 200 µ*M* dNTPs without dTTP, 160 µ*M* dTTP, 40 µ*M* dUTP (*see* **Note 6**), 1.25 units Qiagen HotStar Taq DNA polymerase (5 U/µL), 1 µL 7.7 m*M* PCR primer mixes and 10 ng genomic DNA.
3. PCRs are initiated by a 15-min polymerase activation step at 95°C, and completed by a final 10-min extension step at 72°C. The cycles for PCR are as follows: 25 cycles of 30 s denaturation at 95°C, 30 s annealing at 60°C, and 50 s extension at 72°C; followed by another 10 cycles of 30 s at 95°C, 30 s at 58°C, and 50 s at 72°C.
4. Three microliter aliquots of all amplicons are visualized with ethidium bromide staining under ultraviolet light on a 2.5% agarose gel following electrophoresis in 0.5X TBE buffer.

3.5. PCR Product Purification and Fragmentation

1. The multiplex PCR products are pooled for each individual genomic DNA sample (and negative control), and the total volume is measured.
2. 2.5 volumes of ice-cold 99% ethanol is added, followed by 0.25 volumes of 10 *M* ammonium acetate solution, and the PCR products are precipitated at −20°C overnight.
3. The samples are centrifuged at 20,800*g* and 4°C for 20 min. The supernatant is removed carefully, and the DNA pellet is washed with 400 µL of ice-cold 70% ethanol. The DNA pellet is then dissolved in 15 µL water. Three microliters of the precipitated DNA is checked for quality control by electrophoresis on a 2.5% agarose gel.
4. The remaining precipitated DNA is then fragmented, and unincorporated dNTPs inactivated, by digestion with 1 UNG and 1 U shrimp alkaline phosphatase (SAP) for 1 h at 37°C, followed by enzyme inactivation for 10 min at 95°C, in a 20-µL reaction mixture containing 2 µL 10X digestion buffer. Three microliters of digests are visualized by electrophoresis on a 2.5% agarose gel followed by ethidium bromide staining.

3.6. APEX Reaction

1. The APEX reaction is performed in a total volume of 50 µL by adding (on ice) 17 µL fragmented DNA template, 1 µL of 2 pmol/µL Npg1 Positive control template oligonucleotide, 1 µ*M* of each fluorescently labeled dideoxynucleotide triphosphate (Texas Red-ddATP, Cy3-ddCTP, Cy5-ddGTP, R110-ddUTP), 5 U Thermo Sequenase™ DNA polymerase diluted in its dilution buffer, to 2X Thermo Sequenase reaction buffer (10X—260 m*M* Tris-HCl, 65 m*M* MgCl$_2$, pH 9.5).

2. The reaction mixture is applied to the array of APEX and AS-APEX probes previously printed on the CodeLink slide, that has been washed two times in 95°C water and placed on a ThermoElectron HyPro100 incubation plate set at 58°C.

3. The reaction mixture is covered with a small piece of Parafilm (precut to a size slightly larger than the area of the array grid) and the APEX reaction is allowed to proceed at 58°C with agitation (setting 1 on the HyPro100) for 20 min.

4. Following the incubation period, slides are washed first with warm water to allow the parafilm to dislodge, and then with 95°C water (2 min) to fully remove the template DNA, enzyme and excess ddNTPs. Further washing in 0.3% alcanox (3 min) and 95°C water (2 min) ensures low background on the array images.

5. Following the last hot water wash step, each slide is slowly removed from the water using tweezers, allowing the excess water to drain from the slide surface.

3.7. Microarray Imaging and Spot Channel Intensity Extraction

1. Slide microarrays are imaged using an *arrayWoRx^e Auto* Biochip Reader.

2. Exposure times for each dye are set up to give approx 60–70% pixel saturation for selected "Npg1 Positive control" probe spots. Resolution of the imager is set to 10 µm.

3. Four 16-bit gray-scale TIFF files for each array are obtained (one from each channel—*see* **Fig. 3**) and these are analyzed using *softWoRx* Tracker software V.2.23.02.

4. Spot intensity values ("Cell" background method) and probe name/grid coordinates are exported to Microsoft Excel the intensity values are normalized by setting up to 82 "Npg1" Positive control' spots, widely distributed across each array, to an average value of 20,000 U per channel, with the exported normalized intensity value calculated from the scale factor x [median signal—median background]). If some of the Npg1 Positive control spots have high levels of local background or have not reacted well, they will adversely affect this calculation, and hence should not be included as "positive controls."

3.8. Data Management and Genotyping

1. Excel files containing the *softWoRx* normalized spot intensity values and probe name/grid coordinates for each sample are imported into SNP Chart, where the data for each of the SNPs across each sample can be visualized and genotyped (*see* **Fig. 4**).

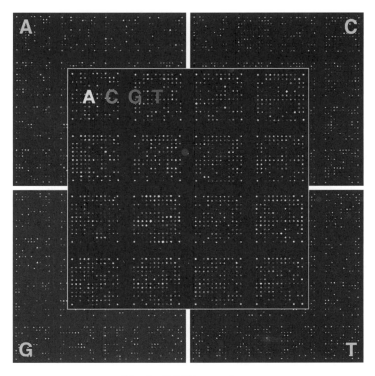

Fig. 3. APEX array images.

2. A four color "blended" TIFF image of the array can also be imported into SNP Chart and this serves a useful spot quality control function (*see* **Fig. 3**).
3. Recently, SNP Chart has been upgraded to allow for automated genotyping and testing of multiple auto-calling algorithms.

4. Notes

1. Plastic containers were used (e.g., GLAD®-type food storage boxes) for all hot water wash steps. For safety purposes use two containers, one inside the other.
2. The price of the APEX and AS-APEX probe oligonucleotides represents a significant start-up cost (due to the required 5′ amino-linker modification). However, due to the very small amount (0.5–1 nL) of each probe deposited as a spot on the array by the printing pin tool, even a 10 nmol oligonucleotide synthesis scale will generate enough probe material to print thousands of arrays.
3. The laboratory has successfully used several other types of glass slides for APEX arrays, including "Genorama® SAL" slides (http://www.asperbio.com), and "Aldehyde-" and "Epoxy-" coated slides (http://www.nuncbrand.com).
4. Unless stated otherwise, all solutions should be prepared in water that has a resistivity of 18.2 MΩ-cm and total organic content of less than five parts per billion. This standard is referred to as "water" in the text.

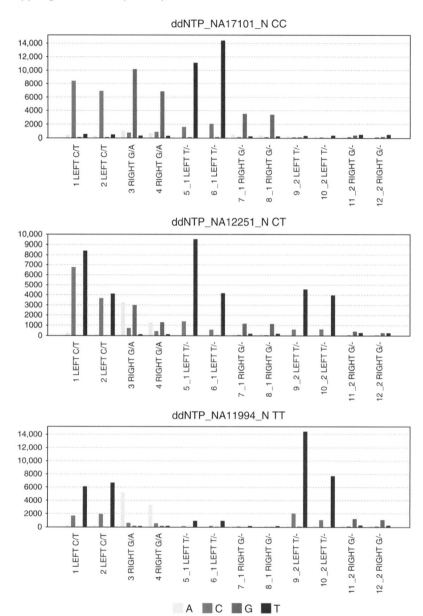

Fig. 4. Examples of SNP charts (rs929215 C/T SNP).

5. Allele-specific single base extension of these AS-APEX probes during the reaction is contingent on the presence of the actual complementary base at the SNP site in the patient template DNA *(15–17)*. This provides a useful redundancy in the probe set that gives considerable robustness in the genotyping, and hence confidence in the data.

6. Incorporation of the dUTP allows the amplified DNA to be enzymatically sheared by UNG to produce a DNA size of about 100 bases, optimal for hybridization to the oligonucleotides on the microarray.

Acknowledgments

The author would like to thank Jian Ruan for technical assistance, and Peter Paré for continued support. The research described has been supported by the National Sanitarium Association (Canada), the British Columbia Lung Association, the Canadian Institutes of Health Research, and the Michael Smith Foundation for Health Research.

References

1. Lander, E. S., Linton, L. M., Birren, B., et al. (2001) Initial sequencing and analysis of the human genome. *Nature* **409,** 860–921.
2. Venter, J. C., Adams, M. D., Myers, E. W., et al. (2001) The sequence of the human genome. *Science* **291,** 1304–1351.
3. Risch, N. and Merikangas, K. (1996) The future of genetic studies of complex human diseases. *Science* **273,** 1516–1517.
4. Lord, P. G. and Papoian, T. (2004) Genomics and drug toxicity. *Science* **306,** 575.
5. Wang, D. G., Fan, J. B., Siao, C. J., et al. (1998) Large-scale identification, mapping, and genotyping of single-nucleotide polymorphisms in the human genome. *Science* **280,** 1077–1082.
6. Ansell, S. M., Ackerman, M. J., Black, J. L., Roberts, L. R., and Tefferi, A. (2003) Primer on medical genomics. Part VI: genomics and molecular genetics in clinical practice. *Mayo Clin. Proc.* **78,** 307–317.
7. Hood, L., Heath, J. R., Phelps, M. E., and Lin, B. (2004) Systems biology and new technologies enable predictive and preventative medicine. *Science* **306,** 640–643.
8. Saiki, R. K., Gelfand, D. H., Stoffel, S., et al. (1988) Primer-directed enzymatic amplification of DNA with a thermostable DNA polymerase. *Science* **239,** 487–491.
9. Hirschhorn, J. N., Sklar, P., Lindblad-Toh, K., et al. (2000) SBE-TAGS: an array-based method for efficient single-nucleotide polymorphism genotyping. *Proc Natl. Acad. Sci. USA* **97,** 12,164–12,169.
10. Oliphant, A., Barker, D. L., Stuelpnagel, J. R., and Chee, M. S. (2002) BeadArray technology: enabling an accurate, cost-effective approach to high-throughput genotyping. *Biotechniques* **56–58,** 60–51.
11. Kennedy, G. C., Matsuzaki, H., Dong, S., et al. (2003) Large-scale genotyping of complex DNA. *Nat. Biotechnol.* **21,** 1233–1237.
12. Shumaker, J. M., Metspalu, A., and Caskey, C. T. (1996) Mutation detection by solid phase primer extension. *Hum. Mutat.* **7,** 346–354.
13. Kurg, A., Tonisson, N., Georgiou, I., et al. (2000) Arrayed primer extension: solid-phase four-color DNA resequencing and mutation detection technology. *Genet Test* **4,** 1–7.

14. Syvanen, A. C. (1999) From gels to chips: "minisequencing" primer extension for analysis of point mutations and single nucleotide polymorphisms. *Hum. Mutat.* **13,** 1–10.
15. Pastinen, T., Raitio, M., Lindroos, K., et al. (2000) A system for specific, high-throughput genotyping by allele-specific primer extension on microarrays. *Genome Res.* **10,** 1031–1042.
16. Gemignani, F., Perra, C., Landi, S., et al. (2002) Reliable detection of β-thalassemia and G6PD mutations by a DNA microarray. *Clin. Chem.* **48,** 2051–2054.
17. Tebbutt, S. J., He, J. Q., Burkett, K. M., et al. (2004) Microarray genotyping resource to determine population stratification in genetic association studies of complex disease. *Biotechniques* **37**, 977–985.
18. Tonisson, N., Zernant, J., Kurg, A., et al. (2002) Evaluating the arrayed primer extension resequencing assay of TP53 tumor suppressor gene. *Proc. Natl. Acad. Sci. USA* **99,** 5503–5508.
19. Tonisson, N., Kurg, A., Kaasik, K., Lohmussaar, E., and Metspalu, A. (2000) Unravelling genetic data by arrayed primer extension. *Clin. Chem. Lab. Med.* **38,** 165–170.
20. Dawson, E., Abecasis, G. R., Bumpstead, S., et al. (2002) A first-generation linkage disequilibrium map of human chromosome 22. *Nature* **418,** 544–548.
21. Tebbutt, S. J., Rogers, H. J., and Lonsdale, D. M. (1994) Characterization of a tobacco gene encoding a pollen-specific polygalacturonase. *Plant. Mol. Biol.* **25,** 283–297.
22. Tebbutt, S. J., Opushnyev, I. V., Tripp, B. W., et al. (2005) SNP Chart: an integrated platform for visualization and interpretation of microarray genotyping data. *Bioinformatics* **21,** 124–127.
23. Devlin, B. and Roeder, K. (1999) Genomic control for association studies. *Biometrics* **55,** 997–1004.
24. Reich, D. E. and Goldstein, D. B. (2001) Detecting association in a case-control study while correcting for population stratification. *Genet Epidemiol.* **20,** 4–16.
25. Freedman, M. L., Reich, D., Penney, K. L., et al. (2004) Assessing the impact of population stratification on genetic association studies. *Nat. Genet.* **36,** 388–393.

11

Protein Chip for Detection of DNA Mutations

Xian-En Zhang and Li-Jun Bi

Summary

A large number of human genetic diseases, bacterial drug resistances, and single-nucleotide polymorphisms are caused by gene mutations. Rapid and high-throughput mutation detection methods are urgently demanded. A protein chip method for detection of single-base mismatches and unpaired bases of DNA was developed using a genetic fusion molecular system Trx-His$_6$-(Ser-Gly)$_6$-Strep tagII-(Ser-Gly)$_6$-MutS (THLSLM). The THLSLM coding sequence was constructed by attaching *Strep tag II* and *mutS* gene to the vector pET32a (+) sequentially with insertion of a (Ser-Gly)$_6$ coding sequence before and behind *Strep tagII* gene, respectively. The fusion protein THLSLM was expressed in *Escherichia coli* AD494 (DE3) and purified using Ni^{2+}-chelation affinity resin. The results of bioactivity assay showed that THLSLM both binds to mismatched DNA and interacts with streptavidin. THLSLM was immobilized on the chip matrix coated with the streptavidin through Strep tagII-streptavidin binding reaction. The resulting protein chip was used to detect the mismatched and unpaired mutations in the synthesized oligonucleotides, as well as a single-base mutation in *rpoB* gene from *Mycobacterium tuberculosis*, with high specificity. The method could potentially serve as a platform to develop the high-throughput technology for screening and analysis of genetic mutations.

Key Words: Detection; fusion molecular system; mutations; protein chip; THLSLM; high-throughput.

1. Introduction

A large number of human genetic diseases, bacterial drug resistances, and single-nucleotide polymorphisms are caused by gene mutations including single-base substitutions and small insertions or deletions of bases in the genome *(1–4)*. The rapid rate of discovery of disease genes increases the need for developing high-throughput mutation detection methods allowing screening of many individuals at multiple loci. Among the number of approaches for detection of DNA mutations, DNA direct sequencing is regarded as being most reliable, but quite time-consuming and costly *(5,6)*. The other widely used methods for

From: *Methods in Molecular Biology, vol. 382: Microarrays: Second Edition: Volume 2*
Edited by: J. B. Rampal © Humana Press Inc., Totowa, NJ

small genetic alteration detection include single-stranded conformation polymorphism *(7)*, denaturing gradient gel electrophoresis *(8)*, enzyme or chemical mismatch cleavage *(9,10)*. These methods are all based on the use of polyacrylamide gel electrophoresis and quite labor intensive, which significantly limits their suitability in automated applications requiring rapid screening of samples.

The MutS protein, a crucial component of the DNA mismatch repair system in many organisms, such as *Escherichia coli*, can specifically recognize and bind all possible single-base mismatches as well as 1–4 base insertion or deletion loops specifically with varying affinities independent of other proteins or cofactors *(11–13)*. The mismatch-binding activity of MutS has been exploited for detection of DNA mutations in several formats *(14–19)*, but none of the MutS-based mutation detection methods have enjoyed widespread application for high throughput, parallel, and multiplex analysis. In recent years, biochip technology has developed rapidly and has shown its great advantage in the parallel analysis of multiple sequences, so it is a point to use this feature in combination with MutS for detection of DNA mutations.

A new biochip format for detecting DNA mutations were proposed, which was essentially a MutS-based protein chip using a genetic fusion molecular system Trx-His$_6$-(Ser-Gly)$_6$-Strep tagII-(Ser-Gly)$_6$-MutS (THLSLM) is proposed. Compared with the DNA chip method, the proposed method may have two main advantages. First, other than labeling MutS protein, which is rather difficult, contrarily only nucleotide were labeled, which is a facile job. Second, unlike the multioligonucleotide chip, the MutS-based protein chip is much easier to prepare because it is a unique biomolecule chip. This advantage was more obvious when an "anchor chain" model *(20)*, a gene manipulation-based technique was introduced, where the orientation of all the protein molecules was well controlled and the homogeneity of the chip could be greatly improved *(21)*.

2. Materials

2.1. Construction of THLSLM Expression System

1. *E. coli* AD494 (DE3) and plasmid pET32a (+) from Novagen (Shanghai, China).
2. PCR primers for cloning *mutS* gene (Sangon Company, Shanghai, China).
3. Ampicillin: 50 mg/mL in Millpore water and store in single use aliquots at –20°C. Working solutions prepared by dilution.
4. Kanamycin sulphate: 60 mg/mL in Millpore water and store in single use aliquots at –20°C. Working solutions prepared by dilution.
5. 5-Bromo-4-chloro-3-indolyl β-D-Galactopyranoside: 50 μg/mL in N, N′-dimethylformamide. Cover with aluminum foil and store in single use aliquots at –20°C. Working solutions prepared by dilution.
6. Restriction enzymes, DNA polymerase, and T4-DNA ligase obtained from Takara (Dalian, China) and Promega (Beijing, China).

7. PCR purification mini kit and gel extraction mini kit purchased from Huaxun.
8. *E. coli* DH5α used for all bacterial transformations and plasmid propagations stored in the laboratory.
9. *E. coli* K-12, its genome as template to amplify *mutS* gene, stored in the laboratory.

2.2. Cell Culture, Lysis, and Purification

1. Isopropylthio,β-D-galactoside (IPTG) dissolved at 0.4 M of the concentration in Millipore water and stored in single use aliquots at −20°C. Working solutions prepared by dilution.
2. Luria Bertanii (LB) medium supplemented with 50 µg/mL ampicillin, 60 µg/mL kanamycin sulphate and 0.6 mM IPTG.
3. ALON metal affinity resins (Novagen).
4. Extracting buffer: 0.5 M NaCl, 20 mM Tris-HCl, pH 7.9. Stored at room temperature.
5. Washing buffer: 5 mM imidazole, 0.5 M NaCl, 20 mM Tris-HCl, pH 7.9. Stored at room temperature.
6. Elution buffer: 250 mM imidazole, 0.5 M NaCl, 20 mM Tris-HCl, pH 7.9. Store at room temperature.
7. Dialyzing buffer: 50 mM Tris-HCl, pH 7.2, 100 mM KCl, 1.0 mM dithiothreitol (DTT), and 1.0 mM ethylenediaminetetra-acetic acid (EDTA). Store at room temperature.

2.3. Native Polyacrylamide Gel Electrophoresis

1. 30% Acrylamide/bisacrylamide solution: it contains 29% (m/v) acrylamide and 1% (m/v) bisacrylamide (this is a neurotoxin when unpolymerized; care should be taken to avoid the exposure to it). Store at 4°C.
2. TBE buffer (5X): 450 mM Tris-HCl, pH 8.0, 10 mM EDTA. Store at room temperature.
3. Ammonium persulfate: prepare 10% solution in water and immediately freeze in single use (200 µL) aliquots at −20°C.
4. N, N, N, N′-Tetramethyl-ethylenediamine (Bio-Rad, Hercules, CA).
5. Low-ionic strength buffer: 7 mM Tris-HCl, 3.3 mM NaAc, 1 mM EDTA, pH 7.9. Store at room temperature.

2.4. Mismatch-Binding Activity Assay

1. 10X assay buffer: 200 mM Tris-HCl, pH 7.6, 50 mM MgCl$_2$, 1 mM DTT, and 0.1 mM EDTA. Store at 4°C.
2. Sucrose: prepare 50% solution in water. Store at 4°C.
3. EB coloring buffer: 0.5 µg/mL EB, 7 mM Tris-HCl, 3.3 mM NaAc, 1 mM EDTA, pH 7.9. Store at room temperature.

2.5. Strep TagII Bioactivity Assay

1. Blocking buffer: 4 mM KH$_2$PO$_4$, 16 mM Na$_2$HPO$_4$, 115 mM NaCl, 3% (w/v) bovine serum albumin (BSA), 0.5% (v/v) Tween-20. Store at 4°C.
2. Phosphate-buffered saline (PBS) buffer: 4 mM KH$_2$PO$_4$, 16 mM Na$_2$HPO$_4$, 115 mM NaCl. Store at 4°C.

3. Washing buffer: 4 mM KH$_2$PO$_4$, 16 mM Na$_2$HPO$_4$, 115 mM NaCl, 0.1% Tween-20. Store at 4°C.
4. Streptavidin–alkaline phosphatase conjugate: 100 µg/mL was prepared and stored at –20°C.
5. Buffer B: 1 mM ZnSO$_4$, 5 mM MgCl$_2$, 1 M Tris-HCl, pH 8.0. Store at 4°C.
6. *p*-Nitrophenyl phosphate (*p*NPP): *p*NPP was prepared at 0.5 mg/mL in the buffer B. Store at 4°C.

2.6. Preparation of the Protein Chip

1. Streptavidin-modified (single-layer) chips (96-well gold coating-well glass slides) gifted by nanoARC Inc. Store at 4°C.
2. 1X assay buffer: 20 mM Tris-HCl, pH 7.6, 5 mM MgCl$_2$, 0.1 mM DTT, and 0.01 mM EDTA. Store at 4°C.
3. Blocking solution: 3% (w/v) BSA, 2 mg/mL salmon sperm DNA (Stratagene), and 0.5% (v/v) Tween-20 solved in 1X assay buffer. Store at 4°C.
4. Washing buffer: 20 mM Tris-HCl, pH 7.6, 5 mM MgCl$_2$, 0.1 mM DTT, 0.01 mM EDTA, and 0.1% Tween-20. Store at 4°C.

2.7. Detection of DNA Mutations

1. Cy3-labeled oligonucleotides 5′-AAT AGT TCT CAG GTX GAC GGA TCT GGA CAC-3′ were mixed with the corresponding oligonucleotides.
2. The first group DNA sample: heteroduplexes containing different mismatches at X site and homoduplexes.
3. The second group DNA sample: the short heteroduplexes with 1–4 unpaired bases, nucleotides A, AG, AGC, and AGCC were inserted, respectively, between positions 15 and 16 of the complementary (nonlabeled) strand.
4. The third group DNA sample: 130-, 320-, and 612-bp *rpoB* gene obtained by PCR, respectively, from wild-type and rifampin-resistant *Mycobacterium tuberculosis* using primers listed in **Table 1**.
5. 1X assay buffer: 20 mM Tris-HCl, pH 7.6, 5 mM MgCl$_2$, 0.1 mM DTT, and 0.01 mM EDTA.
6. Bio Imaging System (Gene Company).
7. GenePix 4000B (Axon Instrument).
8. GenePix Pro 4.0 analysis software (Axon Instruments).

3. Methods

To obtain active fusion protein THLSLM, therefore, it is important to design the components of the fusion protein. The vector was constructed for expression of THLSLM. Coding sequences of thioredoxin (Trx), linker peptide (Ser-Gly)$_6$, Strep tagII (-Trp-Ser-His-Pro-Gln-Phe-Glu-Lys-), linker peptide (Ser-Gly)$_6$ and MutS are fused in sequence by gene splicing in vitro. The function of thioredoxin is to increase the soluble yields of the fusion protein so as to overcome the "inclusion body" problem. This tag is used to facilitate the purification of the

Table 1
PCR Primers Used to Amplify *rpoB* Gene

Oligonucleotide description PCR primers	Sequence
Forward primer (130 bp)	5′-GGC GAT CAA GGA GTT CTT C-3′
Reverse primer (130 bp)	5′-GCA CGC TCA CGT GAC AGA CC-3′
Forward primer (320 bp)	5′-GCG AGC TGA TCC AAA ACC A-3′
Reverse primer (320 bp)	5′-GGT TTC GAT CGG GCA CAT-3′
Forward primer (612 bp)	5′-TGT TGA AAA CTT GTT CTT CA-3′
Reverse primer (612 bp)	5′-AGC CGA TCA GAC CGA TGT-3′

desired protein. The linker peptide made up of serine and glycine was designed to minimize the steric hindrance and provide reasonable space for the components of the fusion protein to retain their native conformation. The mismatch-binding activity of THLSLM was determined by bandshift assays using a short duplex DNA fragment containing a mismatched basepair. Ultimately, it is important to determine whether Strep tagII bioactivity of THLSLM remained. The Strep tagII bioactivity assay was performed in a 96-well microtiter plate based on the nature of Strep tagII/streptavidin binding reaction.

3.1. Construction of THLSLM Expression Vector

1. All oligonucleotides were designed to incorporate proper restriction enzymatic sites for cloning. Primers M-1 (5′–GAA TTC ATG AGT GCA ATA GAA AAT TTC-GAC-3′) and M-2 (5′–AAG CTT TAT TTT TAT TTG ATT CGT CAG TTA T-3′), spanning the *E. coli mutS* gene, were synthesized to amplify *mutS* with *E. coli* K-12 genome as template.
2. Restriction enzymatic sites *Eco*RI and *Hind*III were introduced to the N- and C-terminal primers, respectively.
3. Amplification was performed under the following conditions: after a hot-start step (94°C, 6 min), samples were subjected to 30 cycles of denaturation at 94°C for 1 min, annealing at 55°C for 1 min, extension at 72°C for 3 min, and a final cycle of elongation for 7 min at 72°C.
4. PCR products were purified by a PCR Purification Mini Kit (Promega) and then cloned into the pGEM-T vector.
5. The resulting plasmid pGEM-T-*mutS* was then digested with *Eco*RI and *Hind*III, and MutS encoding fragment was recovered and inserted into the *Eco*RI/*Hind*III site of pET32a (+), yielding the plasmid termed as pET32a-*mutS*.
6. The synthesized oligonucleotides LS1 (5′-GAT CTG AGC GGC TCT GGA TCA GGA TCT GGC AGC GCT TGG AGC CAC CCG CAG TTC GAA AAA GGC GCC GAT-3′) and LS2 (5′-ATC GGC GCC TTT TTC GAA CTG CGG GTG GCT CCA AGC GCT GCC AGA TCC TGA TCC AGA GCC GCT CA-3′) were annealed at 70°C for 5 min to form a dsDNA fragment containing Linker

peptide-Strep tagII coding sequence with *Bgl*II and *Eco*V at its N- and C-termini, respectively. Another linker peptide coding sequence with *Eco*V and *Eco*I was similarly obtained by annealing L1 (5′-ATC AGC GGC TCA GGA TCT GGA TCA GGA TCT GGC G-3′) and L2(5′-AAT TCG CCA GAT CCT GAT CCA GAT CCT GAG CCG CTG AT-3′).

7. The two resulting short fragments were then inserted to replace the corresponding sequence between *Bgl*II and *Eco*RI in pET32a-*mutS*, yielding the final fusion expression vector pET32a-*(Ser-Gly)₆-Strep tagII-(Ser-Gly)₆-mutS*, which was to express the fusion molecular system Trx-His₆-Linker peptide-Strep tagII-Linker peptide-MutS (THLSLM).

3.2. Expression and Purification of THLSLM

1. The competent cells of *E. coli* AD494(DE3) were transformed with the fusion expression vector pET32a-*(Ser-Gly)₆-Strep tagII-(Ser-Gly)₆-mutS* and yielded the recombinant strain *E. coli* AD494(DE3)/ pET32a-*(Ser-Gly)₆-Strep tagII-(Ser-Gly)₆-mutS*.

2. A single colony of the strain was picked up from a fresh LB plate containing ampicillin and kanamycin and inoculated into 10 mL LB liquid medium containing the same antibiotics. The culture was incubated with rotary shaking (300 rev/min) at 37°C for 6–8 h to midexponential phase. Cells were collected by centrifugation at 5000g for 20 min and the pellet was resuspended in 1 mL of fresh LB.

3. One milliliter suspension was inoculated in 1000 mL of LB medium containing same antibiotics. The culture was allowed to grow with rotary shaking (300 rev/min) at 30°C for 4 h before 0.6 mM IPTG were added, and continue to be incubated at 30°C for 5 h.

4. Cultures were then chilled on ice for 10 min, and cells were collected by centrifugation. Cell pellets were washed twice with 100 mL of ice-cold water and stored at –20°C *(22)*.

5. Purification of the fusion protein THLSLM was performed with His-Bind™ resin slurry. First, gently invert the bottle of Ni²⁺-charged chromatography resin to mix the slurry, transfer 5 mL to a glass column slowly and allow the resin to pack under gravity flow. Then, the resin was washed with 10 column-volumes of sterile H₂O. Finally, equilibrate the resin with 10 column-volumes of 1X extraction/wash buffer.

6. The stored cell pellets were thawed on ice and resuspended in 10 mL of 1X extraction buffer per 100 mL of cell culture. The cells were lysed by sonication. The lysates were centrifugalized at 12,000g for 30 min, and the supernatants were loaded onto the column. The column was then washed with 10 column volumes of 1X extraction/wash buffer.

7. The desired protein THLSLM was eluted by washing the resin column with 1X elution buffer. The elution buffer containing the purified fusion protein THLSLM was dialyzed against 1000 mL of dialyzing buffer overnight *(23)*.

3.3. Mismatch-Binding Activity Assay

1. The synthesized oligonucleotides 5′-GTG TCC AGA TCC GTC XAC CTG AGA ACT ATT-3′ were mixed with the corresponding oligonucleotides.

2. The mixture was heated and slowly cooled to room temperature to form heteroduplexes containing GT mismatch at X site and homoduplexes, respectively.

3. The binding reaction mixture contained 1-μL of duplex DNA, 39 μL of Millipore water, 5 μL of 10X assay buffer, and 5 μL of THLSLM of different concentrations. The reaction mixture was held on ice for 30 min before 20 μL of 50% sucrose was added.

4. The resulting mixture was loaded onto 6% nondenaturing polyacrylamide gel in low-ionic strength buffer. The electrophoresis was performed at 10 V/cm at 4°C until the bromophenol blue dye band in the control lane migrated about 6 cm.

5. After the gel was stained by EB, the relative amount of unbound duplexes, an indication of THLSLM bioactivity was quantitated on the Bio Imaging System.

6. The band in the lane with the addition of mismatched DNA was much weaker than those in the other lanes. BSA addition had no influence on the migration rates of both mismatched DNA and complementary DNA (*see* **Note 1**).

7. THLSLM concentration effect the mismatch-binding activity. The amount of unbound mismatched DNA decreased apparently with the increase in THLSLM concentration.

8. The mismatch-binding activity of THLSLM to mismatched DNA increased with the increase of magnesium ions strength. Therefore, 5 mM of magnesium ions and 0.5 mg/mL THLSLM were chosen to perform the binding reaction experiments throughout (*see* **Note 2**).

3.4. Strep TagII Bioactivity Assay

1. A 100-μL aliquot of 100 μg/mL THLSLM solution was first immobilized by incubating at 4°C overnight in the wells of the 96-well microtiter plate by physical absorption with four duplicates. The wells without THLSLM were used as a parallel negative control.

2. After removal of the solution, the wells were blocked with 3% (w/v) BSA and 0.5% (v/v) Tween-20 in PBS buffer for 2.5 h.

3. After washing three times with PBS-Tween-20 (0.1% Tween-20), 100 μL of 100 μg/mL streptavidin-alkaline phosphatase conjugate was added and resultant mixture was incubated for 1 h.

4. Unbound conjugate was removed by washing twice with PBS-Tween-20 and twice with PBS buffer.

5. Finally, 100 μL of 0.5 mg/mL *p*NPP in buffer B was applied to each well. The substrate *p*NPP could be broken down by alkaline phosphate to produce *p*-nitrophenol (yellow color).

6. The Strep tagII bioactivity of THLSLM was evaluated according to the color change. It was found that wells with immobilized THLSLM developed visible color within 20 min, while the wells of negative control showed no detectable signal.

3.5. Preparation of the Protein Chip

1. The diluted THLSLM solution at a concentration of 0.6 mg/mL was added to the wells of the chip coated with streptavidin and incubated at room temperature for 0.5 h.

2. After removal of the solution, the wells were blocked with 2 μL/well blocking solution.

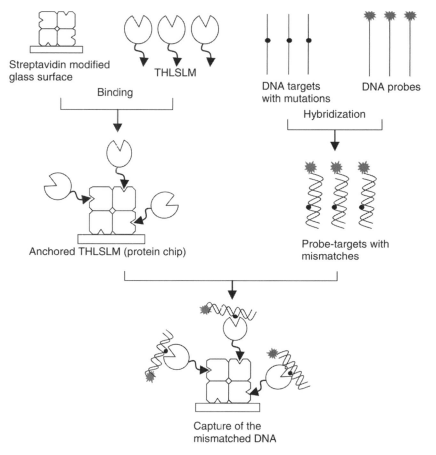

Fig. 1 Schematic diagram of the protein chip for gene mutation assay.

3. The wells were washed three times using 1X assay buffer with the addition of Tween-20 to a final concentration of 0.1% (v/v). The protein chip was then ready for detection of DNA mutations (*see* **Note 3**).

3.6. Mutation Detection of Oligonucleotides

Figure 1 shows schematic diagram of the protein chip for detection of DNA mutations. Cy3-labeled oligonucleotides 5′-AAT AGT TCT CAG GTX GAC GGA TCT GGA CAC-3′ were mixed with the corresponding oligonucleotides. Three groups of DNA samples were prepared. In the first group, the mixture was heated and slowly cooled to room temperature to form heteroduplexes containing different mismatches at X site and homoduplexes. In the second group, nucleotides A, AG, AGC, and AGCC were inserted, respectively, between positions 15 and 16 of the complementary (nonlabeled) strand. The resulting oligonucleotides were annealed to the Cy3-labeled fragments to generate short heteroduplexes with one to four unpaired bases. In the third group, 130-, 320-, and 612-bp *rpoB* gene were obtained by PCR, respectively, from wild-type and rifampin-resistant *M. tuberculosis* using primers. All the forward primers were labeled with Cy3 at their ends. The denatured PCR products from wild-type and mutant were mixed together and incubated at 70°C to produce duplex *rpoB* gene fragments containing a point mutation.

1. A 0.5-μL aliquot of Cy3-labeled oligonucleotides containing mismatched or unpaired bases and Cy3-labeled complementary oligonucleotides were added to the wells of the protein chip. The chip was incubated at room temperature for 30 min to perform the mismatch-binding reaction.
2. The mismatched oligonucleotides were then recognized and bound *in situ* by THLSLM anchored on the chip.
3. The unbound DNA fragments were removed by washing three times with 1X assay buffer with and without 0.1% (v/v) Tween-20 sequentially.
4. A GenePix 4000B fluorescence scanner with the sensitivity of 0.1 fluorophores/μm^2 for Cy3 was used to obtain the Cy3 fluorescence images. All images were analyzed using GenePix Pro 4.0 analysis software.
5. Analysis of the binding reactions show that the protein chip readily detected nearly all the mismatches tested here but bound most strongly to the fragment that contained a GT mismatch (**Table 2**, *see* **Note 4**). In addition, it also bound well to the fragments containing GG, AA, and GA mismatches. The weakest mismatch binding was observed to be a CC mismatch, which had been found to be generally refractory to repair by the *E. coli* mismatch repair system *(24)*.
6. The binding affinity to oligonucleotides containing 1–4 unpaired bases of the protein chip was also studied. It was found that the protein chip recognized and bound duplex DNA fragment containing unpaired bases specifically (**Table 3**). The affinity of the chip to duplex DNA with one or two unpaired bases was determined to be stronger than to those with three or four unpaired bases.

3.7. The Effect of Sequence Context of Oligonucleotides on Mismatch-Binding Activity of the Protein Chip

1. The effect of sequence context of oligonucleotides on mismatch-binding activity of the protein chip was investigated using a series of duplex oligonucleotides containing GT, TG, GA, AG, AC, CA, CT, and TC mismatches.
2. The results demonstrated that the influence of sequence context on the binding reaction varied among different mismatch types. GT and TG, GA and AG, AC and CA are detected equally well using the protein chip, respectively. It seems that well-detected mismatches are easily detected independent of sequence context.
3. The finding that the TC mismatch was better detected than the CT mismatch also suggests that some mismatch recognition depend on the sequence in the vicinity of the mismatch (**Table 4**).

Table 2
Detection of 30 bp Oligonucleotides With Different Mismatches

Sample number	Mismatches in probe-target hybrids	Pixel intensity (532 nm)	Signal-to-noise[a]
1	Full complementary	2083 ± 1129	~1/1
2	TC mismatch	10,575 ± 1014	~5/1
3	GT mismatch	16,715 ± 1152	~8/1
4	GG mismatch	11,432 ± 1036	~6/1
5	GA mismatch	12,044 ± 1195	~6/1
6	AC mismatch	9642 ± 914	~5/1
7	TT mismatch	5623 ± 821	~3/1
8	AA mismatch	10,151 ± 1061	~5/1
9	CC mismatch	4077 ± 989	~2/1

[a]Calculated by the discrimination signal ratio between the mean value of the mismatched DNA and that of the control.

Table 3
Detection of 30 bp Oligonucleotides With 1–4 Unpaired Bases

Sample number	Unpaired bases in probe-target hybrids	Pixel intensity (532 nm)	Signal-to-noise[a]
1	Full complementary	1939 ± 914	~1/1
2	One unpaired base	13,098 ± 1036	~7/1
3	Two unpaired bases	12,679 ± 821	~6/1
4	Three unpaired bases	6742 ± 1095	~3/1
5	Four unpaired bases	5749 ± 1061	~3/1

[a]Calculated by the discrimination signal ratio between the mean value of the unpaired DNA and that of the control.

3.8. Detection of the Mutation in rpoB Gene From Rifampin-Resistant M. tuberculosis Strain

Tuberculosis is one of the most deadly and common infectious diseases, whose global spread is further complicated by the ubiquitous appearance of drug-resistant strains *(25)*. The resistance of *M. tuberculosis* to rifampin has been determined to be the cause of some mutations that occurred in the *rpoB* gene *(26)*. The most frequent mutations are associated with codons 531 and 526. Here, we take the 531 position as a mutation model to investigate the potential application of the protein chip.

1. The 130-, 320-, and 612-bp-long *rpoB* gene fragments were obtained from both wild-type and mutant with A-C substitution at position 531, respectively.

Table 4
**Effect of Sequence Context on the Mismatch Binding
to 30 bp Oligonucleotides**

Sample number	Mismatches in probe-target hybrids	Sequence context	Pixel intensity (532 nm)	Signal-to-noise[a]
1	Full complementary	GGT G GAC	2096 ± 1105	~1/1
2	TC mismatch	GAT T GAG	10,426 ± 1014	~5/1
3	CT mismatch	CCA T CTG	4628 ± 1152	~2/1
4	GT mismatch	GGT G GAC	14,023 ± 896	~7/1
5	TG mismatch	GCA G TTG	14,639 ± 914	~7/1
6	GA mismatch	GGT G GAC	11,082 ± 821	~6/1
7	AG mismatch	CCA G CTG	10,987 ± 1061	~5/1
8	AC mismatch	GCA A TTG	11,523 ± 989	~6/1
9	CA mismatch	CCA A CTG	11,098 ± 1029	~6/1

[a]Calculated by the discrimination signal ratio between the mean value of the impaired or unpaired DNA and that of the control.

Table 5
Detection of Mutations in *rpoB* Gene From *Mycobacterium tuberoculosis*

Sample number	Size of PCR product (bp)	Gene fragments	Pixel intensity (532 nm)	Signal-to-noise[a]
1	130	Full complementary	2210 ± 1062	~1/1
2	130	A/G + T/C	11,880 ± 1014	~5/1
3	320	Full complementary	2676 ± 1095	~1/1
4	320	A/G + T/C	7987 ± 1036	~3/1
5	612	Full complementary	2789 ± 1149	~1/1
6	612	A/G + T/C	6912 ± 896	~2/1

[a]Calculated by the discrimination signal ratio between the mean value of the impaired DNA and that of the control.

2. The PCR products from wild-type and mutant were mixed, denatured, and annealed to form heteroduplexes. Then the diluted heteroduplexes were spotted to the wells of the protein chip. Fluorescent signal was imaged and analyzed (**Table 5**).

3. The assay results, an average of four duplicates, respectively, showed that the pixel intensities of heteroduplexes at 532 nm were 11,880, 7987, and 6912, and that of homoduplexes were 2210, 2676, and 2789 in the same sequence of 130, 320, and 612 bp, respectively (*see* **Notes 5** and **6**).

4. Notes

1. The interaction between THLSLM and mismatched duplex DNA resulted in the formation of a THLSLM-DNA complex that had a very low mobility and hardly

entered the gel matrix. Since it is difficult to estimate the amount of duplex DNA bound to THLSLM directly, the decrease of the unbound duplex DNA was measured to indicate the mismatch-binding characteristic of THLSLM.

2. Nearly no unbound DNA was visualized on the gel once the concentration of THLSLM increased to 0.5 mg/mL, which was threefold over the DNA concentration. Although stoichiometry for DNA with MutS protein is 1:1, the complete binding of mismatches by MutS was at a ratio of 3:1 (MutS to DNA). There are several reasons. First, the existing forms of MutS proteins in solution may be a dimer or tetramer *(27)*, and only the dimer can bind a mismatch; binding kinetics is a dynamic process; the high ratio of MutS to DNA sample is propitious to the reaction toward the complete binding of targets.

3. The fusion molecular system THLSLM, constructed by gene splicing in vitro was used to make a protein chip for specific detection of mutations. The addition of the linker peptides provides spacers to minimize steric hindrance among the functional components of the fusion protein *(28)*. This fusion protein THLSLM could be efficiently immobilized on a streptavidin-modified chip with controlled homogeneity through streptavidin-Strep tagII interaction.

4. The protein chip can recognize and bind nearly all the duplex DNA with mismatched basepair(s). Mismatched DNA duplexes are easily distinguished from complementary DNA duplexes. However, the mismatch types cannot be readily distinguished from one another because most of the mismatches generated approximately the same signal level. The weaker binding to the CC mismatch does not diminish the utility of this method for mutation detection, because every wild-type/mutant pairing gives rise to two different mismatches: CC and GG, the latter can be easily recognized by THLSLM.

5. The fluorescence of heteroduplexes was significantly higher than that of control samples without mismatch. It seems that THLSLM protein chip tended to bind more relatively short mismatched DNA fragments. The signal level from high to low determined for three heteroduplexes was in sequence of 130, 320, and 612 bp. But results showed that the fluorescent signal of homoduplexes increased with increases of DNA length. Nonspecific binding of THLSLM to DNA increased owing to an increase of DNA length. The nonspecific binding, i.e., binding to homoduplexes, especially in the case of a long-length DNA fragment, may be attributed to the possibilities that PCR products formed mismatches when annealed with the genomic DNA from the homologous chromosome or that they produced some secondary structure with mismatches. This nonspecific binding of course will increase the background signal, however, it did not affect the detection result as the nonspecific signal level was much weaker compared with the specific binding reactions under a controlled concentration of THLSLM.

6. The study has demonstrated that the MutS-based protein chip features high specificity, wide range of target DNA size (30–612 bp), ease of handling, and short detection cycle (within 2 h). The results are confident in the range of DNA 90 nM to 90 pM. The sensitivity requirements of genetic analysis vary with the different samples and purposes. For instance, to analyze the human genetic diseases, the

sample could be obtained in ample quantity; sensitivity of the method is not a problem. However, for detection of some trace pathogens (a few hundreds or thousands individuals), especially with single-copy genes in them, the sensitivity becomes a problem. This limitation occurs to all existing detection methods. In this case, an additional gene amplification process is necessary.

Acknowledgment

Supports from the National Science Foundation of China (30270306), Ministry of Science and Technology and the Chinese Academy of Science are acknowledged. We thank NanoARC Inc. for the gift of the streptavidin-coated chip.

References

1. Stewart, P. R., el-Adhami, W., Inglis, B., and Franklin, J. C. (1993) Analysis of an outbreak of variably methicillin-resistant *Staphylococcus aureus* with chromosomal RFLPs and mec region probes. *J. Med. Microbiol.* **38,** 270–277.
2. Meggouh, F., Benomar, A., Rouger, H., et al. (1998) The first de novo mutation of the connexin 32 gene associated with X linked Charcot-Marie-Tooth disease. *J. Med. Genet.* **35,** 251–252.
3. Zoller, B. and Dahlback, B. (1995) Resistance to activated protein C caused by a factor V gene mutation. *Curr. Opin.* **2,** 358–364.
4. Everett, L. A., Glaser, B., Beck, J. C., et al. (1997) Pendred syndrome is caused by mutations in a putative sulphate transporter gene (PDS). *Nat. Genet.* **17,** 411–422.
5. Wong, C., Dowling, C. E., Saiki, R. K., Higuchi, R. G., Erlich, H. A., and Kazazian, H. H., Jr. (1987) Characterization of β-thalassaemia mutations using direct genomic sequencing of amplified single copy DNA. *Nature* **330,** 384–386.
6. Saiki, R. K., Sharf, S., Faloona, F., et al. (1985) Enzymatic amplification of β-globin genomic sequences and restriction site analysis for diagnosis of sickle cell anemia. *Science* **230,** 1350–1354.
7. Sugano, K., Fukayama, N., Ohkura, H., et al. (1995) Single-strand conformation polymorphism analysis by perpendicular temperature-gradient gel electrophoresis. *Electrophoresis* **16,** 8–10.
8. Guldberg, P. and Guttler, F. (1993) A simple method for identification of point mutations using denaturing gradient gel electrophoresis. *Nucleic Acids Res.* **21,** 2261–2262.
9. Deeble, V. J., Roberts, E., Robinson, M. D., Woods, C. G., Bishop, D. T., and Taylor, G. R. (1999) Comparison of enzyme mismatch cleavage and chemical cleavage of mismatch on a defined set of heteroduplexes. *Genetic Testing* **1,** 253–259.
10. Hacia, J. G. (1999) Resequencing and mutational analysis using oligonucleotide microarrays. *Nat. Genet.* **21,** 42–47.
11. Lu, A. L., Clark, S., and Modrich, P. (1983) Methyl-directed repair of DNA base-pair mismatches in vitro. *Proc. Natl. Acad. Sci. USA* **80,** 4639–4643.
12. Lieb, M. (1987) Bacterial genes mutL, mutS, and dcm participate in repair of mismatches at 5-methylcytosine sites. *J. Bacteriol.* **169,** 5241–5246.

13. Jiricny, J., Su, S. S., Wood, S. G., and Modrich, P. (1988) Mismatch-containing oligonucleotide duplexes bound by the *E. coli* mutS-encoded protein. *Nucleic Acids Res.* **25,** 7843–7853.

14. Wagner, R., Debbie, P., and Radman, M. (1995) Mutation detection using immobilized mismatch binding protein (MutS). *Nucleic Acids Res.* **11,** 3944–3948.

15. Ellis, L. A., Taylor, G. R., Banks, R., and Baumberg, S. (1994) MutS binding protects heteroduplex DNA from exonuclease digestion in vitro: a simple method for detecting mutations. *Nucleic Acids Res.* **11,** 2710–2711.

16. Gotoh, M., Hasebe, M., Ohira, T., et al. (1997) Genetic Analysis. *Bimolecular Eng.* **14,** 47–50.

17. Geschwind, D. H., Rhee, R., and Nelson, S. F. (1996) A biotinylated MutS fusion protein and its use in a rapid mutation screening technique. *Genet. Anal.* **13,** 105–111.

18. Nelson, S. F. (1995) Genomic mismatch scanning: current progress and potential applications. *Electrophoresis* **16,** 279–285.

19. Behrensdorf, H. A., Pignot, M., Windhab, N., and Kappel, A. (2002) Rapid parallel mutation scanning of gene fragments using a microelectronic protein-DNA chip format. *Nucleic Acids Res.* **15,** E64.

20. Shao, W. H., Zhang, X. E., Liu, H., Zhang, Z. P., and Cass, A. E. (2000) Anchor-chain molecular system for orientation control in enzyme immobilization. *Bioconjug. Chem.* **11,** 822–826.

21. Shi, J. X., Zhang, X. E., Xie, W. H., et al. (2004) Improvement of homogeneity of analytical biodevices by gene manipulation. *Anal. Chem.* **76,** 632–638.

22. Feng, G. and Winkler, M. E. (1995) Single-step purifications of His6-MutH, His6-MutL and His6-MutS repair proteins of *Escherichia coli* K-12. *Biotechniques* **19,** 956–965.

23. Laurent, G., Celine, B., and Peter, B. (1999) ATP hydrolysis-dependent formation of a dynamic ternary nucleoprotein complex with MutS and MutL. *Nucleic Acids Res.* **27,** 2325–2331.

24. Modrich, P. (1991) Mechanisms and biological effects of mismatch repair. *Annu. Rev. Genet.* **25,** 229–253.

25. Gutierrez, M. C., Galan, J. C., Blazquez, J., Bouvet, E., and Vincent, V. (1999) Molecular markers demonstrate that the first described multidrug-resistant *Mycobacterium bovis* outbreak was due to *Mycobacterium tuberculosis. J. Clin. Microbiol.* **37,** 971–975.

26. Miller, L. P., Crawford, J. T., and Shinnick, T. M. (1994) The *rpoB* gene of *Mycobacterium tuberculosis. Antimicrob. Agents Chemother.* **38,** 805–811.

27. Biswas, I., Ban, C., Fleming, K. G., et al. (1999) Oligomerization of a MutS mismatch repair protein from *Thermus aquaticus. J. Biol. Chem.* **274,** 23,673–23,678.

28. Zhou, Y. F., Zhang, X. E., Liu, H., Zhang, Z. P., Zhang, C. G., and Cass, A. E. (2001) Construction of a fusion enzyme system by gene splicing as a new molecular recognition element for a sequence biosensor. *Bioconjug. Chem.* **12,** 924–931.

12

Screening of cDNA Libraries on Glass Slide Microarrays

Dave K. Berger, Bridget G. Crampton, Ingo Hein, and Wiesner Vos

Summary

A quantitative screening method was developed to evaluate the quality of cDNA libraries constructed by suppression subtraction hybridization (SSH) or other enrichment techniques. The SSH technique was adapted to facilitate screening of the resultant library on a small number of glass slide microarrays. A simple data analysis pipeline named SSHscreen using "linear models for microarray data" (limma) functions in the R computing environment was developed to identify clones in the cDNA libraries that are significantly differentially expressed, and to determine if they were rare or abundant in the original treated sample. This approach facilitates the choice of clones from the cDNA library for further analysis, such as DNA sequencing, Northern blotting, RT-PCR, or detailed expression profiling using a custom cDNA microarray. Furthermore, this strategy is particularly useful for studies of nonmodel organisms for which there is little genome sequence information.

Key Words: cDNA library; microarray; normalization; SSH; suppression subtraction hybridization; limma.

1. Introduction

There are many protocols for the construction of cDNA libraries (*1*); however, few methods have been developed to evaluate the quality of the library in a quantitative manner. For example, this can lead to unnecessary wastage of time and resources in sequencing redundant clones in an expressed sequence tag (ESD) project. A partial solution is to use an enrichment technique in construction of the cDNA library such as suppression subtraction hybridization (SSH) (*2*), or normalization (*3,4*). In addition, most researchers prefer to employ a test for quality control of the resultant library clones, which is often carried out by inverse Northern blot analysis with nylon membranes and radioactively labeled probes. This approach has disadvantages because of the difficulty of normalizing data correctly between two membranes (e.g., treated vs control)

From: *Methods in Molecular Biology, vol. 382: Microarrays: Second Edition: Volume 2*
Edited by: J. B. Rampal © Humana Press Inc., Totowa, NJ

and the use of radioactivity, and therefore, in practice only a qualitative result is obtained *(5)*.

A detailed protocol for the construction of cDNA libraries enriched for differentially expressed genes based on SSH is described; however, the standard protocol has been adapted so that sufficient material is produced to screen the resultant cDNA library clones using a small number of glass slide microarrays. A simple data analysis pipeline named SSHscreen using "linear models for microarray data" (limma) functions in the R computing environment *(6)* has been developed to analyze spot intensity data, thereby screening clones in the cDNA libraries to identify those that are significantly differentially expressed. Furthermore, a calculation is also made to determine whether the clones were derived from rare or abundant transcripts in the treated sample. This can be very useful in determining whether Northern blots or a more sensitive technique, such as real-time RT-PCR, would be necessary to study expression of individual messenger RNAs (mRNAs) in detail. Differentially expressed genes can then be chosen for further investigation such as DNA sequencing or construction of a custom cDNA microarray for detailed expression profiling studies, such as a time-course experiment after treatment. Several investigators have used SSH to generate a cDNA library that was subsequently used for expression profiling on microarrays *(7,8)*. This approach is especially useful for studies of nonmodel organisms for which whole genome sequence is not available, and is illustrated with an example from the cereal crop pearl millet. In addition, the method can be adapted to screen cDNA libraries of differentially expressed genes constructed by methods other than SSH *(3)*.

Experimental design is important so that cDNA libraries are constructed that capture the maximum number of differentially expressed genes between untreated control samples (named "Driver") and treated samples (named "Tester"). Two points should be considered, first, whether the experimental aim requires a wide or narrow subtraction. An example of a narrow subtraction is an investigation of early responses (within 1–5 h) of barley plants to avirulent and virulent races of a fungal pathogen, in which 21 differentially expressed transcripts were identified *(9)*. In contrast, a wide subtraction is the example used here, in which the aim was to identify as many transcripts as possible that are differentially expressed in the general defense response of a plant to a range of pathogens. In this case the treated sample was a pool of RNA from leaves from a disease resistant pearl millet line that had been wounded and treated separately with elicitors of bacterial and fungal origin (Tester) and the control sample was a pool of RNA from untreated leaves from a susceptible pearl millet line (Driver). Furthermore, RNA from a range of different time-points after treatment (5, 14, and 24 h) was pooled for the Tester and at the corresponding times for the Driver. The second issue to consider is whether sufficient material is available to make both forward and reverse cDNA subtraction libraries. A reverse subtraction library is one in which the "Tester" would be the untreated sample and the "Driver" the treated sample. This would capture genes that are downregulated in the biological system under study, which are likely to be as important as upregulated genes *(10)*. In the current example, a reverse subtraction library was also constructed and screened; however, this will not be described because the procedure is the same, but with switched starting materials.

The SSH cDNA library clones are arrayed on glass slide microarrays, and quantitative screening is carried out by calculating three different "SSH enrichment ratios" for each clone. Calculations are based on the intensity values of each spot (cDNA clone) as a result of hybridization by cyanine (Cy) dye-labeled complex cDNA probes. The complex cDNA probes produced during the SSH process are named Unsubtracted Driver (UD), Unsubtracted Tester (UT), and Subtracted Tester (ST). The following contrasts are hybridised to the arrays: ST vs UD, ST vs UT, and UD vs UT. Pairs of complex cDNA probes are each labeled with Cy3 or Cy5 dyes with a dye-swap repeat of each slide to make up a total of six microarray slides for the complete screening experiment. After hybridization, scanning, and normalization of the data to account for intensity biases within and between slides, the SSH enrichment ratios are calculated and a statistical test is applied. The data can be plotted on either of two graphs to determine whether individual clones represent transcripts that are up- or downregulated by the treatment and whether the clones were enriched by the SSH process.

The calculations can be most readily explained using a simple hypothetical dataset (**Table 1**). This dataset contains genes that either change (up- or downregulated) or do not change (housekeeping genes) after a particular treatment. Data values have been assigned to each gene within each of the UD, UT, and ST samples. These data values are the intensity values (after normalization) of each spot (gene) on the microarray after hybridization with the Cy-dye labeled ST, UD, or UT probes, and therefore, can be used to calculate the hypothetical SSH enrichment ratios. Note also that these intensity values correspond directly to the number of cDNA molecules of each gene in the ST, UD, or UT probes, which initially were derived from mRNA transcripts from that gene. SSH enrichment ratio 1 (ER1) values are calculated as $\log_2(ST/UD)$ and give an indication whether cDNAs have been enriched relative to levels in the Driver sample by the SSH process (positive ER1 values). SSH enrichment ratio 2 (ER2) values are calculated as $\log_2(ST/UT)$ and give an indication whether cDNAs have been enriched or reduced relative to levels in the UT sample by the normalization process. Positive ER2 values, therefore, represent cDNAs that were rare in the UT sample and have been enriched by the SSH process, whereas, negative ER2 values represent cDNAs that were abundant in the UT sample and have been

Table 1
Hypothetical Data Set From Microarray Screening of SSH cDNA Library

Code for hypothetical gene	Effect of treatment on expression of each hypothetical gene[a]	Relative amount of transcripts of each hypothetical gene after treatment[b]	UD (intensity[c])	UT (intensity[c])	ST (intensity[c,d])	ER1[e]	ER2[e]	Inverse ER2	ER3[e]
2XupR	Upregulated (twofold)	Rare	125	250	500	2	1	−1	1
4XupA	Upregulated (fourfold)	Abundant	250	1000	500	1	−1	1	2
2XupA	Upregulated (twofold)	Abundant	500	1000	500	0	−1	1	1
2XdR	Downregulated (twofold)	Rare	500	250	500	0	1	−1	−1
4XdR	Downregulated (fourfold)	Rare	1000	250	500	−1	1	−1	−2
2XdA	Downregulated (twofold)	Abundant	2000	1000	500	−2	−1	1	−1
hR	No change (e.g., housekeeping gene)	Rare	250	250	500	1	1	−1	0
hA	No change (e.g., housekeeping gene)	Abundant	1000	1000	500	−1	−1	1	0

[a]Difference in transcript (cDNA) levels for each hypothetical gene between UT and UD.

[b]Abundance of transcripts (cDNA) for each hypothetical gene in unsubstracted tester.

[c]Hypothetical intensity (after normalization) of spot on microarray after hybridization of Cy-dye labeled ST, UD or UT probes. These intensity values correspond directly to the number of transcripts (cDNA) in the ST, UD or UT probes.

[d]The hypothetical intensity values for ST reflect the subtraction and normalization that has occurred by the suppression subtractive hybridization process.

[e]ER1 ($\log_2 ST/UD$), ER2 ($\log_2 ST/UT$), and ER3 ($\log_2 UT/UD$) values are calculated using the hypothetical intensity values for each gene, and used to plot the graphs shown in **Figs. 1** and **2**.

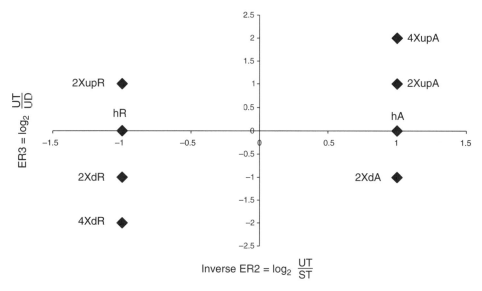

. Fig. 1. Illustration of microarray screen of SSH library clones using Enrichment ratio 3 (ER3) and inverse Enrichment ratio 2 (inverse ER2) values calculated from data from eight hypothetical genes (**Table 1**). Data points in quadrant 1 (e.g., *2XupR*) represent genes that were upregulated and rare in the treated sample UT and have been enriched by the SSH procedure. Data points in quadrant 2 (e.g., *2XupA* and *4XupA*) represent genes that were upregulated and abundant in the treated sample UT, and have been recovered but down-normalized by the SSH procedure. Data points on the *x*-axis represent housekeeping genes that are unchanged by the treatment (e.g., *hR* and *hA*). Data points in quadrant 3 (e.g., *2XdR* and *4XdR*) and quadrant 4 (e.g., *2XdA*) represent genes that are downregulated by the treatment, i.e., escaped the subtraction.

reduced by normalization in the ST sample. SSH enrichment ratio 3 (ER3) values are calculated as $\log_2(\text{UT/UD})$ and give an indication whether cDNAs are up- or downregulated by the treatment (positive or negative ER3 values, respectively).

Figure 1 represents a plot to screen SSH cDNA library clones by plotting ER3 vs inverse ER2. Data points above the *x*-axis of the graph (ER3>0 ~ UT>UD) represent genes upregulated by the treatment (hypothetical genes *2XupR, 2XupA,* and *4XupA*; **Table 1**). These are the expected clones in the SSH library that can be chosen for further study. Data points below the *x*-axis of the graph (ER3<0 ~ UT<UD) represent genes that are downregulated by the treatment (hypothetical genes *2XdA, 2XdR,* and *4XdR*; **Table 1**), which should not be represented in the SSH cDNA library, and most likely escaped the subtraction. These clones can be discarded from further study. The *x*-axis on the graph (ER3 = 0 ~ UD = UT) represents housekeeping genes that have similar

transcript levels in untreated (Driver) and treated (Tester) samples (hypothetical genes *hR* and *hA*; **Table 1**). These should not be present in the SSH cDNA library, because they should have been removed by the subtraction, and can therefore, be discarded from further study.

Normalization also occurs during the SSH procedure, whereby rare transcripts are enriched and abundant transcripts are reduced, which has the advantage of reducing redundant clones in the SSH cDNA library *(11)*. The inverse ER2 values (**Fig. 1**) reflect the level of normalization for each gene, from which one can infer whether it is rare or abundant in the treated UT sample. Data points in quadrant 1 (top left) of **Fig. 1** represent upregulated genes that are rare in the UT sample (negative inverse ER2 value and positive ER3 value, for example, gene *2XupR*). This information is useful since these will have a low level of mRNA transcripts in the treated sample, which may need a sensitive verification method such as real-time RT-PCR. In contrast, data points in quadrant 2 (top right) of **Fig. 1** represent upregulated genes that are abundant in the UT sample (positive inverse ER2 value and positive ER3 value, for example genes *2XupA* and *4XupA*) that should require a less sensitive verification technique such as a Northern blot.

Figure 1, therefore, places the data for genes with different behaviors in each of the four quadrants, which allows for easy interpretation. A good quality SSH cDNA library should have most of the data points from the microarray screen above the *x*-axis when plotted as in **Fig. 1**.

Figure 2 represents an alternative plot as reported in **ref.** *12* that can also be used to screen SSH cDNA library clones by plotting ER1 vs ER2. The diagonal line on the graph (ER1 = ER2 ~ $\log_2$(ST/UD) = $\log_2$(ST/UT) ~ UD = UT) represents housekeeping genes that have similar transcript levels in untreated (Driver) and treated (Tester) samples (genes *hR* and *hA*). Data points above the diagonal line (ER1>ER2 ~ UT>UD) represent genes that are upregulated by the treatment (genes *2XupA, 4XupA,* and *2XupR*). Data points below the diagonal line (ER1<ER2 ~ UT<UD) represent genes that are downregulated by the treatment (genes *2XdA, 2XdR,* and *4XdR*). **Fig. 1** can also be used for identifying genes that are rare and upregulated in the UT sample (positive ER2 value and above the diagonal line, for example, gene *2XupR*) and genes that are abundant and upregulated in the UT sample (negative ER2 value and above the diagonal line, for example genes *2XupA* and *4XupA*).

Two independently constructed SSH cDNA libraries (from banana and pearl millet) were screened using this microarray approach, and the results were verified using an independent technique, inverse Northern blot analysis *(12)*. The implementation of these calculations in the software SSHscreen includes statistical tests of the microarray data to provide further confidence in choice of genes for further study (*see* **Subheading 3.15.**).

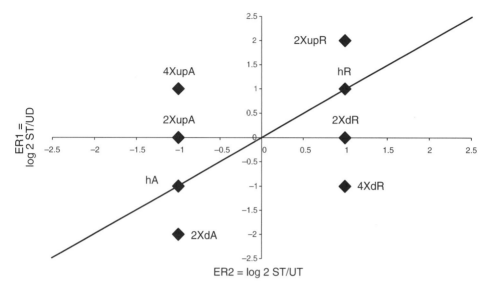

Fig. 2. Illustration of alternative microarray screen of SSH library clones using Enrichment ratio 1 (ER1) and Enrichment ratio 2 (ER2) values calculated from data from eight hypothetical genes (**Table 1**). Data points on the diagonal line (ER1 = ER2 ~ UD = UT) represent housekeeping genes that are unchanged by the treatment (e.g., *hR* and *hA*). Data points above the diagonal line with negative ER2 values (e.g., *2XupA* and *4XupA*) represent genes that are upregulated and abundant in the treated sample UT. Data points above the diagonal line with positive ER2 values (e.g., *2XupR*) represent genes that are upregulated and rare in the treated sample. Data points below the diagonal line (e.g., *2XdA, 2XdR,* and *4XdR*) represent genes that are downregulated by the treatment, for example, and escaped the subtraction.

2. Materials

2.1. Total RNA Isolation

1. RNase Away (Molecular BioProducts, San Diego, CA; cat. no. 7000).
2. RNeasy® Plant Mini Kit (Qiagen, Hilden, Germany; cat. no. 74904).
3. NanoDrop® ND-1000 UV-Vis Spectrophotometer (Nanodrop Technologies, Wilmington, DE).
4. Mortar and Pestle or A11 Basic Grinder (IKA®-Werke GMBH and CO.KG, Staufen, Germany).
5. Simplicity 185 Water purification system (Millipore, Molsheim, France; cat. no. SIMS5V000).

2.2. Messenger RNA Isolation From Total RNA

1. Oligotex mRNA purification kit (Qiagen, cat no. 70022).
2. ActinF primer: 5′ ACCGAAGCCCCTCTTAACCC 3′.
3. ActinR primer: 5′ GTATGGCTGACACCATCACC 3′.

4. dATP, dCTP, dGTP, dTTP (2.5 m*M* each) (Bioline, London, UK; cat. no. BIO-39025).
5. BIOTAQ™ DNA Polymerase 5 U/µL (Bioline, cat. no. BIO-21060).

2.3. cDNA Synthesis

1. cDNA synthesis system (Roche Diagnostics, Basel, Switzerland; cat. no. 1117831).
2. RNeasy MinElute™ Cleanup kit (Qiagen, cat. no. 74204).

2.4. RsaI Digestion

1. *Rsa*I (Roche Diagnostics).

2.5. Adaptor Ligation

1. PCR-Select cDNA subtraction kit (BD Biosciences Clontech, Palo Alto, CA; cat. no. K1804-1).
2. Adaptor 1 (10 µ*M*): 5′-CTAATACGACTCACTATAGGGCTCGAGCGGCCGCC-CGGGCAGGT-3′ 3′-GGCCCGTCCA-5′.
3. Adaptor-2R (10 µ*M*): 5′-CTAATACGACTCACTATAGGGCAGCGTGGTCGCG-GCCGAGGT-3′ 3′-GCCGGCTCCA-5′.
4. PCR Primer 1 (10 µ*M*): 5′-CTAATACGACTCACTATAGGGC-3′.

2.6. Primary and Secondary Hybridization

1. 4X Hybridization buffer (BD Biosciences Clontech).
2. Mineral oil.
3. Dilution buffer (BD Biosciences Clontech).

2.7. Primary PCR Amplification

1. Advantage PCR Polymerase (BD Biosciences Clontech, cat. no. 639201).

2.8. Secondary PCR Amplification

1. Nested PCR primer 1 (10 µ*M*): 5′-TCGAGCGGCCGCCCGGGCAGGT-3′ (BD Biosciences Clontech).
2. Nested PCR primer 2R (10 µ*M*): 5′-AGCGTGGTCGCGGCCGAGGT-3′ (BD Biosciences Clontech).

2.9. Cloning PCR Products Into Vectors

1. pGEM-T Easy vector kit (Promega, Madison, WI; cat. no. A1380).
2. Luria Broth, ampicillin (50 mg/mL stock), laminar flow cabinet.

2.10. Preparation of SSH cDNA Microarrays on Glass Slides

1. 96-Well culture plate (Costar/Corning, NY; cat. no. 612U96).
2. Thermo-Fast® 96-well Semi-Skirted PCR plate (Abgene, Epsom, UK; cat. no. 0900).

3. Silicon sealing mats (Costar/Corning, cat. no. 6555).
4. 96-Pin plate replicator (Amersham Biosciences, Little Chalfont, UK; cat. no. 250520).
5. SP6 and T7 primers.
6. Finnpipet Digital MCP (8 channel) (ThermoLabsystems, Vantea, Finland; cat. no. 4510000).
7. Multiscreen® PCR Purification Plates (Millipore, cat. no. MANU03050).
8. Multiscreen Vacuum Manifold (Millipore, cat. no. MAVM0960R).
9. 96-Microwell storage plate (Nunc, Roskilde, Denmark; cat. no. 267385), 96 well caps sealing mats (Nunc, cat. no. 276000; can be reused after treatment with 10% bleach and autoclaving).
10. SPD111V vacuum centrifuge (Savant, Holbrook, NY).
11. Corning® Gaps II Coated Slides (Corning, cat. no. 40005).
12. Dimethyl sulfoxide (DMSO).
13. 384-Well microplates (Amersham Biosciences, cat. no. RPK0195) (specific requirement for Array Spotter Generation III).
14. Thermo-Sealer (Abgene, cat. no. AB-0384), Easy Peel Thermo-sealer sheets (Abgene, cat. no. AB-0745).
15. Array Spotter Generation III (Molecular Dynamics, Sunnyvale, CA).

2.11. Labeling Complex Probes With Cy Dyes for Microarray Screening

1. Klenow Fragment, exo⁻ (5 U/μL) (Fermentas, Vilnius, Lithuania; cat. no. EPO421).
2. Hexanucleotide mix, 10X (Roche Diagnostics, cat. no. 1277081).
3. QIAquick Gel Extraction Kit (Qiagen, cat. no. 28704).
4. Cy™3-dUTP (1 mM) (cat. no. PA53022), Cy5-dUTP (1 mM) (cat. no. PA55022) (Amersham Biosciences).

2.12. Hybridization of Cy-Labeled Probes to Glass Slide Microarrays

1. Bovine serum albumin Fraction V (Roche Diagnostics, cat. no. 735086).
2. 20X Standard saline citrate (ultrapure) (Invitrogen, Paisley, Scotland; cat. no. 15557-036).
3. 10% Sodium dodecyl sulfate (ultrapure) (Invitrogen, cat. no. 24730-020).
4. 4X Microarray hybridization buffer (Amersham Biosciences, cat. no. RPK0325).
5. Formamide.
6. HybUP hybridization chamber (NB Engineering, Pretoria, South Africa).
7. Cover slips (Marienfield, Bad Mergentheim, Germany; cat. no. MARI5124601).
8. Genepix™ 4000B scanner (Axon Instruments, Foster City, CA).
9. Genepix Pro 5.1 (Axon Instruments) or ArrayVision™ Software (Molecular Dynamics Inc., CA).

2.13. Microarray Data Analysis

1. PC with Windows operating system.
2. R software (2.2.0; download from http://cran.r-project.org).

3. Limma software (linear models for microarray data) (version 2.2.0) (install from within R version 2.2.0).
4. SSHscreen software (download from http://microarray.up.ac.za/SSHscreen/).

3. Methods
3.1. Total RNA Isolation

Total RNA is extracted from 0.5 g fresh weight biological material of the Tester and Driver treatments (*see* **Notes 1** and **2**). Material is frozen in liquid nitrogen and stored at –80°C or immediately ground to a fine powder using a mortar and pestle (preautoclaved) and processed for RNA extraction. A sterile spatula that has been wiped with RNase Away and cooled in liquid nitrogen is used to spoon the frozen powder into a 1.5-mL microfuge tube containing Buffer RLT of the RNeasy Plant Mini Kit. The manufacturer's protocol is followed (*see* **Note 3**). In the final step, the RNeasy column is transferred to a new 1.5-mL Eppendorf tube and RNA is eluted twice into the same collection tube by adding 2×50 µL diethylpyrocarbonate-treated water directly onto the membrane followed by centrifugation for 1 min at 10,000*g*. The expected RNA yield is approx 100 µL at a concentration of approx 200 ng/µL per sample (from 0.5 g starting material). The quantity of extracted RNA is determined with a Nanodrop spectrophotometer. RNA is of sufficient quality with respect to contaminating proteins and organic compounds/carbohydrates if the OD260/280 > 1.8 and OD260/OD230 > 2.0 (*see* **Note 4**).

3.2. Messenger RNA Isolation From Total RNA

In order to enrich mRNA and to decrease the amount of ribosomal RNA (rRNA) and transfer RNA (tRNA) from eukaryotic total RNA samples, poly A^+ mRNA is isolated from 40 µg of RNA starting material (Driver and two Testers; *see* **Note 2**) using an Oligotex mRNA purification kit according to the manufacturers specifications. In the final step, isolated mRNA is eluted with 2X 50 µL elution buffer (10 m*M* Tris-HCl, pH 7.5) and immediately used for downstream applications. The quantity of mRNA is determined with a Nanodrop spectrophotometer. A test can be done to check there is no contaminating genomic DNA by PCR amplification of a "housekeeping gene" (*see* **Note 5**). Primers should be designed so that they flank an intron and the product should not contain an *Rsa*I site. Primers (actinF, actinR) that amplify a fragment from the actin gene in plants were used in the current study. PCR is carried out in a total volume of 20 µL containing 1 µL mRNA, 2 µL 10X Taq buffer, 2 µL dNTPs (stock 2.5 m*M* of each) and 2 µL each of primers actinF (2.5 µ*M* stock) and actinR (2.5 µ*M* stock), and 0.25 µL BIOTAQ™ (5 U/µL). Thermocycler conditions are 94°C for 1 min; 30 cycles of (94°C for 30 s; 55°C for 30 s; 72°C for 1.5 min); 72°C for 5 min.

3.3. cDNA Synthesis

cDNA synthesis of mRNA of the Driver and Tester samples (300 ng to 2 µg of each) is carried out with the cDNA synthesis system kit. RNA secondary structures are eliminated before first-strand cDNA synthesis by preincubation of the reactions for 10 min at 70°C. First strand cDNA is made using AMV Reverse Transcriptase. Following synthesis of the first-strand cDNA, RNase H is used to nick the RNA-strand of the RNA-cDNA heteroduplex. DNA Polymerase I is used with *Escherichia coli* ligase to synthesize second-strand cDNA by nick translation. T_4 DNA Polymerase is added in order to fill in 5′ protruding ends with dNTPs, utilizing the 5′–3′ polymerase-activity of the enzyme, and to generate blunt ends from DNA molecules with 3′ overhangs, using the 3′–5′ exonuclease-activity. The double-stranded cDNA products are then purified using a MinElute Cleanup Kit (*see* **Notes 6** and **7**).

3.4. RsaI Digestion

*Rsa*I digestion of Tester and Driver cDNAs are carried out separately for 4 h at 37°C in total volumes of 40 µL containing 20 µL cDNA (~400 ng cDNA), 1X *Rsa*I buffer, 3 U *Rsa*I, and sterile distilled water (SDW) (*see* **Note 8**). The *Rsa*I digested cDNA is purified using a MinElute Cleanup Kit (*see* **Note 6**). Two reactions of each should be carried out for both the Forward and Reverse libraries. The quantity of the *Rsa*I-digested cDNA is assessed on a Nanodrop spectrophotometer (recovery of >50% is required). Divide the *Rsa*I-digested treated (Tester) and control (Driver) samples into two aliquots each, one Tester/Driver pair for the forward library (protocol described **in Subheading 3.5.–3.11.**) and one Tester/Driver pair for the reverse library (same protocol but the treated sample is Driver and control sample is Tester).

3.5. Adaptor Ligation

Ligation of adaptors to the Tester cDNA is necessary for PCR-based amplification of subtracted material in the first and second rounds of PCR following subtraction (*see* **Note 9**). To ligate adaptors, Tester cDNA has to be subdivided into two portions, and each will be ligated with a different adaptor, either Adaptor 1 or Adaptor 2R. The Driver cDNA will also be subdivided into two portions, one which will be ligated to Adaptors (*see* **Note 10**) and one which will not be ligated to Adaptors (used for hybridizations; *see* **Subheading 3.6.** and **3.7.**).

Purified, *Rsa*I-digested Tester cDNA is diluted by adding 5 µL of sterile water to 1 µL of cDNA. From this, 2 µL of diluted Tester cDNA is placed in a 0.5-mL microfuge tube and mixed with 2-µL Adaptor 1 (10 µ*M*), 3 µL SDW, 2 µL 5X DNA Ligation buffer, 1 µL T_4-DNA ligase (400 U/µL) (termed Tester sample 1-1). An equivalent reaction has to be prepared in order to ligate Adaptor 2R to Tester cDNA (Tester sample 1-2). A third tube denoted Tester sample 1-3 is prepared, made up of 2 μL from sample 1-1 and 2 μL from sample 1-2. Tester sample 1-3 contains Tester cDNA and a mix of Adaptor 1 and Adaptor 2R and will be used to amplify UT material, and also to test that ligation has occurred. A fourth tube denoted Driver sample 1-4 is prepared, made up of 0.8 μL Driver cDNA (diluted by adding 5 μL of sterile water to 1 μL of purified *Rsa*I-digested Driver cDNA), 0.4 μL Adaptor 1 (10 μ*M*), 0.4 μL Adaptor 2R (10 μ*M*), 0.8 μL 5X ligation buffer, 0.4 μL T$_4$ DNA ligase (400 U/μL) and 1.2 μL SDW. Driver sample 1-4 contains Driver cDNA and a mix of Adaptor 1 and Adaptor 2R and will be used to amplify UD material, and also to test that ligation has occurred. Incubate the four ligations overnight at 14°C. Heat the ligations to 72°C for 5 min to inactivate the T$_4$ DNA Ligase.

3.5.1. Test of Adaptor Ligation

The samples 1-3 and 1-4, generated in **Subheading 3.5.**, are used to assess the efficiency of the adaptor ligation. Retain the rest of samples 1-3 and 1-4 for further processing in **Subheading 3.8.** PCR is carried out in a total volume of 25 μL containing 1 μL sample 1-3 (diluted 1 in 4) or 1 μL sample 1-4 (diluted 1 in 4), 2.5 μL 10X Taq buffer, 2 μL dNTPs (stock 2.5 m*M* of each) and 2 μL of PCR Primer 1 (10 μ*M*), and 0.25 μL BIOTAQ™ (5 U/μL). Thermocycler conditions are 94°C for 1 min; 27 cycles of (94°C for 30 s; 66°C for 30 s; 72°C for 1.5 min); 72°C for 5 min. Analyse 7 μL of the PCR-products on a 2% (w/v) agarose gel. Because samples 1-3 and 1-4 each contain both adaptors 1 and 2R, PCR amplification with PCR Primer 1 should result in a smear of products representing different *Rsa*I restriction enzyme products generated in **Subheading 3.4.**

3.6. Primary Hybridization

Ligated Tester samples 1-1 and 1-2 are diluted by adding 9 μL of SDW to 1 μL of each in separate tubes (*see* **Note 11**). Two reactions are setup in 0.2-mL microfuge tubes, each in a reaction volume of 4 μL. The first contains 1.5 μL of Tester 1-1 ligation (1:10 diluted), 1.5 μL of *Rsa*I digested cDNA (Driver) (without adaptors) and 1 μL of 4X hybridization buffer (Clontech) (*see* **Note 12**). The second contains 1.5 μL of Tester 1-2 ligation (1:10 diluted), 1.5 μL of *Rsa*I digested cDNA (Driver) (without adaptors) and 1 μL of 4X hybridization buffer. Once the hybridizations are set up, overlay the samples with 10 μL of mineral oil and then incubate at 98°C in a Thermocycler for 1 min and 68°C > 6 h (do not exceed 12 h incubation). Retain the tubes at 68°C for the next step.

3.7. Secondary Hybridization

In preparation for the secondary hybridization reaction, a master mix is made containing 1 μL of *Rsa*I digested cDNA (Driver) (without adaptors), 1 μL of 4X hybridization buffer and 2 μL SDW (*see* **Notes 13** and **14**). One microliter of the above master mix is placed into a clean tube, overlaid with approx 7 μL of mineral oil, and incubated at 98°C for 1.5 min (to denature sample). At the same time, set a pipet to 30 μL and use this to remove all the sample (4 μL) from the primary hybridization of the Tester 1-2 ligation reaction into a pipet tip (try not to take too much mineral oil) and then withdraw some air into the tip and then remove the entire freshly denatured Driver (4 μL) into the same tip. Transfer all of this into the primary hybridization Tester 1-1 ligation reaction and mix up and down with a pipet. Incubate at 68°C overnight. Add 100 μL of Dilution Buffer (Clontech) to the secondary hybridization mix. Heat to 68°C for 7 min. Store reaction in 20-μL aliquots at –20°C.

3.8. Primary PCR Amplification

The primary PCR is carried out in a total volume of 25 μL containing 1 μL of the secondary hybridization mix (template), 2.5 μL 10X PCR buffer, 0.5 μL dNTPs (10 mM), 1 μL of PCR Primer 1 (10 μM), and 0.5 μL 50X Advantage cDNA polymerase mix (*see* **Note 15**). Thermocycler conditions are 1 cycle of 75°C for 5 min; 94°C for 1 min, followed by 30 cycles in total of 94°C for 30 s; 66°C for 30 s; 72°C for 1.5 min; and a final cycle at 72°C for 5 min. Take 3 μL samples after 20, 25, and 30 cycles to optimise PCR conditions (*see* **Subheading 3.9.**). Set up three additional PCR reactions to 30 cycles with different templates, namely 1 μL of the 1-3 Tester ligation (1:100 dilution), 1 μL of the 1-4 Driver ligation (1:100 dilution), or 1 μL SDW.

3.9. Secondary PCR Amplification

The secondary PCR amplification is carried out with the same reaction conditions and cycles as the primary PCR, except that 1 μL each of the Nested PCR primer 1 (Clontech 10 μM) and Nested PCR primer 2R (Clontech 10 μM) is added instead of PCR primer 1, and the first cycle is 94°C for 1 min (*see* **Note 16**). Six PCR reactions are prepared, each containing different templates consisting of 1 μL of the products of the primary PCR reactions that have been diluted 1:10. The templates (from reactions carried out in **Subheading 3.8.**) are hybridization/primary PCR after 20, 25 and 30 cycles, the 1-3 Tester ligation/primary PCR (30 cycles), 1-4 Driver ligation/primary PCR (30 cycles) and SDW control/primary PCR (30 cycles). Take 5-μL samples after 20, 25, and 30 cycles from only the reactions containing the hybridization/primary PCR as template to check for optimal PCR conditions (*see* **Note 16**) and analyze on a 2% (w/v) agarose gel. The other reactions should be analyzed after 30 cycles only.

3.10. Bulking Up of Secondary PCR Products

Repeat the secondary PCR amplification using the PCR conditions optimized in **Subheadings 3.8.** and **3.9.** Set up four or more reactions of each template

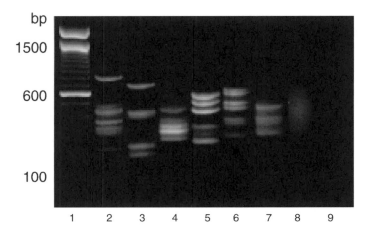

Fig. 3. Example of secondary PCR amplification of SSH products (from pearl millet). Six individual secondary PCR reactions were set up in order to bulk up primary PCR SSH products (*see* **Subheading 3.10.** and **Note 17**). **Lane 1**, 100 bp DNA ladder (Invitrogen); **lanes 2–7**, secondary amplification of subtracted tester cDNAs (sample 1-1); **lane 8**, amplification of UT cDNAs (sample 1-3); **lane 9**, sterile distilled water negative control.

using "hybridization/primary PCR" as template (subtracted Tester material) or Tester 1-3 ligation/primary PCR as template (UT material) or Driver 1-4 ligation/ primary PCR as template (UD material) (*see* **Fig. 3**). Include a negative control PCR reaction with SDW as template. Purify secondary PCR products of the subtracted and the unsubtracted materials using a MinElute kit in 50 μL amounts following the protocol described in **Subheading 3.3.** (*see* **Notes 17–20**) but elute twice with 10 μL water. The pooled and purified secondary PCR products from the Tester "hybridization/primary PCR" are named ST, whereas, the secondary PCR products from Tester 1-3 are named UT, and from Driver 1-4 are named UD. The quantity of cDNA is determined spectrophoto- metrically on a Nanodrop (*see* **Notes 17–20**).

3.11. Cloning PCR Products Into Vectors

The Forward SSH cDNA library is constructed by ligating purified ST insert DNA (*see* **Subheading 3.10.**) with pGEM-T Easy vector. The pGEM-T Easy vector kit has been optimised using a 1:1 molar ratio of insert DNA to vector (*see* **Notes 21** and **22**). Reactions are carried out in 10 μL reaction volume con- taining 10–50 ng ST insert DNA (usually 3 μL), 50 ng pGEM-T Easy vector (50 ng/μL), 1X rapid ligation buffer (30 mM Tris-HCl—pH 7.8, 10 mM MgCl$_2$, 10 mM DTT, 1 mM ATP, 5% [w/v] polyethylene glycol MW 8000) and 3 Weiss units T$_4$ DNA Ligase. Samples are incubated at 14°C for 16 h and stored at −20°C. Ligation products are transformed into *E. coli* JM109 cells using standard procedures. At least 1920 clones from the Forward SSH cDNA library (and 1920 clones from the Reverse SSH cDNA library if this subtraction is also done) can be stored as *E. coli* glycerol stocks in 96-well culture plates sealed with their own lids at −80°C (*see* **Note 23**).

3.12. Preparation of SSH cDNA Microarrays on Glass Slides

Aliquot 75 µL Luria Bertanii medium (with 50 µg/µL ampicillin) per well into 96-well culture plates. Inoculate the culture plates with the cDNA clones stored as *E. coli* glycerol stocks at −80°C using a 96-pin plate replicator. Work in a laminar flow cabinet and flame-sterilize the replicator before inoculating each plate. Incubate plates at 37°C overnight with shaking at 150 rpm. A PCR master mix for 96 reactions is prepared and 100-µL aliquots are added to each well of a Thermo-Fast® 96-well PCR plate. Each PCR reaction contains 1X Taq buffer, 2.5 mM MgCl$_2$, 0.2 mM of each dNTP and 0.3 µM of SP6 and T7 primers each (which flank the insert in the pGEM-T Easy vector), and 1 U BIOTAQ™. The sterilized 96-pin plate replicator is inserted into the 96-well overnight culture, and then inserted into the 96-well plate containing the PCR reaction mixes, thereby transferring a small amount of each culture. Each PCR plate is sealed with a silicon mat. Thermocycler conditions are 94°C for 5 min; 30 cycles of (94°C for 30 s; 50°C for 30 s; 72°C for 1 min.); 72°C for 5 min. PCR reaction products are purified by loading into 96-well Multiscreen purification plates on a vacuum manifold. A vacuum (24 in Hg = ± 800 mbar) is applied for 10 min. The purification plate is removed from the manifold and blotted dry from underneath with paper towels. SDW (50 µL) is added to each well and the plate is shaken vigorously for 5 min. The purified PCR products are removed from each well with a multichannel pipet and transferred to a 96-well storage plate. The samples are stored at −20°C. In preparation for arraying, the samples in the 96-well storage plates are dried in a vacuum centrifuge and resuspended in 15 µL SDW. A 2-µL sample of each reaction is visualised on a 1% agarose Electro-Fast® Stretch gel to check that a single band is amplified in each reaction and to estimate the yield (~2.5 µg of each PCR product is required for arraying). An equal volume of 100% DMSO is added to each sample to obtain at least 16 µL of each PCR product at a concentration of >125 ng/µL in 50% DMSO, which is required for arraying. The samples are transferred into 384-well plates before arraying onto Corning Gaps II Slides using an Array Spotter (*see* **Note 24**). Each sample is arrayed in duplicate on each slide. The slides are stored in a desiccator at room temperature. The 384-well plates containing the remaining samples are sealed with a Thermo-sealer and stored at −20°C.

3.13. Labeling Complex Probes With Cy Dyes for Microarray Screening

Complex probes need to be generated from ST, UT, and UD to screen the SSH library on glass slide microarrays, but common sequences such as the adaptors have to be removed by restriction digest with *Rsa*I. After restriction digestion with *Rsa*I using reaction conditions as described in **Subheading 3.4.**, digested samples are purified by agarose gel electrophoresis followed by excision of the cDNA fragments and discarding the removed adaptors, using the QIAquick Gel Extraction Kit. The quantity of each cDNA complex probe is determined spectrophotometrically on a Nanodrop.

Complex cDNA probes of ST, UT, or UD with the adaptors removed are labeled by incorporation of Cy™5 or Cy™3 dUTP using Klenow enzyme. To set up two labeling reactions (for hybridization to one slide), two tubes each containing 14 µL with 200 ng of template cDNA (ST, UT, or UD) are denatured by heating at 95°C for 5 min. and immediately placed on ice. A master dNTP mix for two labeling reactions (for hybridization to one slide) is made up of 1.6 µL of dATP, dGTP, and dCTP (2.5 m*M* each) and 2.2 µL of 2.5 m*M* dTTP. This is split into two equal volumes of 1.9 µL per tube. Add 0.3 µL Cy3-dUTP (1 m*M*) to one tube (Cy™3-dNTP mix) and 0.3 µL Cy5-dUTP (1 m*M*) to the other tube (Cy™5-dNTP mix). Labeling reactions are carried out in a total volume of 20 µL containing 200 ng denatured ST cDNA, 2 µL of 10X Klenow buffer, 2 µL Cy3-dNTP mix or Cy5-dNTP mix, 2 µL hexanucleotide mix (10X), 1 µL Klenow enzyme. After incubation at 37°C for 2–20 h, the reaction is stopped with 2 µL 0.5 *M* EDTA pH 8.0. Complex probes are purified using a QIAquick column and eluted in 45 µL SDW. A Nanodrop is used to determine the yield of Cy-dye labeled probe and Cy-dye incorporation of each sample (require >50 pmoles of probe per sample, with a Cy-dye molecule incorporated at a frequency of >15 and <50 per unlabeled nucleotide).

3.14. Hybridization of Cy-Labeled Probes to Glass Slide Microarrays

Six glass slides arrayed with the SSH cDNA Forward library (*see* **Note 25**) are initially incubated in filtered pretreatment solution (3.5X standard saline citrate [SSC]; 0.2% sodium dodecyl sulfate [SDS]; 1% bovine serum albumin) without probe at 60°C for 20 min (*see* **Note 26**). Before hybridization, equimolar amounts (>50 pmoles) of each pair of appropriate probes labeled with the different Cy dyes are mixed in a 0.5-mL tube, and dried in a vacuum centrifuge. The following probe combinations are prepared: Cy™3-labeled ST and Cy5-labeled UD; dye swap: Cy™3-labeled UD and Cy™5-labeled ST probe; Cy3-labeled ST and Cy™5-labeled UT; dye swap: Cy™3-labeled UT and Cy5-labeled ST probe; Cy3-labeled UT and Cy™5-labeled UD; dye swap: Cy3-labeled UD and Cy5-labeled UT. The Cy-labeled probes are resuspended in 40 µL hybridization solution made up of 50% formamide, 25% microarray hybridization buffer and 25% SDW. The probe mixtures are placed at 92°C for 2 min to denature the probes, and then immediately placed on ice for 1 min. The glass slide is placed in a manual hybridization chamber (*see* **Note 27**) and the denatured probe mixture is applied and overlaid with a cover slip. The chamber is sealed and submerged in a water bath at 42°C for 16 h. After hybridization, slides are washed for 4 min at 42°C with 1X SSC/0.2% SDS, 0.1X SSC/0.2% SDS (twice) followed by three washes in 0.1X SSC for 1 min at room temperature. The slides are rinsed with distilled water, placed in a 50-mL Falcon tube, dried by centrifugation at 1000g for 2 min. in an Eppendorf (5810R) centrifuge, and scanned with a Genepix 4000B scanner. Genepix Pro 5.1 or ArrayVision software is used to localize and integrate every spot on the array.

3.15. Analysis of Microarray Data

We present a simple data analysis pipeline in the R computing environment in the form of the add-on package SSHscreen. It provides an analysis tool for performing quantitative screening of SSH cDNA library clones using cDNA microarray data. SSHscreen is built around an existing R package limma (linear models for microarray data), which provides the functionality for importing and analyzing gene expression microarray data. Limma provides the tools for assessing designed experiments and assessing differential expression through the use of linear models. It has features which make the analyses stable even for experiments with a small number of arrays—this is achieved by borrowing information across genes. The theoretical details underlying limma are outlined in **ref. 6**. We illustrate the functionality of SSHscreen by means of an example analysis performed using the pearl millet data.

3.15.1. Installing R

Currently, SSHscreen is designed for the Windows platform. Download the R (V2.2.0) installation file for Windows from http://cran.r-project.org. The correct version can be found under.....CRAN.........Mirrors (choose the CRAN mirror site that is geographically closest) (e.g., if you are in UK choose http://cran.uk.r-project.org/).......Precompiled Binary Distributions......... Windows (95 and later)......base......R-2.2.0-win32.exe (Setup program: ~25 megabytes). To start the installation just double-click the file and follow the instructions. Choose a working directory for R. If you installed manually, make a shortcut to R-2.2.0\bin\Rgui.exe on your desktop or somewhere on the Start menu file tree. Right-click the shortcut, select Properties...and change the "Start in" field to your working directory.

3.15.2. Installing the Limma Package in R

We need to install the limma package, because the SSHscreen package is dependent on limma. Launch R and select......Packages>Select repositories .....from the drop-down menu. Make sure that the Bioconductor repository is selected. Next select Packages>Set CRAN mirror to select the CRAN mirror site closest to your location and click OK. Finally, select Packages>Install package(s) from the drop-down menu and select the limma package (version 2.2.0) from the list of packages and click OK (*see* **Note 28**). In subsequent sessions, load the installed package by typing library(limma) at the R command prompt.

3.15.3. Installing and Loading the SSHscreen Package in R

Download the zip file for the SSHscreen package from http://microarray. up.ac.za/SSHscreen/. In R, select Packages>Install package(s) from local zip files… from the drop-down menu and specify the path to the downloaded zip file. Load the installed package by typing library(SSHscreen) at the command prompt.

3.15.4. Data Preparation

Data from several image analysis programs can be read directly into SSHscreen. This includes output from GenePix Pro, ArrayVision, Imagene, Spot, and other software. Data from image analysis software not directly supported by SSHscreen can also be read in. These data files should be saved as tab-delimited text files and should contain columns named "SpotLabel," "GeneList," "Rf," "Gf," "Rb," "Gb," which contain the spot labels (i.e., label that corresponds to the spot position on the array), the names of the SSH cDNA clones, and the red (Cy5) and green (Cy3) foreground and background, respectively. Optional columns named "SNR," "SNG," and "flag" that contain the signal to noise ratios and flags for the different spots can also be provided. Example data files for the pearl millet data (captured in ArrayVision and prepared manually) can be downloaded from the SSHscreen website (ER1vsER2.zip; ER3vsER2inv.zip). These data files are in the general tab-delimited format described above, and can be prepared with any spreadsheet programme such as Microsoft® Excel. However, it is recommended that output files generated by any of the supported image analysis programs are read directly into SSHscreen. The source of the data files should be specified through the "source" argument to the SSHscreen function (e.g., source = "other" for tab-delimited data files prepared as described above; source = "genepix" for ".gpr" data files from Genepix Pro 5.1, and so on).

3.15.5. Preparing the "Targets" File

The targets file describes which SSH cDNA sample was hybridized to each channel of each array. It should be in tab-delimited text format and should contain a row for each microarray in the experiment. The file must be named Targets.txt. The targets file must contain a FileName column, giving the names of the files containing the expression data (as described earlier), a Cy3 column giving the cDNA type labeled with Cy3 dye for that slide and a Cy5 column giving the cDNA type labeled with Cy5 dye for that slide. Other columns are optional (*see* **Note 29**).

3.15.6. Preparing the "Spot Types" File

The spot types file identifies special array elements such as control spots. The tab-delimited file is used to set the control status of each spot on the arrays, so that plots may highlight different types of spots in an appropriate way. The file should have a SpotType column giving the names of the different spot types. One or more other columns should have the same names as columns in the gene list and should contain patterns or standard terms sufficient to identify the spot-type. Any other columns are assumed to contain plotting attributes, such as colors or symbols, to be associated with the spot-types. This file must be named SpotTypes.txt. Further details for preparing targets and spot type files with examples are provided in the limma user's guide available at http://bioinf.wehi.edu.au/limma/usersguide.pdf. The targets and spot types files can be prepared using any text editor but spreadsheet programs such as Microsoft® Excel are convenient (*see* **Note 29**). Move the data, targets and spot types files for the "ER1vsER2" analysis to a single directory (e.g., /ER1vsER2/) and the corresponding files for the "ER3 vs inverse ER2" analysis into a different directory (e.g., /ER3vsER2inv/). If the data is in Genepix Pro 5.1 format (i.e., ".gpr" files), provide the ".gal" file in the same directories so that the limma software can associate the genelist with spot positions on each array.

3.15.7. Using the SSHscreen Function

For details on how to use SSHscreen type help (SSHscreen) at the R command line (assuming that the package has been loaded). This provides detailed information on how to use the function, which can be used to perform either the "ER1 vs ER2" or "ER3 vs inverseER2" SSH screening procedures. The required procedure can be specified through the "method" argument. The path to the directory containing the data, targets and spot type files is specified through the "path" argument. The "ndups" argument specifies the number of duplicate spots on the arrays, while the "spacing" argument specifies the spacing between the rows of the array corresponding to duplicate spots. The argument "irregular" = TRUE, "ndups" = 2 (number of replicates), "spacing" = 1 is commonly used when the user wishes SSHscreen to count the spacing between replicate spots automatically based on the gene IDs. The default is "irregular" = FALSE and in this case it is important to provide the spacing of the replicate cDNA spots after subtraction of the number of control and blank spots between each duplicate cDNA spot. The above help file provides further details and options. The SSHscreen function includes implementation of limma calculations for within- and between-slide normalization of the microarray data (*see* **Note 30**).

3.15.8. Analyzing the Pearl Millet Data as an Example

We perform an example "ER3 vs inverseER2" analysis using the pearl millet data. Download the zipped data from http://microarray.up.ac.za/SSHscreen/ and unzip it to any directory, say C:\data\ER3vsER2inv. Execute SSHscreen using this data with following command:

SSHscreen(path = "C:\\data\\ER3vsER2inv", source= "other", ndups = 2, spacing = 32, method = "ER3").

An "ER1vsER2 analysis" as in **ref. *12*** can also be done with the sample pearl millet data that can be downloaded from http://microarray.up.ac.za/SSHscreen/. Execute SSHscreen using this data with following command:

SSHscreen(path = "C:\\data\\ER1 vs ER2," source = "other," ndups = 2, spacing = 32, method = "ER1").

The same procedures can be followed with the users' own data provided the input files are correctly prepared. For example, suppose we want to perform an "ER3 vs inverse ER2" analysis and we have GenePix Pro 5.1 output files UT_UD.gpr, UD_UT.gpr, ST_UT.gpr, and UT_ST.gpr along with suitably prepared "targets," "spot types" and ".gal" files in the same directory, say C:\data\ER3vsER2inv\gpr\. We can execute SSHscreen using this data with following command (assuming the data is spotted as duplicate spots on each array):

SSHscreen(path = "C:\\data\\ER3vsER2inv\\gpr\\", source = "genepix", irregular = TRUE, ndups = 2, spacing = 1, method = "ER3,"norm.plot = TRUE).

The SSHscreen help file contains many more options and details.

3.15.9. Interpreting the Output From the "ER3 vs Inverse ER2" Analysis

Figure 4 shows the "Up/Down Regulation" plot of ER3 vs inverse ER2 produced for the pearl millet data, which corresponds to **Fig. 1**. The clones that are statistically significantly up- or downregulated, as determined by the linear model, are also indicated. For this example data, two tables of output are returned to the screen. The first table is named "tt.ud" and lists the 100 most significant up- and downregulated clones. This table includes, amongst other things, the *t*-statistics and corresponding *p*-values outputted by limma. Clones with positive *t*-values are upregulated, while clones with negative *t*-values are downregulated. In this example, *p*-values are not corrected for multiple testing by controlling the false discovery rate (FDR). This can be specified by the argument "adjust." This adjustment is recommended when the user applies the above screening procedure to their own data, with the user deciding upon an acceptable FDR (*see* **Note 31**). The argument "toplist" can be used to control the number of genes that are returned. In this example of

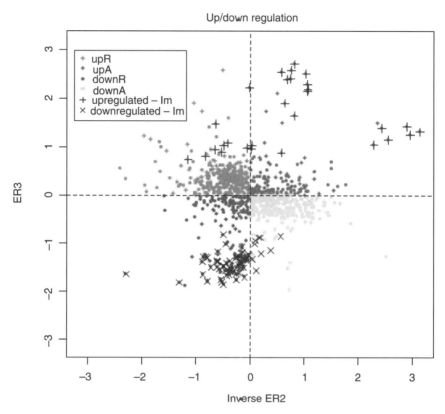

Fig. 4. ER3 vs Inverse ER2 plot produced by SSHscreen for the pearl millet data. A black and white image is shown. The standard SSHscreen plot shows data points in color as follows: upregulated/rare (upR) transcripts (quadrant 1; blue), upregulated/abundant (upA) transcripts (quadrant 2; green), downregulated/rare (downR) transcripts (quadrant 3; purple), and downregulated/abundant (downA) transcripts (quadrant 4; yellow).

"ER3 vs inverse ER2" analysis for the pearl millet data, a second plot is generated (named "Abundance") that indicates which clones are rare or abundant, according to the statistical linear model (figure not shown). The second table returned to the screen is named "tt.ar" and lists the 100 most significant abundant and rare clones, according to the statistical linear model. In addition, it is advisable to display data in various ways as a quality check and to check for unexpected effects. SSHscreen produces MA-plots *(13)* for the raw and normalized data for each slide when the argument "norm.plot" is specified as TRUE (the default is FALSE). A MA–plot is a scatter plot of log intensity ratios $M = \log_2(R/G)$ vs average log intensities $A = \log_2(R \times G)/2$, where, R and G represent the fluorescence intensities in the red and green channels, respectively.

4. Notes

1. To minimize RNA degradation by RNases, all solutions are autoclaved and glassware, mortars, and pestles are baked at 200°C for 16 h before use. Sterile, disposable pipet tips, and microfuge tubes are used and gloves are always worn and frequently changed. Diethylpyrocarbonate treated distilled water is used for RNA extraction steps. SDW (18.2 MΩ) is prepared by a Simplicity 185 water purification system. A mortar and pestle is sufficient for grinding leaf material; however, an A11 Basic Grinder is recommended for grinding woody or fibrous plant tissue for RNA extraction.

2. In this example, the Tester samples are pearl millet leaves treated separately by wounding and fungal elicitor (100 mg/mL chitin) or wounding and bacterial elicitor (100 mg/mL flagellin). RNA was extracted independently in triplicate samples of 0.5 g leaves per sample harvested at 5, 14, and 24 h after each treatment. Two pools of Tester RNA are made, one Tester (chitin treatment) by pooling 15 μg of RNA from each of the three time-points (45 μg); and the other Tester (flagellin treatment) by pooling 15 μg of RNA from each of the three time points (45 μg). Driver samples are pearl millet leaves treated with sterile distilled water (SDW), and RNA was pooled from six replicate extractions (45 μg).

3. RNeasy technology is based on silica-gel-based membranes with selective nucleic acid binding properties. A high-salt buffer system allows binding of up to 100 μg of RNA (>200 bases) to the membrane. The presence of highly denaturing guanidine isothiocyanate ensures the isolation of intact RNA by inactivating RNases released in the disrupted plant material.

4. In addition, 2 μg RNA can be analyzed by electrophoresis under nondenaturing conditions on 1.2% (w/v) agarose gels. Although RNA forms secondary structures under these conditions, the presence of distinct ribosomal bands and the absence of genomic DNA or degraded RNA may be used as indicators of the quality of RNA.

5. The expected result is that no fragment corresponding to the size of the genomic actin fragment should be amplified. If DNA contamination is present, this should be removed by one of two methods, either treating the mRNA with DNaseI on RNAeasy purification columns or subjecting the mRNA sample to a second purification with the Oligotex mRNA purification kit.

6. This protocol is designed to purify double-stranded DNA fragments from PCR reactions resulting in high concentrations of DNA. Fragments ranging from 70 bp to 4 kbp are purified from primers, nucleotides, polymerases, and salts using MinElute spin columns in a microcentrifuge. Ensure that the elution buffer is dispensed directly onto the center of the membrane for complete elution of bound DNA. The average eluate volume is 9 μL from 10 μL elution buffer volume. Maximum elution efficiency is achieved between pH 7.0 and 8.5. When using water, make sure that the pH value is within this range, and store DNA at –20°C as DNA may degrade in the absence of a buffering agent. The purified DNA can also be eluted in TE (10 m*M* Tris-HCl, 1 m*M* EDTA, pH 8.0), but the EDTA may inhibit subsequent enzymatic reactions.

7. Second strand cDNA synthesis product quantity is determined spectrophotometrically on a Nanodrop (expected yield of 400 ng cDNA from 40 µg RNA starting material). The quality is determined by PCR for a "housekeeping gene" cDNA (actin is commonly used in plants as described in **Subheading 3.2.**). The expected result is amplification of an actin cDNA fragment of the correct size (i.e., smaller than the genomic actin fragment because the intron should have been spliced out).

8. In order to create Tester and Driver DNA fragments of less than 2 kb in size with blunt ends, purified cDNA has to be digested with *Rsa*I, a four-base-cutting restriction enzyme (GT/AC). The Driver used in this example was 400 ng Driver cDNA, and the Tester cDNA was 200 ng Tester (chitin treatment) pooled with 200 ng Tester (flagellin treatment). Note that these amounts are sufficient for a Forward library, but twice this amount should be digested with *Rsa*I for additional construction of a Reverse library.

9. The protocol for construction of a Forward library only is described from this point onward, because construction of a Reverse library follows the same steps but with reversed starting material.

10. The protocol deviates from the standard protocol for the PCR-Select cDNA subtraction kit, in which adaptors are not ligated to the Driver cDNA. In the protocol described in this chapter, a sample of Driver cDNA with ligated adaptors is necessary for later screening of the cDNA library on glass slide microarrays.

11. In the primary hybridization, an excess of Driver cDNA (without adaptors) is added to each Tester cDNA (1-1 and 1-2), and the samples are heat denatured and allowed to anneal. Nontarget cDNAs, present in both Tester and Driver, form hybrids and are removed from the subtraction. Remaining single-stranded cDNAs are available for the second hybridization and contain an enrichment of differentially expressed genes. The Tester must be diluted relative to Driver material before subtraction. After adaptor ligation, the Tester has already been diluted 1:6, followed by 1:5 (30-fold dilution in total). The Tester should be diluted a further 10-fold by adding water (total of 300-fold dilution) before hybridization with Driver material (the dilution level can be varied to increase or decrease the stringency of the subtraction).

12. Make sure that the hybridization buffer has fully dissolved by placing it at room temperature (25°C) for approx 15 min or heat to 37°C for 10 min.

13. In the secondary hybridization, both samples from the primary hybridization containing Adaptor 1- or Adaptor 2R-ligated Tester hybridised with an excess of Driver, are mixed together, and freshly denatured *Rsa*I-digested Driver cDNA (without adaptors) is added to further enrich for differentially expressed sequences. Single stranded DNAs, not annealed in the first hybridization, form new hybrid molecules, carrying different adaptors on each end.

14. It is important to keep all the samples at 68°C during the procedure outlined in **Subheading 3.7.**

15. Differentially expressed cDNAs, enriched through hybridization, are amplified using PCR. In the primary suppression PCR only double stranded cDNAs with different adaptors at the 5′- and 3′-ends are exponentially amplified, whereas,

those with the same adaptors form secondary structures and amplification is suppressed. It is important to use a hot-start Taq polymerase such as the Advantage cDNA polymerase (made up of KlenTaq1, a proofreading polymerase and TaqStart Antibody), for the primary and secondary PCR reactions.

16. The secondary, nested PCR amplification is used to further reduce background and to enrich for differentially expressed sequences. Optimized PCR conditions are the combination of cycles in the primary and secondary PCR that yield the maximum number of PCR product bands with similar intensities from the reactions containing the hybridization/primary PCR as template. Nonoptimal reactions are those that yield a smear of products or individual products of high intensity.

17. Carry out sufficient PCR reactions of each to have enough material for Cy dye labeling and screening on microarray slides (1 μg of each sample per library). It is not uncommon for different products to be observed when samples of the secondary PCR products are analyzed by agarose gel electrophoresis (*see* **Fig. 3**). The reason for this is that template molecules are highly diluted before amplification in the primary and secondary PCR reactions.

18. Purification of secondary PCR products is required for probe generation for Southern analysis, microarray screening and cloning of PCR products into the pGEMT-easy Vector. Glycerol and high salt concentrations from PCR reactions are liable to inhibit subsequent applications and have to be removed.

19. Control PCR reactions can be set up using the "hybridization/primary PCR" as template (subtracted material) or Tester 1-3 or Driver 1-4 ligation/primary PCR as template (unsubtracted material) with primers designed to a plant "housekeeping gene" such as actin using the protocol described in **Subheading 3.2**. Take 3 samples after 20, 25, and 30 cycles and analyse on a 2% (w/v) agarose gel. The expected result is that the actin cDNA PCR product should decrease as the cycles increase when the template is the subtracted material, whereas the same product should increase with the cycles when the template is the unsubtracted material 1-3 or 1-4 *(12)*.

20. Southern hybridization analysis can be used to assess the level of enrichment following the SSH procedure and in addition to evaluate the level of background *(9)*. PCR products from ST, UT, and UD are separated by gel electrophoresis, transferred to a positively charged nylon membrane and hybridized with complex probes derived from ST, UT, or UD after adaptor removal.

21. An optional step is to size fractionate the ST insert DNA on an agarose gel and purify two sets of fragments (large fragments >500 bp and small fragments >100 bp and <500 bp). These are then ligated separately into the pGEM-T Easy vector. This step aims to reduce the proportion of smaller fragments in the SSH cDNA library, which are often preferentially inserted into cloning vectors.

22. pGEM-T Easy vector contains T7 and SP6 RNA polymerase promoters, flanking a multiple cloning region within the α-peptide coding region of the enzyme β-galactosidase. Insertion of DNA leads to inactivation of the α-peptide and allows identification of recombinant clones by color screening on indicator plates. The pGEM-T Easy vector contains a 3′-terminal deoxythymidine supporting

ligation of PCR products with a 3′-terminal deoxyadenosine, added during PCR amplification by nonproofreading enzymes such as *Taq* DNA polymerase. Alternative PCR product cloning vectors are available from a range of other commercial suppliers.

23. Two tests can be carried out to determine if the SSH cDNA library is of sufficient quality to continue with screening on glass slide microarrays. Standard PCR reactions can be carried out on 20 random clones from the library using primers flanking the multiple cloning region such as T7 and SP6 primers, and analyzed by agarose gel electrophoresis. At least 80% of the clones should contain single inserts and if these are of different sizes, this will indicate that the library has the desired low level of redundancy. The second test is to determine the DNA sequence of the inserts in the random selection of 20 clones. DNA sequence analysis is used to confirm the presence of expected adaptor sequences at the ends of each insert, and that each insert codes for an open reading frame. In addition, one can identify *Rsa*I restriction sites within the insert (which may indicate ligation of noncontiguous *Rsa*I fragments). A BLAST search *(14)* can determine whether the redundancy is low and whether the inserts represent fragments of genes that are likely to be involved in the biological response under investigation.

24. This step was carried out at the ACGT Microarray Facility (http://microarray.up. ac.za/); however, it can be done at any microarray core facility that is setup to spot cDNAs on glass slide microarrays. The reverse subtraction SSH cDNA library can be arrayed on the same glass slide if it was constructed from the same material at the same time as the forward subtraction SSH cDNA library. Thereby, both libraries can be screened at the same time.

25. The protocol is given for six slides to illustrate both methods of screening the SSH cDNA library. However, four slides are the minimum number required for a screen, by either hybridizing the contrasts UT vs UD and ST vs UT with a dye swap of each (*see* **Fig. 1**; "ER3 vs inverse ER2" analysis), or hybridizing the contrasts ST vs UD and ST vs UT with a dye swap of each (*see* **Fig. 2**; "ER1 vs ER2" analysis and *[12]*). In addition, greater statistical power can be gained by technical replicates of each contrast (e.g., four slides in duplicate). If both analyses are done, it is useful to identify those clones that are common to both the toptable of the "ER1 vs ER2" analysis and the toptable of the "ER3 vs inverse ER2" analysis (tt.ud).

26. Prehybridization with unlabeled adaptors can be carried out to reduce background.

27. Manual hybridization was carried out in a HybUP hybridization chamber (NB Engineering) designed to process five glass slides at a time; however, there are a range of alternative commercially available manual or automated hybridization chambers.

28. If the direct link to limma is blocked by your local Internet firewall, download the limma zip file from http://cran.r-project.org and install it in R by selecting Packages>Install package(s) from local zip files…from the drop-down menu.

29. Different "targets" and "spot types" files need to be prepared for the "ER1 vs ER2" analysis and "ER3 vs ER2inv" analysis. Examples of these files are provided within the millet example analysis procedure. The "targets" and "spot types" files

must be placed in the same directory as the data files. The path to this directory should be specified through the "path" argument in the SSHscreen function call.

30. The SSHscreen software implements both within- and between-array normalizations. The wide variety of within-array and between-array normalization procedures supported in limma is available in SSHscreen. These methods and general issues involved in the normalization of two-color cDNA microarray data are discussed in **ref. *15***. Questions about the SSHscreen software can be directed to dave.berger@up.ac.za.

31. The FDR is an error rate that controls the expected number of unchanged genes among those that are changed (differentially expressed). Statistical methods controlling the FDR are relatively new and rely on some assumptions. However, they tend to be more suited for studies of an exploratory, rather than confirmatory nature. The nature and aims of a particular study usually guides users as to what FDR they are willing to accept. For more details on the control of error rates in a microarray context readers can refer to **ref. *16***.

Acknowledgments

This work was supported by the National Research Foundation and the African Center for Gene Technologies (ACGT), South Africa.

References

1. Ying, S. -Y. (2003) Complementary DNA libraries: an overview, in *Generation of cDNA Libraries: Methods and Protocols in Molecular Biology, vol. 221,* (Ying, S. -Y. ed.), Humana Press, Totowa, NJ, pp. 1–12.

2. Diatchenko, L., Lukyanov, S., Lau, Y. F., and Siebert, P. D. (1999) Suppression subtractive hybridization: a versatile method for identifying differentially expressed genes. *Methods Enzymol.* **303,** 349–380.

3. Bonaldo, M. F., Lennon, G., and Soares, M. B. (1996) Normalization and subtraction: two approaches to facilitate gene discovery. *Genome Res.* **6,** 791–806.

4. Zhulidov, P. A., Bogdanova, E. A., Shcheglov, A. S., et al. (2004) Simple cDNA normalization using kamchatka crab duplex-specific nuclease. *Nucleic Acids Res.* **32,** E37.

5. Mahalingam, R., Gomez-Buitrago, A., Eckardt, N., et al. (2003) Characterizing the stress/defense transcriptome of *Arabidopsis*. *Genome Biol.* **4,** R20.1–R20.14.

6. Smyth, G. K. (2004) Linear models and empirical Bayes methods for assessing differential expression in microarray experiments. *Statist. Appl. Genet. Mol. Biol.* **3,** 1–26.

7. Yang, G. P., Ross, D. T., Kuang, W. W., Brown, P. O., and Weigel, R. J. (1999) Combining SSH and cDNA microarrays for rapid identification of differentially expressed genes. *Nucleic Acids Res.* **27,** 1517–1523.

8. Vallee, M., Gravel, C., Palin, M. F., et al. (2005) Identification of novel and known oocyte-specific genes using complementary DNA subtraction and microarray analysis in three different species. *Biol. Reprod.* **73,** 63–71.

9. Hein, I., Campbell, E. I., Woodhead, M., et al. (2004) Characterisation of early transcriptional changes involving multiple signalling pathways in the a13 barley interaction with powdery mildew (*Blumeria graminis* f. sp. *hordei*). *Planta* **218,** 803–813.

10. Cao, W., Epstein, C., Liu, H., et al. (2004) Comparing gene discovery from Affymetrix GeneChip microarrays and Clontech PCR-select cDNA subtraction: a case study. *BMC Genomics* **5,** 26.

11. Diatchenko, L., Lau, Y. F., Campbell, A. P., et al. (1996) Suppression subtractive hybridization: a method for generating differentially regulated or tissue-specific cDNA probes and libraries. *Proc. Natl. Acad. Sci. USA* **93,** 6025–6030.

12. Van den Berg, N., Crampton, B. G., Hein, I., Birch, P. R., and Berger, D. K. (2004) High-throughput screening of suppression subtractive hybridization cDNA libraries using DNA microarray analysis. *Biotechniques* **37,** 818–824.

13. Dudoit, S., Yang, Y. H., Callow, M. J., and Speed, T. P. (2002) Statistical methods for identifying differentially expressed genes in replicated cDNA microarray experiments. *Statistica Sinica* **12,** 111–139.

14. Altschul, S. F., Gish, W., Miller, W., Myers, E. W., and Lipman, D. J. (1990) Basic local alignment search tool. *J. Mol. Biol.* **215,** 403–410.

15. Yang, Y. H. and Thorne, N. P. (2003) In Goldstein, D. R. (ed.), Science and Statistics: A Festschrift for Terry Speed. pp. 403–418.

16. Wit, E. and McClure, J. (eds.) (2004) *Statistics for Microarrays: Design, Analysis and Inference*. Wiley, Chichester, UK.

13

ArrayPlex SA

A Turn-Key Automated Gene Expression Target Preparation System

Handy Yowanto

Summary

Automated target preparation for gene expression analysis eliminates the time-consuming and labor-intensive manual process, which is error prone and diverts scientists from value added activities. Target preparation methods were developed, on the fully integrated ArrayPlex SA® system, based on the field proven Biomek FX®, which streamlined the target preparation procedure allowing up to 96 samples to be processed in less than 36 h. The process is comprised of three functional methods, cDNA synthesis, in vitro transcription, and fragmentation, providing the users with the ability to consolidate runs for optimal use of instrument time and minimize reagent cost. Starting with sets of eight tRNA samples, the cDNA synthesis method synthesizes the first and second strand DNA followed by a cDNA clean-up step using an ultrafiltration plate. The in vitro transcription method then amplifies and biotin-labels the cDNA to cRNA in 6 h at 37°C, and purifies the product using a solid support extraction plate. Finally, the fragmentation method quantifies the cRNA, adjusts the concentration to the recommended 0.625 µg/mL and fragments the cRNA prior to an off-line hybridization. Universal human reference RNA with concentration ranging from 1 to 7.5 µg were prepared on the ArrayPlex SA, compared against a manual method and scanned using Affymetrix human genome U133 Plus 2.0 array GeneChip® cartridges. Nested analysis of variance was then performed to identify sources of variability between the automated and manual methods.

Key Words: Affymetrix GeneChip; ArrayPlex SA; automated target preparation; Beckman coulter Biomek FX; gene expression; in vitro transcription.

1. Introduction

Expression microarrays provide a vehicle for exploring the gene expression in a manner that is rapid, sensitive, systematic, and comprehensive. In recent years, it has evolved to be a powerful technique dealing with biological

From: *Methods in Molecular Biology, vol. 382: Microarrays: Second Edition: Volume 2*
Edited by: J. B. Rampal © Humana Press Inc., Totowa, NJ

questions that are involved in the transcriptional state of oncology, development, and drug discovery *(1–6)*.

Knowledge discovery relies on the ability of users to prepare the target RNA with high degree of reproducibility, in order to reduce the need to replicate experiments and the data reliability that allow researchers to make comparisons between different laboratories. However, manual target preparations of total RNA samples are tedious, time-consuming and can be error prone. In order to address the variability and consistency issues associated with manual target preparation, we have developed validated gene expression target preparation methods on the ArrayPlex SA designed specifically for the Affymetrix GeneChip cartridges that incorporates the ENZO RNA Transcript Labeling reaction.

Based on the field proven Biomek FX, the ArrayPlex SA offers a turn-key solution that focus on the quality, quantity, and integrity of the cRNA in the Affymetrix gene expression target preparation. The system is capable of processing up to 96 samples in multiple of eight samples under 36 h. Beside the small footprint, the instrument features an on-deck thermocycler, thermoshaking alp to keep the reaction mixture cool during processing, a peltier unit that maintains the reagent master mixes at 4°C and a solid phase extraction alp for oligonucleotides purification and clean-up.

Universal human reference (UHR) RNA containing total RNA from 10 different cell lines was used in the study to provide a better representation of the instrument performance. Two plates containing 8 and 96 UHR RNA at 1, 2, 5, and 7.5 µg starting materials were processed using the ArrayPlex SA and randomly selected samples were compared against manually prepared target cRNA at 2, 5, and 7.5 µg starting materials processed by two different technicians.

2. Materials

2.1. ArrayPlex Stand-Alone System

System parts are from Beckman Coulter Inc., Fullerton, CA, unless otherwise indicated.

1. Biomek FX Dual Arm with Multi-channel-L & Span-8 pods.
2. Biomek 96-Channel disposable tip pipetting head (200 µL).
3. Biomek system software v3.2.
4. Biomek disposable tip loader.
5. Standard single position ALP.
6. Span-8 waste ALP.
7. Biomek FX 4 × 3 ALP (Beckman Coulter Inc., Fullerton, CA).
8. Windows XP automation controller.
9. 15 inch Flat screen monitor, black.

10. Biomek FX filtration system for Promega Multi-Plate Wizards with vacuum valve.
11. Biomek Span-8 disposable tip kit.
12. Custom thermoshaking ALP.
13. Custom peltier unit.
14. PTC-0200 DNA engine Chassis (MJ Research).
15. RAD-0200 Remote Alpha Dock System (MJ Research, Watertown, MA).
16. RPS-0200 Remote Alpha Dock Power Supply (MJ Research).
17. ALP-2296 Moto Alpha Unit, 96 wells (MJ Research).
18. MJ Remote Alpha Dock Device Integration Kit.

2.2. Labwares

1. P+ compression lid arched (MJ Research, cat. no. MLS-2022).
2. Plate separator (Array Automation, Fullerton, CA; cat. no. AA2673961).
3. Filter collar (Array Automation, cat. no. AA3246721).
4. 96-Well plate lid (Phenix Technologies, Hayward, CA; cat. no. ML-5009).
5. Labware lid cover (Array Automation, cat. no. AA5426719).
6. 96-Well optical plate (Costar, Corning, Acton, MA; cat. no. 3635).
7. HardShell 96-well plate (MJ Research, Watertown, MA; cat. no. HSP-9601).
8. Cooling block (Array Automation, cat. no. AA6320313).
9. Cooling block insulator (Array Automation, cat. no. AA6573457).
10. 96-Well polypropylene V bottom plate (E & K Scientific, Santa Clara, CA; cat. no. EK-21201).
11. 96-Well polystyrene U-bottom plate (E & K Scientific, cat. no. EK-20101).
12. 2-mL Centrifuge tubes (VWR, West Chester, PA; cat. no. 20170-237).
13. 96-Deep well plate (Abgene, Rochester, NY; cat. no. AB-0932).
14. Pyramid reservoir (Innovative Microplate, Chicopee, MA; cat. no. S30014).
15. 150 mL Full module reservoir (Beckman Coulter, cat. no. 372784).
16. 75 mL Half module reservoir (Beckman Coulter, cat. no. 372786).
17. 40 mL Quarter module reservoir (Beckman Coulter, cat. no. 372790).
18. Span 8 P250 sterile tips (Beckman Coulter, cat. no. 379502).
19. Span 8 P20, sterile tips (Beckman Coulter, cat. no. 379505).
20. Multichannel AP96 P250 Sterile Tips (Beckman Coulter, cat. no. 717252).
21. Multichannel AP96 P20 Sterile Tips (Beckman Coulter, cat. no. 394627).

2.3. cDNA Synthesis

1. T7-oligo(dT) primer, 5′-GGCCAGTGAATTGTAATACGACTCACTATAGGGAG-GCGG-(dT)$_{24}$-3′ (*see* **Note 1**).
2. First Strand cDNA reagent master mix (*see* **Note 2**).
3. Second Strand cDNA reagent master mix (*see* **Note 3**).
4. T4 DNA polymerase reagent master mix (*see* **Note 4**).

2.4. In Vitro Transcription Synthesis and Fragmentation

1. T7 RNA Polymerase reagent master mix (*see* **Note 5**).
2. 5X Fragmentation buffer (*see* **Note 6**).

2.5. Purification and Clean-Up

1. RNEasy™ 96 Kit (Qiagen GmbH, Hilden, Germany; cat. no. 74181).
2. MinElute™ 96 Ultrafiltration PCR purification kit (Qiagen GmbH, cat. no. 28053).

2.6. Hybridization

1. Acetylated bovine serum albumin (BSA) solution (50 mg/mL) (Invitrogen Life Technologies, Carlsbad, CA).
2. Herring sperm DNA (Promega Corporation, Madison, WI).
3. GeneChip eukaryotic hybridization control kit (Affymetrix, Santa Clara, CA).
4. Control oligonucleotides B2, 3 n*M* (Affymetrix).
5. 5 *M* NaCl, RNase-free, DNAse-free (Ambion, Austin, TX).
6. 12X Stock sodium MES buffer (*see* **Note 7**).
7. 2X Hybridization buffer (*see* **Note 8**).

2.7. Washing and Staining

1. Nonstringent wash A buffer (*see* **Note 9**).
2. Stringent wash B buffer (*see* **Note 10**).
3. 2X Stain buffer (*see* **Note 11**).
4. 10 mg/mL Goat IgG stock (*see* **Note 12**).

2.8. Miscellaneous

1. Eukaryotic Poly-A RNA control (Affymetrix).
2. Diethylpyro-carbonate (DEPC) treated water, nuclease-free, 0.2 μm filtered, autoclaved (Ambion, Austin, TX).
3. RNA 6000 Nano reagents and supplies (Agilent Technologies, Palo Alto, CA).
4. RNA 6000 Ladder (Ambion, Austin, TX).
5. Human genome U133 Plus 2.0 GeneChip Arrays (Affymetrix).
6. UHR RNA (Stratagene, La Jolla, CA).

3. Methods

The most commonly employed technique for RNA amplification is a T7 based linear amplification methods *(7)*. A synthetic oligo(dT) primer containing the phage T7 RNA polymerase promoter is used in the method to prime synthesis the first strand cDNA by reverse transcription of the poly(A)+ RNA component of the total RNA. Second strand cDNA is then synthesized by degrading the poly(A)+ RNA strand with RNase H, followed by second strand synthesis with *Escherichia coli* DNA polymerase 1. During in vitro transcription (IVT) of the double stranded cDNA template using T7 RNA polymerase, amplified antisense RNA (cRNA) is generated. Based on this technique, several protocols have been developed and used in microarray analysis *(8–11)*.

The validated gene expression target preparation method on the ArrayPlex SA system is made up of three functional methods complete with pipetting templates and labware definitions. These functional methods are designed to maximize the use of instrument time and reduce reagent consumptions by allowing samples to be consolidated during subsequent runs.

Five microliters of good quality isolated total RNA or UHR RNA are loaded onto a sterile MJ Research HardShell reaction plate in sets of eight (*see* **Note 13**). We have tested the methods using samples ranging between 1 and 7.5 µg and observed no significant performance variability between varying amounts of UHR RNA starting materials. Although 1 µg of UHR RNA starting material has been used successfully, we recommend at least 2 µg of good quality total RNA as starting materials to ensure that sufficient amount of cRNA are produced. As quality control measure, 2 µL of commercially available Poly-Adenylated RNA controls can also be added to monitor the amplification process quality (*see* **Note 13**).

The cDNA synthesis method will reverse transcribe good quality total RNA into double stranded DNA and purify the double stranded DNA using an ultra filtration plate. User will then have the option to either freeze the samples or proceed with the IVT synthesis (*see* **Note 14**) (**Fig. 1**). During the 6–8 h of IVT linear amplification, biotinylated labels are incorporated into the single stranded cRNA; before purifying the cRNA a solid phase extraction plate method. Once amplified and labeled, the fragmentation method will quantify, adjust the cRNA concentration to 0.625 µg/µL, verify to ensure good dilution accuracy and then fragments the full length cRNA to 35–200 base fragments by metal-induced hydrolysis.

3.1. Preparation of Samples and Poly A Controls

1. UHR RNA is provided in a solution of 70% methanol and 0.1 *M* sodium acetate.
2. Centrifuge the tube at 12,000*g* for 15 min at 4°C.
3. Wash the pellet in 70% ethanol.
4. Centrifuge the tube at 12,000*g* for another 15 min at 4°C.
5. Carefully remove the supernatant and air-dry the pellet at room temperature for 30 min to remove retained ethanol.
6. Resuspend the pellet in DEPC-treated water to 2 µg/µL concentration.
7. Prepare a serial dilution of 1:20, 1:50, and 1:100 from the stock Poly-A control by diluting it using Poly-A dilution buffer.

3.2. cDNA Synthesis

1. Perform BioAnalyzer RNA electropherogram to determine the quality of the total RNA and measure the total RNA concentration using NIST calibrated UV-Vis instrument.
2. Load 5 µL of samples into the MJ Research Hard Shell reaction plate. If less than 5 µL of samples are used, adjust the volume by adding the appropriate amount of water to attain 5 µL for each samples. Add 2 µL of Poly-A RNA control (1:10,000 dilution) into each well.

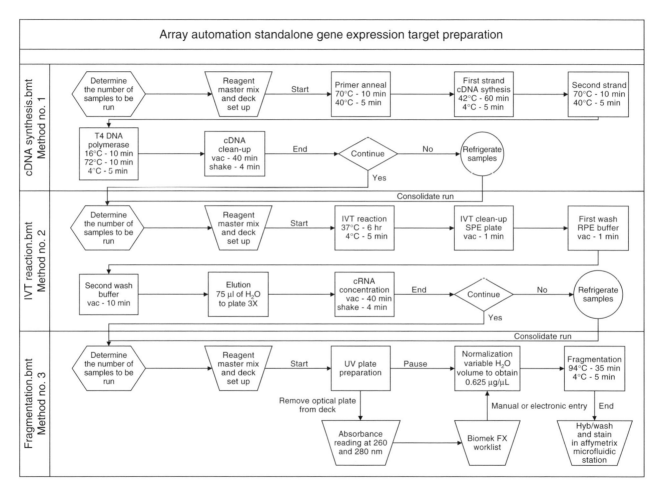

Fig. 1. Automated gene expression target preparation.

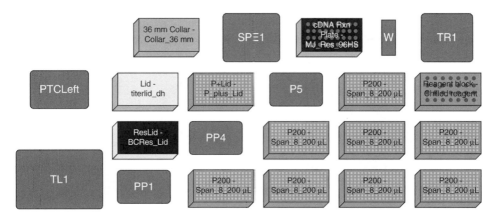

Fig. 2. cDNA synthesis method instrument deck set up configuration.

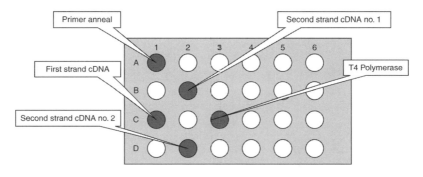

Fig. 3. cDNA synthesis method reagent master mix cold block location map.

3. Reaction plate must be prepared just before to the actual run and keep the samples at 4°C to minimize degradation. Prepare the reagent master mixes and place the tubes in the cooling block (**Fig. 3**).

4. Place a new compression pad on the arch stainless steel back and wipe the compression pad surface using a fresh moist RNAse Zap towelletes and quickly remove the RNAse Zap solution using a clean Kim Wipes. Wipe the static alp, span8 probes, the special lid, collar and the reservoir holder using RNase Zap towelletes.

5. Load all of the necessary labware according to the cDNA synthesis deck layout configuration (**Fig. 2**), and enter the number of samples to be process in a set of eight samples. Verify to make sure that the peltier and the thermoshaking alp are at 4°C before starting the run.

6. Add 20 mL of RNase-free water into the half module reagent reservoir (**Fig. 4**).

7. Continue with the IVT amplification/labeling method or flash freeze the cDNA samples using dry ice for future use.

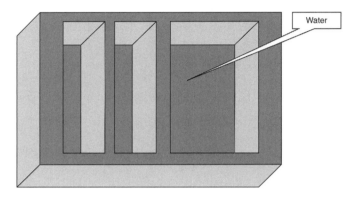

Fig. 4. cDNA synthesis method modular reagent reservoir component location map.

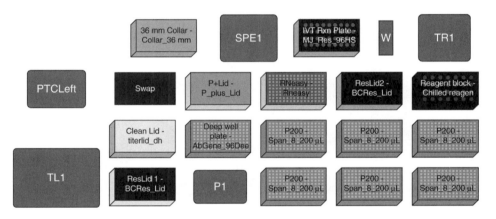

Fig. 5. In vitro transcription synthesis method instrument deck set up configuration.

3.3. In Vitro Transcription Synthesis

1. Using a multichannel pipetor, transfer 25–28 µL of cDNA from the U-bottom plate into a fresh MJ Research Hard Shell reaction plate.
2. Prepare the IVT reagent master mix and load the reagent master mix on the cooling block (**Fig. 6**).
3. Seal the RNEasy plate columns that will not be used using an adhesive back aluminum seal and load all the necessary labwares according to the IVT synthesis deck layout configuration (**Fig. 5**).
4. Enter the number of samples to be process in a set of eight samples and enter the correct number of RNEasy purification plate columns that have been spent. Enter "0" if you are using a new RNEasy purification plate.
5. Add 36.5 mL of RLT buffer into the first quarter module reservoir, 29 mL of 70%

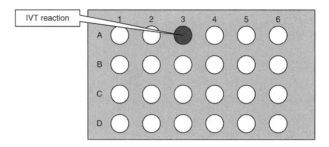

Fig. 6. In vitro transcription synthesis method reagent master mix cold block location map.

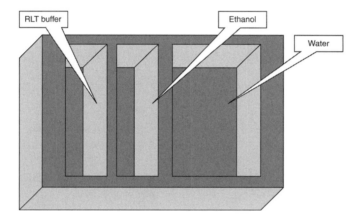

Fig. 7. In vitro synthesis method modular reagent reservoir component location map.

ethanol into the second quarter module reservoir, and 38.5 mL of water into the half module reagent reservoir (**Fig. 7**).

6. Fill the full module reagent reservoir with 216 mL of RPE buffer (**Fig. 8**).
7. Verify to make sure that the peltier and the thermoshaking alp are at 4°C before starting the run.
8. Continue with the fragmentation method once the IVT synthesis has been completed or flash freeze the cRNA samples using dry ice for future use.

3.4. Quantitation, Normalization, and Fragmentation of Samples

1. Prepare the fragmentation buffer and load the buffer on the reagent block (**Fig. 10**).
2. Enter the number of samples to be process in set of eight samples.
3. Verify to ensure that the correct labwares have been loaded on the proper location on the instrument deck before starting the run (**Fig. 9**).
4. Blank the optical plate on the first instrument pause using a calibrated plate reader and return the optical plate back to the deck once it has been blanked and resume the run.

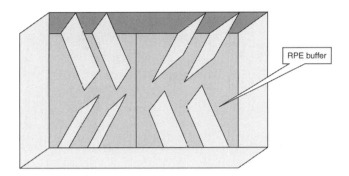

Fig. 8. In vitro synthesis method reagent reservoir.

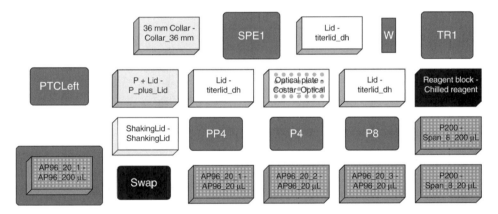

Fig. 9. Fragmentation method instrument deck set up configuration.

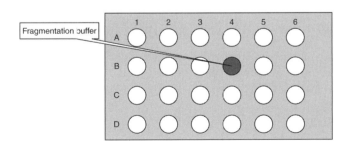

Fig. 10. Fragmentation method reagent master mix cold block location map.

5. The instrument will prepare the optical plate by aspirating 2 µL of sample into each of the optical plate well and then manually transfer the optical plate to the plate reader to be scanned at 260 and 280 nm wavelengths.

6. The amount of water volume will be quantified and entered into the Biomek FX work list file automatically. Resume the instrument run to normalize the samples to the desired concentration of 0.625 µg/µL.

7. Blank and scan the second optical plate to verify that the samples have been diluted accurately.

8. Once sample fragmentation has been completed, transfer 30 µL of the cRNA sample and mix it with an appropriate amount of hybridization buffer. Proceed with the GeneChip hybridization, labeling and scan as described by the Affymetrix user guide.

9. Flash freeze any processed samples using dry ice, if they are not being hybridized, immediately to minimize degradation.

3.5. Sample Analysis Using BioAnalyzer Gel Electrophoresis

1. Allow the reagent to equilibrate to room temperature for 30 min before use.

2. Spin filter 550 µL of RNA 6000 Nano gel matrix in a microcentrifuge for 10 min at 1500g and then aliquot 65 µL of filtered gel into 0.5-mL microfuge tubes.

3. Add 1 µL of dye into the 65 µL filtered gel and vortex thoroughly.

4. Spin the tube for 10 min at room temperature at 13,000g.

5. Take a new RNA Nano chip and place the chip on the Chip Priming Station.

6. Dispense 9 µL of the gel-dye mix into the bottom of the well with a circle "G" marking.

7. Pressurized the well by pushing the syringe plunger on the Chip Priming Station for 30 s before releasing the plunger and then dispense 9 µL of the gel-dye mix into each of the wells marked "G."

8. Pipet 5 µL of the RNA 6000 Nano Marker into the ladder well and into each of the 12 sample wells. Load 1 µL of the RNA 6000 ladder from Ambion into the ladder well and 1 µL of each sample into each of the 12 sample wells.

9. Vortex the chip on the IKA vortexer for 1 min at 2400 rpm.

10. Scan the chip on the Agilent 2100 Bioanalyzer **Figs. 11** and **12**.

3.6. Hybridization

1. Aliquot 30 µL of the fragmented samples and then add 5 µL of control oligonucleotides B2 (3 nM), 15 µL of 20X eukaryotic hybridization controls, 3 µL of Herring sperm DNA (10 mg/mL), 3 µL of Acetylated BSA (50 mg/mL), 150 µL of 2X hybridization buffer and 24 µL of water.

2. Heat the hybridization cocktail to 99°C for 5 min, cool the mixture to 45°C for 5 min and then spin the hybridization cocktail at maximum speed in a microcentrifuge for 5 min to remove any insoluble materials.

3. After equilibrating the probe array to room temperature for at least 30 min, wet the array by filling it with 1X hybridization buffer using a micropipetor.

4. Incubate the 1X hybridization buffer containing probe array at 45°C for 10 min with rotation.

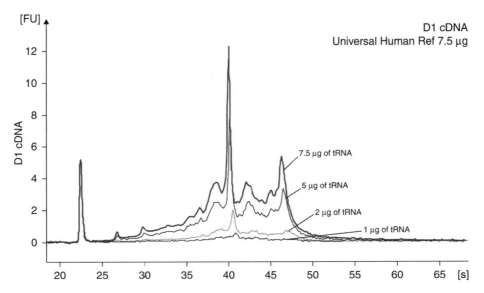

Fig. 11. BioAnalyzer scans of Universal Human Reference cDNA at various starting total RNA amounts. Samples were scanned after it has been processed by the cDNA synthesis method.

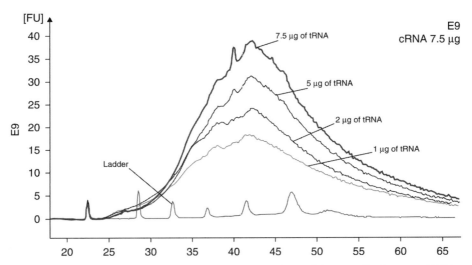

Fig. 12. BioAnalyzer scans of biotin labeled Universal Human Reference cRNA at various starting total RNA amount.

5. Remove the buffer solution from the probe array cartridge and fill with 250 µL of hybridization cocktail.

6. Incubate the probe array in the hybridization oven at 45°C for 16 h.

3.7. Washing, Staining, and Scanning

1. Prepare the streptavidin phycoerythrin (SAPE) by mixing 600 µL of 2X MES stain buffer, 48 µL of acetylated BSA (50 mg/mL), 12 µL of SAPE, and 540 µL of water in an amber 0.5-mL tube.

2. Prepare the antibody solution by mixing 300 µL of 2X MES stain buffer, 24 µL of 50 mg/mL of acetylated BSA, 6 µL of 10 mg/mL normal Goat IgG, 3.6 µL of 0.5 mg/mL biotinylated antibody, and 266.4 µL of water.

3. On the fluidic station controller, Select the antibody amplification stain for eukaryotic targets from the menu and click "Run."

4. Insert the probe array into the designated module in the fluidic station (FS-450/250) and turn the cartridge lever up to the engage position.

5. When prompted, load the first vial containing 600 µL of the SAPE solution mix in sample holder 1, the second vial containing 600 µL antistreptavidin biotinylated antibody in sample holder 2 and another vial containing 600 µL of SAPE solution in the sample holder 3.

6. Press down the needle lever to snap needle into position and start the run.

7. Once the wash and stain procedure has been completed, scan the probe array on the GeneArray Scanner or the GeneChip Scanner 3000.

4. Notes

1. Prepare the primer anneal reagent master mix for each reaction well use 1.1 µL of 100 µM T7—$(dT)_{24}$ primer (Integrated DNA Technologies) and 4.4 µL of water.

2. Prepare the first strand cDNA reagent master mix for each reaction well by mixing 4.2 µL of 5X first strand cDNA buffer (Invitrogen), 2.1 µL of 0.1 M DTT (Invitrogen), 1.1 µL of 10 mM dNTP mix (Invitrogen), 1.1 µL of Super Script II RNase H - Reverse Transcriptase (200 U/µL) (Invitrogen) and 2.1 µL of water in a sterile 2-mL centrifuge tube.

3. Prepare the Second Strand cDNA reagent master mix for each reaction by adding 30.4 µL of 5X second strand cDNA buffer (Invitrogen), 3 µL of 10 mM dNTP mix (Invitrogen), 1 µL of DNA ligase (10 U/µL) (Invitrogen), 4.1 µL of DNA Polymerase I (10 U/µL) (Invitrogen), and 1 µL of RNase H (2 U/µL) (Invitrogen) into two sterile 2-mL centrifuge tubes. Four of the instrument probes will aspirate into one of the Second Strand cDNA reagent master mix tube, while the next four probes will aspirate from the second centrifuge tube.

4. Prepare the T4 DNA Polymerase reagent master mix for each reaction well by mixing 2.3 µL of T4-DNA polymerase (5 U/µL) (Invitrogen), 0.6 µL of 5X T4 DNA polymerase buffer and 1.7 µL of water in a sterile 2-mL centrifuge tube.

5. Prepare the T7 RNA polymerase reagent master mix for each reaction well by mixing 2.2 µL of 20X T7 RNA polymerase, 4.4 µL of 10X buffer, 4.4 µL of 10X Biotin-labeled ribonucleotides, 4.4 µL of 10X DTT, and 4.4 µL 10X RNase

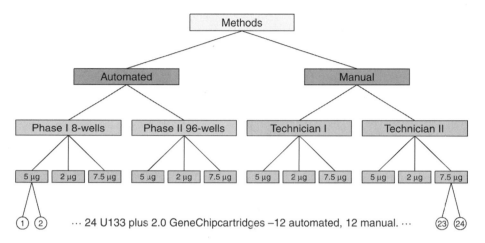

Fig. 13. Experiment design to determine the sources of variation between automated and manual Gene Expression Target Preparation.

Manual

Starting conc. (technician)	Background	P (%)	Average signal (all)	Actin 3'/5' ratio	GAPDH 3'/5' ratio
2 µg tRNA (I)	43.46	56.50	801.2	1.26	0.85
2 µg tRNA (I)	44.28	56.20	803.1	1.16	0.91
2 µg tRNA (II)	47.69	55.10	812.3	1.26	0.89
2 µg tRNA (II)	41.67	57.70	787.5	1	0.87
5 µg tRNA (I)	75.43	50.30	802	1.13	0.86
5 µg tRNA (I)	53.73	53.70	807.6	1.16	0.83
5 µg tRNA (II)	50.00	55.80	801.1	1.03	0.86
5 µg tRNA (II)	45.90	57	813.6	0.96	0.9
7.5 µg tRNA (I)	55.72	51	802.6	1.17	0.83
7.5 µg tRNA (I)	50.74	54.80	797.7	1.14	0.87
7.5 µg tRNA (II)	56.59	54.10	797.9	1.02	0.87
7.5 µg tRNA (II)	43.72	57.20	799.8	0.92	0.99
	ave	54.95			
	SD	0.0236			
	%CV	4.29			

Fig. 14. Affymetrix Gene Expression Human U133 plus 2.0 GeneChip results generated by GeneChip Operating System v1.2.0.037. Target was prepared manually according to the Affymetix GeneChip Expression Analysis Technical manual.

Automated

Starting conc. (technician)	Background	P (%)	Average signal (all)	Actin 3'/5' ratio	GAPDH 3'/5' ratio
1 µg tRNA (II)	44.97	56.10	817.2	1.06	0.87
1 µg tRNA (II)	42.73	57.20	793.5	1.02	0.90
2 µg tRNA (I)	42.09	58.50	780.7	1.01	0.88
2 µg tRNA (I)	46.38	57.10	799.1	1	0.90
2 µg tRNA (II)	58.12	55.80	794.9	1	0.86
2 µg tRNA (II)	50.93	58.30	787.3	0.98	0.88
5 µg tRNA (I)	49.54	56.20	796.8	1.04	0.84
5 µg tRNA (I)	46.50	57.30	787.4	0.99	0.85
5 µg tRNA (II)	59.95	54.90	797	1.02	0.86
5 µg tRNA (II)	51.97	57.40	798.1	0.98	0.87
7.5 µg tRNA (I)	45.54	58.20	773.3	1.04	0.86
7.5 µg tRNA (I)	53.70	57.10	771.6	1.07	0.85
7.5 µg tRNA (II)	56.31	57.20	780.2	1.09	0.88
7.5 µg tRNA (II)	57.46	55.40	790.2	1.05	0.85

	Ave	56.91	
	SD	0.0109	
	%CV	1.92	

Fig. 15. Affymetrix Gene Expression Human U133 plus 2.0 GeneChip results generated by GeneChip Operating System version 1.2.0.037. Target was prepared using the ArrayPlex SA Automated Gene Expression methods.

inhibitor. Reagent for the in vitro Transcription Synthesis can be purchase as a kit from Enzo Life Sciences, BioArray™, High Yield™, RNA Transcript Labeling Kit (T7) (cat. no. 42655-20).

6. The 5X fragmentation buffer is prepared by mixing 0.64 g MgOAc (Sigma) and 0.98 g KOAc (Sigma) into 4 mL of 1 *M* Tris acetate (Sigma), pH 8.1, and adjusting the volume using DEPC treated water to 20 mL. Filter purify the solution using a 0.2-µm filter. For each of the reaction well, load 27 µL of 5X fragmentation buffer.

7. 1.22 *M* MES, 0.89 *M* sodium buffer (12X) stock is prepared by dissolving 70.4 g MES free acid monohydrate (Sigma-Aldrich), and 193.3 g MES sodium salt (Sigma Aldrich) in water adjusted to 1000 mL at pH between 6.5 and 6.7. Filter the solution through 0.2-µm filter.

8. Hybridization buffer (2X) is prepared by mixing 8.3 mL of 12X MES Stock solution, 17.7 mL of 5 *M* NaCl, 4 mL of 0.5 *M* EDTA (Sigma Aldrich), 0.1 mL of 10% Tween-20 and 19.9 mL of water. Store the buffer at 4°C and protect it from light.

9. Wash A, Nonstringent Wash Buffer: mix 300 mL of 20X SSPE (BioWhittaker Molecular Application / Cambrex), 1 mL of 10% Tween-20 (Pierce Chemical) and 699 mL of water. Filter the solution through a 0.2-µm filter.

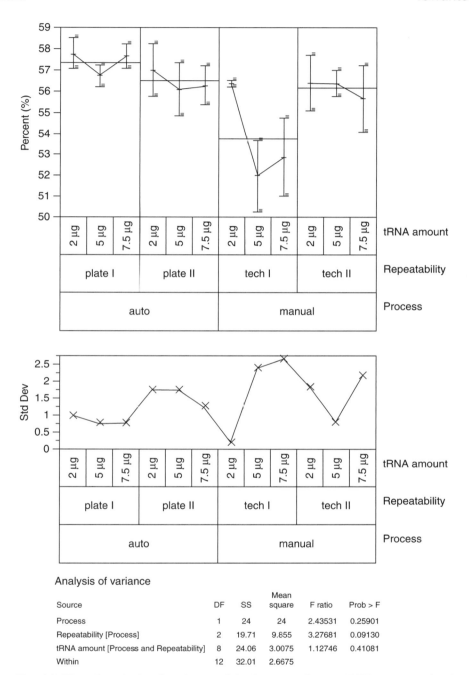

Fig. 16. Nested analysis of variance of the human reference RNA processed using automated and manual target preparation. The results showed higher variation between replicates where samples were processed manually by technicians, when compared to automated target preparation.

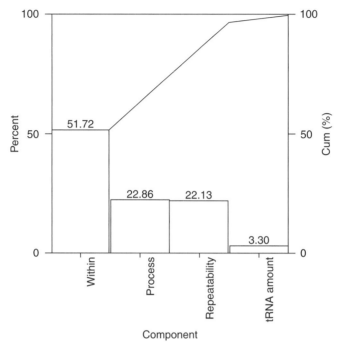

Fig. 17. Pareto chart of Nested ANOVA showed that 74.58% of the variability was associated with the manual target preparation.

10. Wash B, Stringent Wash Buffer: mix 83.3 mL of 12X MES Stock Buffer, 5.2 mL of 5 *M* NaCl, 1 mL of 10% Tween-20, and 910.5 mL of water. Filter through a 0.2-μm filter and store the solution away from light at 4°C.
11. Mix 41.7 mL 12X MES stock buffer, 92.5 mL 5 *M* NaCl, 2.5 mL 10% Tween-20 and 113.3 mL water to make a 250 mL solution. Filter the solution through a 0.2-μm filter and store the solution away from light at 4°C.
12. Resuspend 50 mg of Goat IgG (Sigma Aldrich) in 5 mL of 150 m*M* NaCl.
13. Place all samples on dry ice until ready to be used, to minimize degradation to the RNA samples.
14. Flash freeze samples immediately using dry ice after each process before storing them in the −70°C freezer.
15. Wipe all work surfaces and manual pipetors using RNase Zap towelletes to reduce or eliminate RNAse contamination.

Acknowledgments

The author would like to thank David Helphrey from the Advance Technology Center, Beckman Coulter Inc. and Dr. Denis Heck, at DNA and Protein Microarray Facility, University of California, Irvine (DMAF) for their valuable technical assistance; and the DMAF staffs, Kim T. Nguyen, Sriti Misra, Ching Lan "Amanda" Lim, Cherryl Nugas-Shellby, and Lana Bordcosh for the manual target preparation.

References

1. Lockhart, D. J., Dong, H., Byrne, M. C., et al. (1996) Expression monitoring by hybridization to high-density oligonucleotide arrays. *Nat. Biotechnol.* **14,** 1675–1680.
2. Eisen, M. B., Spellman, P. T., Brown, P. O., and Bortstein, D. (1998) Cluster analysis and display of genome-wide expression patterns. *Proc. Natl. Acad. Sci. USA* **95,** 14,863–14,868.
3. Golub, T. R., Slonim, D. K., Tamayo, P., et al. (1999) Molecular classification of cancer: class discovery and class prediction by gene expression monitoring. *Science* **286,** 531–537.
4. Schena, M., Shalon, D., Davis, R. W., and Brown, P. O. (1995) Quantitative monitoring of gene expression patterns with a complementary DNA microarray. *Science* **270,** 467–470.
5. Schena, M., Shalon, D., Heller, R., Chai, A., Brown, P. O., and Davis, R. W. (1996) Parallel human genome analysis: microarray-based expression monitoring of 1000 genes. *Proc. Natl. Acad. Sci. USA* **93,** 10,614–10,619.
6. Schref, U., Ross, D. T., Waltham, M., et al. (2000) A gene expression database for the molecular pharmacology of cancer. *Nat. Genet.* **24,** 236–244.
7. Phillips, J. and Eberwine, J. H. (1996) Antisense RNA amplification: a linear amplification method for analyzing the mRNA population from single living cells. *Methods* **10,** 283–288.
8. Mahadevappa, M. and Warrington, J. (1999) A high-density prove array sample preparation method using 10- to 100-fold fewer cells. *Nat. Biotechnol.* **17,** 1134–1136.
9. Wang, E., Miller, L. D., Ohnmacht, G. A., Lie, E. T., and Marincola, F. M. (2000) High-fidelity mRNA amplification for gene profiling. *Nat. Biotechnol.* **18,** 457–459.
10. Affymetrix GeneChip Expression Analysis Technical Manual 701021 Rev. 5, Affymetrix Inc. (http://www.affymetrix.com/Download/manuals/expression_manual.pdf).
11. Zhao, H., Hastie, T., Whitfield, M. L., Borrsen-Dale, A. L., and Jeffery, S. S. (2002) Optimization and evaluation of T7 based RNA linear amplification protocols for cDNA microarray analysis. *BMC Genomics* **3,** 31.
12. Nimgaonkar, A., Sanoudou, D., Butte, A., et al. (2003) Reproducibility of gene expression across generations of Affymetrix microarrays. *BMC Bioinforma.* **4,** 27.

14

Tumor–Stroma Interactions of Metastatic Prostate Cancer Cell Lines
Analyses Using Microarrays

Nicolas Wernert, Annette Kaminski, El-Mustapha Haddouti, and Jens Claus Hahne

Summary

Tumor–stroma interactions are of great importance not only for the development and progression of primary prostate carcinoma but probably also for the establishment of metastasis. Fibroblasts are an important stromal cell type encountered by metastatic tumor cells at different sites. In previous investigations, we had found that media conditioned by three metastatic prostate cancer cell lines (LNCaP, PC-3, and DU-145) induced cultured nonprostatic fibroblasts to proliferate or to express matrix-metalloproteinase-1 considered important for tumor invasion. Fibroblast-conditioned media in turn stimulate proliferation of DU-145 cells and migration of PC-3 cells. Both tumor cells and fibroblasts secrete VEGF suggesting that not only metastatic but also stromal cells at metastatic sites contribute to the vascularization of metastasis necessary for continuous growth.

In order to better understand the reciprocal tumor–stroma cross-talk in molecular terms we used the mRNA extracted from stimulated and unstimulated neoplastic and fibroblastic stromal cells for cDNA array hybridization using Affymetrix® chips. The three prostate cell lines influenced the fibroblasts nearly in the same manner. In particular proteins involved in cell adhesion, cell–cell contact, and cell cycle regulation were downregulated in stimulated fibroblasts. In contrast, fibroblasts affected every prostate cancer cell line in different ways, which may be because of the different origin of the metastatic prostate cancer cell lines.

Key Words: Microarray; prostate cancer metastasis; tumor–stroma interactions; fibroblasts; prostate cancer cell lines; RNA extraction.

1. Introduction

Prostate carcinoma is among most frequent cancers in the Western world. Considerable research efforts are therefore, undertaken to unravel its development and progression. Androgen dependence of this tumor is long known *(1–10)*,

From: *Methods in Molecular Biology, vol. 382: Microarrays: Second Edition: Volume 2*
Edited by: J. B. Rampal © Humana Press Inc., Totowa, NJ

for review *see* **ref.** *11* and epidemiological studies point to factors in Western life style *(12–14)*. Linkage analyses suggest hereditary prostate cancer genes *(15–19)* and polymorphisms of modifier genes (related to hormone response, cell protection or DNA repair) can increase tumor risk *(20–24)*. Several proto-oncogenes (such as *myc [25,26]*, *EIF3S3 [27]*, *bcl2 [28]*, or growth factor encoding genes *[29]*) have been found activated and tumor suppressor genes such as *TP53 (30)* and *PTEN (31)* to be inactivated. Further oncogenes and tumor suppressor genes are suggested by comparative genomic hybridization (CGH) or loss of heterozygosity (LOH) analyses *(32)*, for review *see* **ref.** *33–39*. The activity of prostate cancer relevant genes may be modified through epigenetic DNA methylation and histone acetylation *(40–43)*. Signalling pathways such as Wnt signalling can be deregulated *(44)*. We also found that alterations of the proteasome pathway might play a role for prostate carcinoma development by modifying the degradation of proteins relevant for growth or apoptosis *(45)*.

For a long time epithelial stromal interactions have been considered important for development and progression of prostate cancer likewise (for review *see* **refs.** *46*, and *47*).

Carcinomas, including prostate cancer, are not only made up of neoplastic epithelial cells but also of a supporting connective tissue, the tumor stroma, which is involved in several pivotal processes such as tumor vascularization (necessary for a continuous tumor growth) or invasion through the secretion of matrix-degrading proteases by stromal fibroblasts *(48–54)*. An epithelial–stromal intercommunication is already involved in embryonic development of the prostate gland, during which the mesenchyme of the urogenital sinus induces epithelial morphogenesis as well as expression of prostate-secreted proteins *(46,55–57)*. Androgens are supposed to act primarily on prostate stroma, which induces epithelial proliferation and differentiation through paracrine secretion of growth factors such as keratinocyte growth factor, epidermal growth factor, or basic fibroblast growth factor *(55–61)*. This mesenchymal-epithelial cross-talk is maintained throughout life *(62)* with important implications also for prostate cancer development and progression *(63)* (for review *see* **refs.** *46* and *64*). Prostate fibroblasts can modify prostate carcinoma cell growth by modifying proliferation or apoptosis *(65–69)* through paracrine mechanisms *(67,70)*. A previous investigation demonstrated a number of differences at the proteome level between normal and neoplastic epithelial, and between normal and peritumoral stromal cells of prostate carcinomas using SELDI-TOF (SAX-2 Protein Chips, Ciphergen Biosystems Ltd. Guildford, UK) mass spectrometry *(71)*.

A significant proportion of prostate cancers metastasizes through the lymph or blood stream *(72,73)*. Blood-born metastases (most frequently to the skeleton) come about in about 35% of patients resulting in uncontrollable disease *(74,75)*. The metastatic process is made up of many sequential steps including detachment of neoplastic cells from the primary tumor, their migration, degradation of the extracellular matrix, vascular transport of metastatic tumor cells, and finally attachment of cells to metastatic sites by tissue-specific adhesion *(76,77)*. This can be increased by soluble factors from target tissues *(78)*. Invasion and proliferation in the new microenvironment and induction of angiogenesis permitting continuous tumor growth at the metastatic site are the last steps of the metastatic cascade *(79)*.

Tumor–stroma interactions at metastatic sites are as important for successful establishment of metastases as they are for the development of the primary tumor. However, little is known about this cross-talk at metastatic sites of prostate cancer.

In a previous study, aspects of the reciprocal cross-talk between metastatic prostate cancer cell lines and nonprostatic fibroblasts were analysed, which are encountered by metatstatic cells at different sites. This issue will be addressed in cell culture systems because dynamic interactions can not be directly analysed in vivo.

The media conditioned by three metastatic prostate cancer cell lines (LnCaP, PC-3, and DU-145 derived from lymph node, brain, and bone metastases *[80–82]*) induced cultured fibroblasts human foreskin fibroblasts (HFF) to proliferate, which can be translated in vivo to fibrous stroma induction *(83)*. Using ELISA assays, identified in prostate cancer cell conditioned media basic fibroblast growth factor and platelet-derived growth factor (PDGF) as factors, which might be responsible for induction of fibroblast proliferation. Fibroblast conditioned media in turn stimulated proliferation of DU-145 metastatic prostate cancer cells while exhibiting no significant effect on proliferation of the two other neoplastic cell lines. In fibroblast conditioned media keratinocyte growth factor and epidermal growth factor could induce tumor cell proliferation and growth. Both factors are also secreted by the three metastatic prostate cancer cell lines suggesting autocrine effects on tumor cell proliferation.

Vascular endothelial growth factor (VEGF), the most important factor for tumor vascularization, was secreted at highest levels by DU-145 cells followed by PC-3 and LNCaP cells. Fibroblasts secreted VEGF at about half the concentration compared to tumor cells.

The fibroblast conditioned medium increased PC-3 cell migration, which is an important part of tumor cell invasion at metastatic sites. Stromal fibroblasts are themselves known to participate to tumor invasion by secreting various proteases. Quantitative light cycler analyses revealed that media conditioned by DU-145 cells induced in fibroblasts expression of matrix-metalloproteinase-1 mRNA in contrast to LNCaP and PC-3 conditioned media.

At metastatic sites prostate cancer cells must proliferate, induce blood vessels and invade the target tissues. Our results provide evidence that a reciprocal intercellular cross-talk between prostate cancer cells and their fibroblastic microenvironment is involved in each of these steps.

In order to better understand cell behavior in molecular terms, large scale analysis of gene expression at the level of the transcriptome using cDNA microarrays is a powerful approach *(84–95)*. In the present study, therefore, again stimulated cultured fibroblasts with media that had been conditioned by the three metastatic prostate cancer cell lines and the latter with media that had been conditioned by the fibroblasts. Then extracted the mRNA from the cells, reverse transcribed it into cDNA, which used for the hybridization of Affimetrix® chips.

2. Materials
2.1. Cell Culture and Preparation of Conditioned Media

1. Cultivation of HFF cells in Dulbecco's modified Eagle's medium (GIBCO/Invitrogen, Karlsruhe, Germany) supplemented with 0.5% penicillin–streptomycin (GIBCO/Invitrogen) and 0.5% fetal bovine serum (FCS) (GIBCO/ Invitrogen) (*see* **Note 1**).
2. Cultivation of prostate cancer cells in Roswell Park Memorial Institute (RPMI) 1640 medium (Invitrogen) supplemented with 0.5% penicillin–streptomycin (GIBCO/Invitrogen) and 0.5% FCS (GIBCO/Invitrogen) (*see* **Note 1**).
3. D-phosphate-buffered saline (PBS) solution without Ca^{2+} and Mg^{2+} (GIBCO/ Invitrogen).
4. Solution of trypsine (0.25%) and ethylenediamine tetra-acetic acid (EDTA) (1 mM) (GIBCO/Invitrogen).
5. Cell flasks, 24-well plates and 15-mL capillary tubes from Greiner Bio-One (Frickenhausen, Germany).
6. Trypane blue solution for determination of the percentage and the total number of viable cells: 0.36% (w/v) trypane blue and 0.9% (w/v) NaCl in deionized water.
7. Graduated counting chambers (Fast-Read 102®) (Madaus Diagnostik, Köln, Germany).

2.2. RNA Isolation

1. RNaseAway (Molecular Bio Products, San Diego, CA).
2. Sterile, disposable polypropylene tubes (for example: Sarstedt, Nümbrecht, Germany—or equivalent).
3. Glassware should be treated before use to ensure that it is RNase-free. Glassware used for RNA work should be cleaned and oven backed at 240°C for at least 4 h before use (*see* **Note 2**).
4. Cell lysis buffer RLT (*see* **Note 3**) from RNeasy® Mini Kit (Qiagen, Hilden, Germany). RLT-buffer contains thiocyanate, wear gloves when handling and take appropriate safety measures.
5. QIAshredder spin column (Qiagen).
6. 70% (v/v) Ethanol, store at room temperature.
7. RNeasy mini column (Qiagen).

8. Washing buffer RW 1 (Qiagen). RW1-buffer contains thiocyanate, wear gloves when handling and take appropriate safety measures.
9. Washing buffer RPE (Qiagen) (*see* **Note 4**).
10. RNase-free water: suspend deionized water with 0.1% (v/v) diethyl pyrocarbonate (DEPC), allow shaking overnight at room temperature, and then autoclave or heating to 100°C for 15 min to eliminate residual DEPC. Store at room temperature.

DEPC is suspected to be carcinogenic and should be handled with great care, wear gloves and use fume hood when using DEPC.

2.3. DNA Chips

This protocol assumes the use of the Affymetrix GeneChip platform (Affymetrix, Santa Clara, CA) for microarray analysis.

1. T7-oligo-d(T)$_{24}$ primer containing the T7 RNA polymerase binding site [5′-GGCCAGTGAATTGTAATACGACTCACTATAGGGAGGCGG-(dT$_{24}$)-3′] (MWG Biotech, München, Germany).
2. 0.1 *M* Dithiothreitol and store in aliquots at –20°C.
3. First strand cDNA synthesis with Superscript II (Invitrogen).
4. *Escherichia coli* DNA polymerase I (Invitrogen).
5. *E. coli* DNA ligase (TaKaRa, Gennevilliers, France), RNaseH (TaKaRa) and T4 DNA polymerase I (TaKaRa).
6. 0.5 *M* EDTA, dispense the solution into aliquots and sterilize by autoclaving and store at room temperature (*see* **Note 5**).
7. Phenol:chloroform (1:1 [v/v]) solution for extraction of double-stranded cDNA (*see* **Note 6**). Phenol is highly corrosive and can cause severe burns, wear gloves when handling and take appropriate safety measures. All manipulations should be carried out in a chemical hood.
8. Phase-lock gel™ separation tube (Eppendorf, Hamburg, Germany).
9. 3 *M* Sodium acetate (pH 5.0) for cDNA precipitation; dispense the solution into aliquots and sterilize by autoclaving; store at room temperature.
10. RNase-free water (*see* **Subheading 2.2.**, **item 10**).
11. BioArray High Yield RNA Transcript Labeling Kit (Enzo Diagnostics, NY) for synthesis of biotinylated cRNA.
12. RNeasy columns (Qiagen) for purification of cRNA.
13. U133A2.0 microarrays (Affymetrix, Santa Clara, CA).
14. GeneArray scanner 2500 (Agilent, Palo Alto, CA).
15. Affymetrix Microarray Suite 5.0 software (MAS 5.0; statistical algorithm).

3. Methods

A powerful approach to obtain informations about cell behavior in molecular terms is the large scale analysis of gene expression at the level of the transcriptome using cDNA microarrays *(84–95)*. It is in particular possible to compare the results from different microarray experiments. In the present study, we compared the transcriptome of unstimulated cultured fibroblasts to that of fibroblasts that had been previously stimulated with media conditioned by three metastatic prostate cancer cell lines (DU-145, PC-3, and LNCaP, respectively). In reverse experiments the transcriptomes of unstimulated cancer cell lines to those after stimulation with fibroblast conditioned media were compared.

It is important to analyse results obtained by cDNA microarray hybridization with great care. In addition to a critical evaluation of control hybridizations included in all microarray experiments (for determination of hybridization efficiency and possible "technical problems") it is pivotal to perform statistical analyses in order to find significant signal differences between different microarrays. It is recommended to proof expression differences observed in microarray experiments by another method, such as real-time PCR or (at the proteome level) by Western blotting or immunohistochemical staining.

3.1. Preparation of Conditioned Media

1. The HFF and the prostate cancer cell lines (DU-145, PC-3, and LNCaP) are passaged when approaching confluence using trypsin/EDTA to provide new maintenance cultures on 75-cm^2 flasks (*see* **Note 7**) and experimental cultures on 24-well plates. For the experimental cultures, the cell density is of particular importance factor, therefore, cells are counted.
2. Ten microliters of the respective cell culture suspension are mixed with 10-µL trypane blue solution. From this mixture 10 µL are filled into the Fast-Read 102 graduated counting chamber (*see* **Note 8**).
3. Count living cells under a microscope with 10-fold magnification. Living cells appear white, whereas dead cells absorb the blue staining and appear blue (*see* **Note 9**).
4. Calculation of cell density: cell number per field $\times 2 \times 10^4$ = cells/mL.
5. Cell density for experimental cultures is adjusted to 1000 cells/mL.
6. One millilter of the adjusted experimental culture is given in each well of the 24-well plate (1000 cells/well). The plates are incubated at 37°C with 5% CO_2 in a humidified atmosphere for 24 h.
7. Supernatants from wells of the same cell line are pooled in a 15 mL capillary tube and centrifuged at 4000 rpm (1789g) for 10 min at room temperature. Clear supernatants are used for stimulation experiments at once or stored in aliquots at –20°C (*see* **Note 10**).

3.2. Conditioning of Cells

1. Provide from a maintenance culture when approaching confluence a working culture on a 25-cm^2 flask by passaging with trypsin/EDTA. A 1:5 split will provide working cultures that are approaching confluence after 48 h.
2. The medium from working cultures is removed by aspiration. Cultures are then washed twice with 5-mL PBS to eliminate traces of medium. Cells are thereafter overlaid with the supernatants prepared as described above (*see* **Subheading 3.1.**, **step 7**). HFF cells are incubated with the supernatants derived from DU-145, PC-3,

and LNCap, respectively. Tumor cell lines are incubated with supernatant derived from HFF cells. Cultures are incubated at 37°C with 5% CO_2 in a humidified atmosphere for 24 h.

3. Media are removed from cultures by aspiration. Cultures are washed twice with 5 mL PBS to eliminate traces of medium and detached from the flasks by the use of trypsin/EDTA. Cells are resuspended in 5-mL PBS and centrifuged at 1600 rpm (286g) for 5 min at room temperature.

4. The clear supernatant is discarded and the cell pellet is resuspended in 350 µL RLT lysis buffer (*see* **Note 11**).

5. Continue at once with the RNA isolation or store the solved solution at –20°C until use (*see* **Note 12**).

3.3. RNA Isolation and Quantification

1. Make sure that the bench and all tools (e.g., pipets) are RNase-free. Clean bench and pipets with RNaseAway and wear always gloves during all steps when handling RNA.

2. For homogenization, pipet the lysate directly onto a QIAshredder spin column placed into a 2-mL collection tube and centrifuge at 13,000 rpm (15,700g) for 2 min.

3. Add 350 µL of 70% ethanol to the homogenized lysate and mix well by pipetting.

4. Apply the whole sample (700 µL) to a RNeasy mini column placed in a 2-mL collection tube. Close the tube gently and centrifuge at 13,000 rpm (15,700g) for 15 s. Discard the flow-through.

5. Add 700 µL RW 1-buffer to the column. Close the tube gently and centrifuge at 13,000 rpm (15,700g) for 15 s. Discard the flow-through.

6. Transfer the column into a new 2-mL collection tube. Pipet 500 µL RPE-buffer onto and inject into the column. Close the tube gently and centrifuge at 13,000 rpm (15,700g) for 15 s to wash the column. Discard the flow-through.

7. Add another 500 µL RPE-buffer onto the column. Close the tube gently and centrifuge at 13,000 rpm (15,700g) for 2 min to dry the silica-gel membrane in the column. Discard the flow-through (*see* **Note 13**).

8. Transfer the column into a new 1.5-mL collection tube for RNA elution. Pipet 30 µL of RNase-free water and inject directly onto the silica-gel membrane. Close the tube gently and centrifuge at 13,000 rpm (15,700g) for 1 min (*see* **Note 14**).

9. Measure RNA concentration in a 1:100 dilution at wavelengths 260 and 280 nm using an UV-VIS spectrophotometer (Backman Instruments Inc., cat. no. DU-530) (*see* **Note 15**).

10. Store the RNA solution at –80°C until use.

3.4. DNA Chips

1. For first-strand cDNA synthesis use 10 µg total RNA (*see* **Note 16**), add 1 µL of 3′ polyadenylated control RNA mixture and 1 µL 100 µM T7-oligo-d(T)$_{24}$ primer.

2. Incubate the mixture at 70°C for 10 min and put it on ice immediately.

3. Add 4-µL of 5X first-strand buffer, 2 µL 0.1 M dithiothreitol, and 1 µL 10 mM dNTPs, and preincubate the mixture at 42°C for 2 min.

4. Add 2 µL Superscript II and incubate at 42°C for 1 h.

5. For second-strand synthesis add 30 μL 5X second-strand buffer, 91 μL RNase-free water, 3 μL 10 m*M* dNTPs, 4 μL *E. coli* DNA polymerase I, 1 μL *E. coli* DNA ligase, and 1 μL RNaseH.

6. Incubate the reaction mixture at 16°C for 2 h.

7. Add 2.5 μL T4 DNA polymerase I and incubate at 16°C for 5 min.

8. Stop the reaction by the addition of 10 μL 0.5 *M* EDTA.

9. Transfer the solution into a phase-lock gel™ tube.

10. Add an equal volume of phenol:chloroform (1:1 [v/v]) to the mixture.

11. Mix the contents of the tube until an emulsion forms.

12. Centrifuge the mixture at 12,000g for 15 s (*see* **Note 17**).

13. Decant the entire water phase (containing the cDNA) into a new tube.

14. Precipitate cDNA by addition of 0.1 vol 3 *M* sodium acetate and 3 vol ice cold ethanol.

15. Mix the contents well and incubate the mixture on ice for 15 min (*see* **Note 18**).

16. Centrifuge the mixture at maximal speed for 15 min in a precooled centrifuge at 4°C.

17. Discard the supernatant and resolve the cDNA pellet in 12 μL RNase-free water.

18. Perform the in vitro transcription of 5 μL cDNA using the BioArray High Yield RNA Transcript Labeling Kit to form biotin-labeled cRNA.

19. Purify the labeled cRNA on RNeasy columns as described previously (*see* **Subheading 3.3., steps 4–8**).

20. Measure cRNA concentration in a 1:100 dilution at the wavelengths 260 and 280 nm using an UV-VIS spectrophotometer (*see* **Notes 15** and **19**).

21. Hybridize 15–20 μg cRNA after fragmentation to U133A2.0 microarrays at 45°C for 16 h.

22. Wash and stain the microarrays using the Fluidics Station 400 according to the manufacturer's recommendation.

23. Scan the microarrays in a GeneArray scanner 2500.

24. Process the arrays images with the Affymetrix Microarray Suite 5.0 software to determine signals and detection calls (present, absent, and marginal).

25. Identify with Affymetrix DataMining Tool 3.0 the probesets exhibiting a signal $\log_2$ ratio >1.32 and a change *p*-value <0.001 or a signal $\log_2$ ratio <−1.32 and a change *p*-value >0.999 (corresponding to 2.5-fold upregulation or downregulation).

4. Notes

1. Before using FCS, it should be heated and inactivated at 56°C for 30 min (**96**). The precipitates can be removed by centrifugation at 5000 rpm (2795*g*) for 5 min at room temperature.

2. If there is no possibility to oven bake the glassware, then treat it with DEPC. Fill glassware with 0.1% (v/v) DEPC, will allow shaking overnight at room temperature, and then autoclave or heating to 100°C for 15 min to eliminate residual DEPC. DEPC is suspected to be carcinogenic and should be handled with great care. Wear gloves and use fume hood when using DEPC.

3. Before use β-mercaptoethanol must be added to the RLT-buffer. Add 10 μL of 14.3 *M* β-mercaptoethanol per 1 mL RLT-buffer. The solution is stable for 1 mo

after addition of β-mercaptoethanol at room temperature. β-mercaptoethanol is toxic, work in a fume hood and wear appropriate protective clothing.

4. RPE-buffer must be diluted with ethanol before to use. Add four volumes of ethanol (96–100% [v/v]) to obtain a working solution.

5. The disodium salt of EDTA will not go into solution until the pH of the solution is adjusted to 8.0 by addition of NaOH.

6. For preparation of the phenol:chloroform solution mix equal volumes of phenol and chloroform. Equilibrate the mixture by extracting several times with 0.1 M Tris-HCl (pH 7.6). Store the equilibrated mixture under an equal volume of 0.01 M Tris-HCl (pH 7.6) at 4°C in a dark glass bottle.

7. The LNCaP cells grow markedly better in coated cell flasks (CellBind^{TM} Surface; Corning, NY). Therefore, the maintenance cultures of LNCap were cultivated in such coated flasks.

8. Air bubbles are undesirable in the counting chamber. Pipetting slowly and carefully minimizes their appearance.

9. To minimize the error of this method it is recommended to count cells in three or more fields of the counting chamber. The cell number per field is then calculated by using the mean value.

10. If you store conditioned media at –20°C avoid multiple freeze and thaw cycles. Thaw the conditioned media only once directly before use on ice.

11. It is important to solve the whole cell pellet, if this is not possible by pipetting up and down vortex the suspension instead.

12. If you store cell lysates in RLT-buffer at –20°C avoid multiple freeze and thaw cycles. Thaw cell lysates only once directly before use on ice and take care that no cell clumps are visible before continuing.

13. It is important to dry the gel membrane completely since residual ethanol may interfere with downstream reactions. If the column contacts the flow-through (even shortly by removing from the collection tube), transfer the column in another 2-mL collection tube and centrifuge at 13,000 rpm (15,700g) for 1 min.

14. Elution of RNA will result in higher amount if you incubate the column for 5 min at room temperature after addition of RNase-free water and then centrifuge.

15. Because of the salts in your RNA preparation you have to use Tris-HCl buffer (10 mM Tris-HCl, pH 7.5) for blanking the spectrophotometer. The ratio of 260 to 280 nm should be between 1.8 and 2.1.

16. RNA concentration should be 1 µg/µL. If RNA concentration is less you have to add 0.1% (v/v) 4 M LiCl and 2.5% (v/v) ethanol (absolute) to the RNA solution. Incubate the mixture on ice for 2 h. Discard the supernatant carefully and add 500 µL 70% (v/v) ethanol to the pellet. Do not resuspend the pellet. After centrifugation at 15,000 rpm (17,000g) for 15 min at 4°C discard the supernatant carefully. Add 200 µL 70% (v/v) ethanol to the pellet. Do not resuspend the pellet. After centrifugation at 15,000 rpm (17,000g) for 15 min at 4°C discard the supernatant carefully. Dry the pellet in a speed vac for 3 min. Resuspend the pellet in RNase-free water and measure RNA concentration as previously described (*see* **Subheading 3.3.**, **step 9**).

17. During centrifugation the Phase Lock Gel™ builds a barrier between the water and the organic phase, whereby organic phase and interphase are effectively separated.
18. Place reaction mixture at −20°C freezer, to avoid prolong time for DNA precipitation.
19. Store the cRNA at −80°C until use.

Acknowledgments

This work has been supported by German Cancer Association ("Deutsche Krebshilfe" grant no. 10-1877-We 2).

References

1. Denmeade, S. R. and Isaacs, J. T. (2002) A history of prostate cancer treatment. *Nat. Rev. Cancer* **2,** 389–396.
2. Isaacs, W., De Marzo, A., and Nelson, W. G. (2002) Focus on prostate cancer. *Cancer Cell* **2,** 113–116.
3. Jarrard, D. F., Kinoshita, H., Shi, Y., et al. (1998) Methylation of the androgen receptor promoter CpG island is associated with loss of androgen receptor expression in prostate cancer cells. *Cancer Res.* **58,** 5310–5314.
4. Culig, Z., Hobisch, A., Cronauer, M. V., et al. (1994) Androgen receptor activation in prostatic tumor cell lines by insulin-like growth factor-I, keratinocyte growth factor, and epidermal growth factor. *Cancer Res.* **54,** 5474–5478.
5. Hobisch, A., Eder, I. E., Putz, T., et al. (1998) Interleukin-6 regulates prostate-specific protein expression in prostate carcinoma cells by activation of the androgen receptor. *Cancer Res.* **58,** 4640–4645.
6. Veldscholte, J., Ris-Stalpers, C., Kuiper, G. G., et al. (1990) A mutation in the ligand binding domain of the androgen receptor of human LNCaP cells affects steroid binding characteristics and response to anti-androgens. *Biochem. Biophys. Res. Commun.* **173,** 534–540.
7. Visakorpi, T., Kallioniemi, A. H., Syvanen, A. C., et al. (1995) Genetic changes in primary and recurrent prostate cancer by comparative genomic hybridization. *Cancer Res.* **55,** 342–347.
8. Yeh, S. and Chang, C. (1996) Cloning and characterization of a specific coactivator, ARA70, for the androgen receptor in human prostate cells. *Proc. Natl. Acad. Sci. USA* **93,** 5517–5521.
9. Muller, J. M., Isele, U., Metzger, E., et al. (2000) FHL2, a novel tissue-specific coactivator of the androgen receptor. *Embo J.* **19,** 359–369.
10. Gregory, C. W., He, B., Johnson, R. T., et al. (2001) A mechanism for androgen receptor-mediated prostate cancer recurrence after androgen deprivation therapy. *Cancer Res.* **61,** 4315–4319.
11. McKenna, N. J. and O'Malley, B. W. (2001) Nuclear receptor coactivators—an update. *Endocrinology* **143,** 2461–2465.
12. Giovannucci, E. (1999) Nutritional factors in human cancers. *Adv. Exp. Med. Biol.* **472,** 29–42.
13. Schmitz-Drager, B. J., Eichholzer, M., Beiche, B., and Ebert, T. (2001) Nutrition and prostate cancer. *Urol. Int.* **67,** 1–11.

14. Jankevicius, F., Miller, S. M., and Ackermann, R. (2002) Nutrition and risk of prostate cancer. *Urol. Int.* **68,** 69–80.

15. Bratt, O. (2002) Hereditary prostate cancer: clinical aspects. *J. Urol.* **168,** 906–913.

16. Nwosu, V., Carpten, J., Trent, J. M., and Sheridan, R. (2001) Heterogeneity of genetic alterations in prostate cancer: evidence of the complex nature of the disease. *Hum. Mol. Genet.* **10,** 2313–2318.

17. Simard, J., Dumont, M., Soucy, P., and Labrie, F. (2002) Perspective: prostate cancer susceptibility genes. *Endocrinology* **143,** 2029–2040.

18. Slager, S. L., Schaid, D. J., Cunningham, J. M., et al. (2003) Confirmation of linkage of prostate cancer aggressiveness with chromosome 19q. *Am. J. Hum. Genet.* **72,** 759–762.

19. Paiss, T., Worner, S., Kurtz, F., et al. (2003) Linkage of aggressive prostate cancer to chromosome 7q31-33 in German prostate cancer families. *Eur. J. Hum. Genet.* **11,** 17–22.

20. Simard, J., Dumont, M., Soucy, P., and Labrie, F. (2002) Prostate cancer susceptibility genes. *Endocrinology* **143,** 2029–2040.

21. Ingles, S. A., Ross, R. K., Yu, M. C., et al. (1997) Association of prostate cancer risk with genetic polymorphisms in vitamin D receptor and androgen receptor. *J. Natl. Cancer Inst.* **89,** 166–170.

22. Correa-Cerro, L., Berthon, P., Haussler, J., et al. (1999) Vitamin D receptor polymorphisms as markers in prostate cancer. *Hum. Genet.* **105,** 281–287.

23. Makridakis, N. M., Ross, R. K., Pike, M. C., et al. (1999) Association of missense substitution in SRD5A2 gene with prostate cancer in African-American and Hispanic men in Los Angeles, USA. *Lancet* **354,** 975–978.

24. Steinhoff, C., Franke, K. H., Golka, K., et al. (2000) Glutathione transferase isozyme genotypes in patients with prostate and bladder carcinoma. *Arch. Toxicol.* **74,** 521–526.

25. Jenkins, R. B., Qian, J., Lieber, M. M., and Bostwick, D. G. (1997) Detection of c-myc oncogene amplification and chromosomal anomalies in metastatic prostatic carcinoma by fluorescence in situ hybridization. *Cancer Res.* **57,** 524–531.

26. Devi, G. R., Oldenkamp, J. R., London, C. A., and Iversen, P. L. (2002) Inhibition of human chorionic gonadotropin β-subunit modulates the mitogenic effect of c-myc in human prostate cancer cells. *Prostate* **53,** 200–210.

27. Saramaki, O., Willi, N., Bratt, O., et al. (2001) Amplification of EIF3S3 gene is associated with advanced stage in prostate cancer. *Am. J. Pathol.* **159,** 2089–2094.

28. McDonnell, T. J., Troncoso, P., Brisbay, S. M., et al. (1992) Expression of the protooncogene bcl-2 in the prostate and its association with emergence of androgen-independent prostate cancer. *Cancer Res.* **52,** 6940–6944.

29. Djakiew, D. (2000) Dysregulated expression of growth factors and their receptors in the development of prostate cancer. *Prostate* **42,** 150–160.

30. Navone, N. M., Troncoso, P., Pisters, L. L., et al. (1993) p53 protein accumulation and gene mutation in the progression of human prostate carcinoma. *J. Natl. Cancer Inst.* **85,** 1657–1669.

31. Li, J., Yen, C., Liaw, D., et al. (1997) PTEN, a putative protein tyrosine phosphatase gene mutated in human brain, breast, and prostate cancer. *Science* **275,** 1943–1947.

32. Hügel, A. and Wernert, N. (1999) Loss of heterozygosity, malignancy grade and clonality in microdissected prostate cancer. *Br. J. Cancer* **79,** 551–557.

33. Lijovic, M. and Frauman, A. G. (2003) Toward an understanding of the molecular genetics of prostate cancer progression. *J. Environ Pathol. Toxicol. Oncol.* **22,** 1–15.

34. Elo, J. P. and Visakorpi, T. (2001) Molecular genetics of prostate cancer. *Ann. Med.* **33,** 130–141.

35. Srivastava, M., Bubendorf, L., Srikantan, V., et al. (2001) ANX7, a candidate tumor suppressor gene for prostate cancer. *Proc. Natl. Acad. Sci. USA* **98,** 4575–4580.

36. Banham, A. H., Beasley, N., Campo, E., et al. (2001) The FOXP1 winged helix transcription factor is a novel candidate tumor suppressor gene on chromosome 3p. *Cancer Res.* **61,** 8820–8829.

37. Kuzmin, I., Gillespie, J. W., Protopopov, A., et al. (2002) The RASSF1A tumor suppressor gene is inactivated in prostate tumors and suppresses growth of prostate carcinoma cells. *Cancer Res.* **62,** 3498–3502.

38. Abate-Shen, C. and Shen, M. M. (2000) Molecular genetics of prostate cancer. *Genes Dev.* **14,** 2410–2434.

39. Dong, J. T. (2001) Chromosomal deletions and tumor suppressor genes in prostate cancer. *Cancer Metastasis Rev.* **20,** 173–193.

40. Maier, S., Reich, E., Martin, R., et al. (2000) Tributyrin induces differentiation, growth arrest and apoptosis in androgen-sensitive and androgen-resistant human prostate cancer cell lines. *Int. J. Cancer* **88,** 245–251.

41. Adorjan, P., Distler, J., Lipscher, E., et al. (2002) Tumor class prediction and discovery by microarray-based DNA methylation analysis. *Nucl. Acids Res.* **30,** e21.

42. Maruyama, R., Toyooka, S., Toyooka, K. O., et al. (2002) Aberrant promoter methylation profile of prostate cancers and its relationship to clinicopathological features. *Clin. Cancer Res.* **8,** 514–519.

43. Schulz, W. A., Elo, J. P., Florl, A. R., et al. (2002) Genomewide DNA hypomethylation is associated with alterations on chromosome 8 in prostate carcinoma. *Genes Chromosomes Cancer* **35,** 58–65.

44. Gerstein, A. V., Almeida, T. A., Zhao, G., et al. (2002) APC/CTNNB1 (β-catenin) pathway alterations in human prostate cancers. *Genes Chromosomes Cancer* **34,** 9–16.

45. Ko, Y., Hahn, T., Lu, H., et al. (2005) A novel component of the ubiquitin pathway, ubiquitin carboxyl extension protein 1 is overexpressed in prostate cancer. *Int. J. Mol. Med.* **15,** 183–196.

46. Cunha, G. R., Hayward, S. W., and Wang, Y. Z. (2002) Role of stroma in carcinogenesis of the prostate. *Differentiation* **70,** 473–485.

47. Sung, S. Y. and Chung, L. W. (2002) Prostate tumor-stroma interaction: molecular mechanisms and opportunities for therapeutic targeting. *Differentiation* **70,** 506–521.

48. Wernert, N. (1997) The multiple roles of tumor stroma. *Virch. Arch. A* **430,** 433–443.
49. Wernert, N., Locherbach, C., Wellmann, A., Behrens, P., and Hugel, A. (2001) Presence of genetic alterations in microdissected stroma of human colon and breast cancers. *Anticancer Res.* **21,** 2259–2264.
50. Park, C. C., Bissell, M. J., and Barcellos-Hoff, M. H. (2000) The influence of the microenvironment on the malignant phenotype. *Mol. Med. Today* **6,** 324–329.
51. Liotta, L. A. and Kohn, E. C. (2001) The microenvironment of the tumor-host interface. *Nature* **411,** 375–379.
52. McCawley, L. J. and Matrisian, L. M. (2001) Tumor progression: defining the soil round the tumor seed. *Curr. Biol.* **11,** R25–R27.
53. Chrenek, M. A., Wong, P., and Weaver, V. M. (2001) Tumor-stromal interactions. Integrins and cell adhesions as modulators of mammary cell survival and transformation. *Breast Cancer Res.* **3,** 224–229.
54. Bissell, M. J. and Radisky, D. (2001) Putting tumors in context. *Nat. Rev. Cancer* **1,** 46–54.
55. Cunha, G. R. (1994) Role of mesenchymal-epithelial interactions in normal and abnormal development of the mammary gland and prostate. *Cancer* **74,** 1030–1044.
56. Cunha, G. R. (1996) Growth factors as mediators of androgen action during male urogenital development. *Prostate Suppl.* **6,** 22–25.
57. Cunha, G. R., Hayward, S. W., Dahiya, R., and Foster, B. A. (1996) Smooth muscle-epithelial interactions in normal and neoplastic prostatic development. *Acta Anat. (Basel)* **155,** 63–72.
58. Cunha, G. R., Foster, B., Thomson, A., et al. (1995) Growth factors as mediators of androgen action during the development of the male urogenital tract. *World J. Urol.* **13,** 264–276.
59. Sugimura, Y., Foster, B. A., Hom, Y. K., et al. (1996) Keratinocyte growth factor (KGF) can replace testosterone in the ductal branching morphogenesis of the rat ventral prostate. *Int. J. Dev. Biol.* **40,** 941–951.
60. Thomson, A. A., Foster, B. A., and Cunha, G. R. (1997) Analysis of growth factor and receptor mRNA levels during development of the rat seminal vesicle and prostate. *Development* **124,** 2431–2439.
61. Foster, B. A. and Cunha, G. R. (1999) Efficacy of various natural and synthetic androgens to induce ductal branching morphogenesis in the developing anterior rat prostate. *Endocinology* **140,** 318–328.
62. Hayward, S. W., Haughney, P. C., Rosen, M. A., et al. (1998) Interactions between adult human prostatic epithelium and rat urogenital sinus mesenchyme in a tissue recombination model. *Differentiation* **63,** 131–140.
63. Hayward, S. W., Rosen, M. A., and Cunha, G. R. (1997) Stromal-epithelial interactions in the normal and neoplastic prostate. *Br. J. Urol.* **79,** 18–26.
64. Chung, L. W. and Davies, R. (1996) Prostate epithelial differentiation is dictated by its surrounding stroma. *Mol. Biol. Rep.* **23,** 13–19.
65. Condon, M. S. and Bosland, M. C. (1999) The role of stromal cells in prostate cancer development and progression. *In Vivo* **13,** 61–65.

66. Kooistra, A., Romijn, J. C., and Schroder, F. H. (1997) Stromal inhibition of epithelial cell growth in the prostate, overview of an experimental study. *Urol. Res.* **25,** S97–S105.

67. Gleave, M. E., Hsieh, J. T., von Eschenbach, A. C., and Chung, L. W. K. (1992) Prostate and bone fibroblasts induce human prostate cancer growth *in vivo*: implications for bidirectional tumor-stromal cell interactions in prostate carcinoma growth and metastasis. *J. Urol.* **147,** 1151–1159.

68. Olumi, A. F., Dazin, P., and Tlsty, T. D. (1998) A novel coculture technique demonstrates that normal human prostatic fibroblasts contribute to tumor formation of LNCaP cells by retarding cell death. *Cancer Res.* **58,** 4525–4530.

69. Camps, J. L., Chang, S. M., Hsu, T. C., et al. (1990) Fibroblast mediated acceleration of human epithelial tumor growth *in vivo*. *Proc. Natl. Acad. Sci. USA* **87,** 75–79.

70. Kooistra, A., Van den Eijnden-van Raaij, A. J., Klaij, I. A., Romijn, J. C., and Schroder, F. H. (1995) Stromal inhibiton of prostatic epithelial cell proliferation not mediated by transforming growth factor beta. *Br. J. Cancer* **72,** 427–434.

71. Wellmann, A., Wollscheid, V., Lu, H., et al. (2002) Analysis of microdissected prostate tissue with ProteinChip arrays—a way to new insights into carcinogenesis and to diagnostic tools. *Int. J. Mol. Med.* **9,** 341–347.

72. Straub, B., Muller, M., Krause, H., Schrader, M., and Miller, K. (2003) Quantitative real-time RT-PCR for detection of circulating prostate-specific antigen mRNA using sequence-specific oligonucleotide hybridization probes in prostate cancer patients. *Oncology* **65,** 12–17.

73. Schmidt, U., Bilkenroth, U., Linne, C., et al. (2004) Quantification of disseminated tumor cells in the bloodstream of patients with hormone-refractory prostate carcinoma undergoing cytotoxic chemotherapy. *Int. J. Oncol.* **24,** 1393–1399.

74. Bubendorf, L., Schopfer, A., Wagner, U., et al. (2000) Metastatic patterns of prostate cancer: an autopsy study of 1,589 patients. *Hum. Pathol.* **31,** 578–583.

75. Rubin, M. A., Putzi, M., Mucci, N., et al. (2000) Rapid ("warm") autopsy study for procurement of metastatic prostate cancer. *Clin. Cancer Res.* **6,** 1038–1045.

76. Cooper, C. R., McLean, L., Walsh, M., et al. (2000) Preferential adhesion of prostate cancer cells to bone is mediated by binding to bone marrow endothelial cells as compared to extracellular matrix components in vitro. *Clin. Cancer Res.* **6,** 4839–4847.

77. Lehr, J. E. and Pienta, K. J. (1998) Preferential adhesion of prostate cancer cells to a human bone marrow endothelial cell line. *J. Natl. Cancer Inst.* **90,** 118–123.

78. Cooper, C. R., Chay, C. H., Gendernalik, J. D., et al. (2003) Stromal factors involved in prostate carcinoma metastasis to bone. *Cancer* **97,** 739–747.

79. Chay, C. H., Cooper, C. R., Gendernalik, J. D., et al. (2002) A functional thrombin receptor (PAR1) is expressed on bone-derived prostate cancer cell lines. *Urology* **60,** 760–765.

80. Kaighn, M. E., Narayan, K. S., Ohnuki, Y., Lechner, J. F., and Jones, L. W. (1979) Establishment and characterization of a human prostatic carcinoma cell line (PC-3). *Invest. Urol.* **17,** 16–23.

81. Stone, K. R., Mickey, D. D., Wunderli, H., Mickey, G. H., and Paulson, D. F. (1978) Isolation of a human prostate carcinoma cell line (DU 145). *Int. J. Cancer* **21,** 274–281.

82. Horoszewicz, J. S., Leong, S. S., Chu, T. M., et al. (1980) The LNCaP cell line—a new model for studies on human prostatic carcinoma. *Prog. Clin. Biol. Res.* **37,** 115–132.

83. Wernert, N., Gilles, F., Fafeur, V., et al. (1994) Stromal expression of c-ets 1 transcription factor correlates with tumor invasion. *Cancer Res.* **54,** 5683–5688.

84. Bourdeau, I. (2004) Clinical and molecular genetic studies of bilateral adrenal hyperplasias. *Endocr. Res.* **30,** 575–583.

85. Galvin, J. E. and Ginsberg, S. D. (2004) Expression profiling and pharmacotherapeutic development in the central nervous system. *Alzheimer Dis. Assoc. Disord.* **18,** 264–269.

86. Kawakami, Y., Fujita, T., Matsuzaki, Y., et al. (2004) Identification of human tumor antigens and its implications for diagnosis and treatment of cancer. *Cancer Sci.* **95,** 784–791.

87. Haraguchi, N., Inoue, H., Mimori, K., et al. (2004) Analysis of gastric cancer with cDNA microarray. *Cancer Chemother. Pharmacol.* **54,** S21–S24.

88. Sipos, F., Galamb, O., Molnar, B., and Tulassay, Z. (2004) Use of DNA-chips technology in colorectal cancer. *Orv. Hetil.* **145,** 993–999.

89. Mycko, M. P., Papoian, R., Boschert, U., Raine, C. S., and Selmaj, K. W. (2004) Microarray gene expression profiling of chronic active and inactive lesions in multiple sclerosis. *Clin. Neurol. Neurosurg.* **106,** 223–229.

90. Aigner, T., Finger, F., Zien, A., and Bartnik, E. (2004) cDNA-microarrays in cartilage research—functional genomics of osteoarthritis. *Z. Orthop.* **142,** 241–247.

91. Lapillonne, A., Clarke, S. D., and Heird, W. C. (2004) Polyunsaturated fatty acids and gene expression. *Curr. Opin. Clin. Nutr. Metab. Care* **7,** 151–156.

92. Kiefer, J., Alexander, A., and Farach-Carson, M. C. (2004) Type I collagen-mediated changes in gene expression and function of prostate cancer cells. *Cancer Treat. Res.* **118,** 101–124.

93. Goldsmith, Z. G. and Dhanasekaran, N. (2004) The microrevolution: applications and impacts of microarray technology on molecular biology and medicine (review). *Int. J. Mol. Med.* **13,** 483–495.

94. Fargiano, A. A., Desai, K. V., and Green, J. E. (2003) Interrogating mouse mammary cancer models: insights from gene expression profiling. *J. Mammary Gland Biol. Neoplasia* **8,** 321–334.

95. Schwaenen, C., Wessendorf, S., Kestler, H. A., Dohner, H., Lichter, P., and Bentz, M. (2003) DNA microarray analysis in malignant lymphomas. *Ann. Hematol.* **82,** 323–332.

96. Giard, D. J. (1987) Routine heat inactivation of serum reduces its capacity to promote cell attachment. *In Vitro Cell Dev. Biol.* **23,** 691–697.

15

Identification of Small Molecule Targets on Functional Protein Microarrays

Michael Salcius, Gregory A. Michaud, Barry Schweitzer, and Paul F. Predki

Summary

Small molecules, such as metabolites and hormones, interact with proteins to regulate numerous biological pathways, which are often aberrant in disease. Small molecule drugs have been successfully exploited to specifically perturb such processes and thereby, decrease and even eliminate disease progression. Although there are compelling reasons to fully characterize interactions of small molecules with all proteins from an organism for which an intended drug regimen is planned, currently available technologies are not yet up to this task. High-content functional protein microarrays, containing hundreds to thousands of proteins, are new tools that show great potential for meeting this need. In this chapter, we review examples and methods for profiling small molecules on high-content functional protein arrays and discuss considerations for troubleshooting.

Key Words: Drug, kinase; protein microarray; small molecule.

1. Introduction

The use of DNA microarrays to study the expression of thousands of genes in a single experiment has become a highly accepted method in both academia and in the private industry for identifying changes in gene expression owing to external stimuli or disease *(1–4)*. Even though this approach has led to the identification of large numbers of candidate drug targets, the utility of gene expression data for developing small molecule drugs has yet to be proven *(5)*. One complicating factor in identifying drug targets through the use of DNA microarrays is the fact that protein abundance is often poorly correlated with amount of RNA *(6)*. Screening for drug targets by directly assessing the effects of small molecules at the protein level are anticipated to identify leads with a better chance of validation.

From: *Methods in Molecular Biology, vol. 382: Microarrays: Second Edition: Volume 2*
Edited by: J. B. Rampal © Humana Press Inc., Totowa, NJ

More recently, protein microarrays containing hundreds or thousands of proteins have been used for a variety of applications including both molecular interaction and assessment of posttranslational modification assays. Protein microarrays can generally be separated into two classes: antibody arrays and functional protein arrays. Antibody protein arrays usually contain numbers of different antibodies and have been used to measure protein abundance in a variety of samples. In contrast, functional protein microarrays contain many different classes of proteins. A number of articles have been published in the last few years that describe the utility of functional protein arrays for identifying protein–protein, protein–DNA, protein–lipid, and protein–small molecule interactions. In addition, high-content functional protein arrays have been validated for characterizing antibody specificity *(7)* and profiling serum antibodies to identify candidate biomarkers *(8)*.

One application of functional protein arrays that has great potential is the ability to identify novel targets for small molecules. High throughput small molecule screening assays have been developed for this purpose, but these assays require knowledge of the protein/class being targeted. Therefore, one obvious advantage of high-content protein arrays is the ability to identify targets for molecules having an unknown mechanism of action. Alternatively, interactions of molecules with known targets can be assessed in the context of thousands of other proteins to identify unanticipated targets that might explain or predict drug toxicity. Therefore, this approach has the potential to supplement and perhaps eventually decrease the need for animal studies that aim to address the effect of drugs on the body.

Several proof-of-principle experiments for characterizing small molecule protein interactions on functional protein microarrays have been documented in the literature. It is described that interactions of small molecules of varying affinities with a known target protein, PKBP12, on a modified glass microarray *(9)*. The G protein-coupled receptors (GPCR) arrayed on glass slides displayed expected binding to their small molecule ligands *(10)*. The fact that the GPCRs displayed membrane-like properties on the solid state support provides further support that binding conditions similar to those found in vivo can be replicated on functional protein microarrays. The binding of radiolabeled GTP to several GTP-binding proteins on microarrays and were able to derive an equilibrium binding constant similar to the published value *(11)*. In another example, specific binding of a (^{125}I)Triiodothyronine (T3) to its receptor on an array containing a variety of receptors was demonstrated *(12)*. Thus, it seems clear that small molecules can be expected to bind to their targets in protein arrays with similar specificity and affinity as observed with more conventional techniques.

Next, methods for demonstrating interactions of small molecules with proteins immobilized on glass slides are reviewed. At the present time, small molecules must have some type of detectable label for use in this application. It is expected that improvements in this approach will come from enhancements in (1) small molecule labeling efficiencies, (2) enhanced sensitivity of detection of labels or, alternatively, the development of label-free detection technologies, (3) improved slide chemistries, and (4) improved methods of protein deposition.

2. Materials

2.1. Microarray Probing

1. Probing buffer: 50 mM Tris-HCl, pH 7.5, 5 mM MgSO$_4$, 0.1% Brij-35 (keep at 4°C or on ice). A stock solution of Brij-35 at 10% should be used to make the buffer.
2. ProtoArray™ Protein Microarrays (Invitrogen) on modified glass. Store at –20°C until use (*see* **Note 1**). Go to http://www.invitrogen.com/protoarray for additional information on these protein array products.
3. Streptavidin conjugated to AlexaFluor 647 diluted to 750 ng/mL in Probing buffer (sensitive to light, keep in dark at all times. Make up fresh just before use.)
4. Biotinylated (or fluorophore-labeled) probe of interest diluted to 1 nM–10 μM in probing buffer (make up fresh just before use). One microliter is a good starting point.
5. Hybrislips™ (Grace Bio-Labs) used to cover the array to allow low volume probing without drying of the array.
6. Incubation chamber (Evergreen Scientific) used to probe arrays.
7. 50-mL Conical tubes used to perform probings.
8. Glass staining tray (for blocking a large number of arrays).
9. Slide boxes to spin down and store arrays.

2.2. Scanning and Data Analysis

1. Genepix 4000B Fluorescent Microarray Scanner (Molecular Devices) or any scanner that can detect AlexaFluor 647 dye can be used (excitation 647 nm, emission 666 nm). Wavelength is specified by excitation/emission spectrum of detection reagent or fluorophore attached to the small molecule.
2. Genepix Pro 5.1 (Molecular Devices) suggested or other microarray analysis software.
3. A "GAL" file or other file containing information about the location of proteins on the microarray (Gal files for Invitrogen products can be downloaded from http://www.invitrogen.com/protoarray).
4. ProtoArray™ Prospector provides automated data analysis for Invitrogen microarray products (can be downloaded from http://www.invitrogen.com/protoarray).

3. Methods

3.1. Microarray Probing (for a Flowchart of the Protocol, see Fig. 1)

1. Remove the arrays from the freezer and allow to equilibrate for 15 min at 4°C in the packaging provided (*see* **Note 2** and **Fig. 1**).

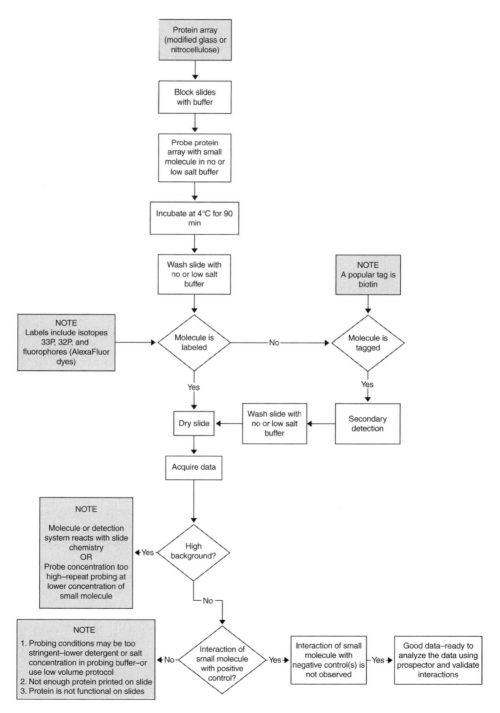

Fig. 1. Flowchart for small molecule probing on protein arrays.

2. Place one array in each incubation chamber (or use the slide mailer provided with some arrays) and add 30 mL of probing buffer to the tube. Cap tightly, and incubate at 4–6°C for 60 min with shaking or rotation. This step blocks the unprinted array surface and equilibrates the array in preparation for probing.

3. Prepare a dilution of the small molecule probe in the probing buffer, 125 μL is suficient volume to probe an array under a Hybrislip (*see* **Note 3**—for a discussion of assay limitations and means of increasing sensitivity *see* **Note 4**).

4. Remove the array from the blocking solution, tap it on a laboratory wipe to remove any excess liquid and apply the probe evenly to the array—move on to the next step immediately—do not let the array dry out (*see* **Note 5**).

5. Remove the plastic covering from each side of the hybrislip, and touching only the edges, place the Hybrislip on the array (*see* **Note 6**).

6. Place the array in a 50-mL conical tube with the printed side up, cap, and secure such that the array surface will be level to prevent the loss of liquid. Incubate the array this way at 4–6°C for 90 min.

7. Remove the array and place in an incubation chamber (array will not fit on rails in the chamber with a Hybrislip, so place it diagonally).

8. Pipet 25 mL of probing buffer along the side of the incubation chamber. This should cause the Hybrislip to release from the array surface.

9. Using forceps carefully lift up the array and remove the Hybrislip. Do not touch the array surface. Place the array back into the incubation chamber along the rails.

10. Incubate the array for 1–2 min on ice (or at 4–6°C), decant the buffer and invert the incubation chamber on a paper towel to remove residual liquid.

11. Pipet 25 mL of buffer along the side of the incubation chamber (*see* **Note 7**).

12. Incubate the array for 1–2 min on ice (or at 4–6°C), decant the buffer, and invert the incubation chamber.

13. Repeat **steps 11–12** once more, for a total of three washes.

14. If the small molecule used is fluorescently labeled, skip to **step 20**.

15. Prepare 25 mL of streptavidin conjugated to Alexa-Fluor 647 at 0.75 μg/mL for each array. Add to the incubation chamber, avoiding pipetting onto the array surface.

16. Incubate chamber on ice (or at 4–6°C) in the dark for 30 min.

17. Decant the solution and invert the incubation chamber on a paper towel to remove excess liquid. Add 25 mL of buffer, pipetting down the side of the chamber.

18. Incubate on ice (or at 4–6°C) for 1–2 min. Decant the solution and invert the incubation chamber on a paper towel to remove excess liquid.

19. Repeat **steps 16–17** twice more for a total of three washes.

20. Remove the array from the incubation chamber and tap one edge gently on a laboratory wipe to remove any buffer.

21. Place the array in a slide holder (or a 50-mL conical tube if a slide holder is unavailable). Ensure the array is properly seated to prevent damage.

22. Centrifuge the array at 800*g* for 1 min at room temperature.

23. Place the array in a slide box with the lid open in the dark for 15–30 min to allow the array to dry (*see* **Note 8**).

24. The array is ready to be scanned when it is completely dry.

3.2. Scanning and Data Analysis

1. Start the Genepix software.
2. Slide open the door on top of the Genepix 4000B scanner, open the slide holder and place the slide array side down with the barcode toward the front of the machine. Close the slide holder, and slide the door closed.
3. Check the settings for the scanner and set accordingly—the 635 laser should be selected—a good starting point is a PMT gain of 600, laser power 100%, pixel size 10 µm, lines to average 1, focus position 0 µm.
4. Scan the array using preview mode (Alt+P), watching for any saturated signals—if signals are saturated, lower PMT until no longer saturated. If signals are too weak, increase until the signals become stronger. It is suggested the PMT gain stay within a range of 400–900 PMT.
5. Click the "view scan area" button (Ctrl V) to define the area of the array to be scanned in detail. Click and drag the box around the area to be scanned.
6. Click on the "data scan" button to start the scan (Alt+D). Again, watch for saturated pixels, and if any appear, stop the scan, lower the PMT, and rescan.
7. Once the scan is done, load the array list file to analyze the array. Click on the "File…" button, then the "load array list…" (Alt+Y), navigate to the location of the array list file and select it. A grid will appear on the screen.
8. Using the "fiduciary" spots, align the grid to the spots on the array. Every Invitrogen Protoarray™ contains dye-labeled antibody in each subarray, and these signals can be used to align the grid on the array.
9. Now go to each subarray and carefully align each subarray grid to the dye-labeled antibody (*see* **Note 9**).
10. Use the "align blocks" button and select "align all features in all blocks" (Shift F5). Now carefully view each subarray and ensure the program has accurately found each feature (*see* **Note 10**). If it is necessary to change the size of the feature found, press the "Feature Mode" button (F), then hold down the "CTRL" key and use the up/down arrows to adjust the size of the feature circle.
11. Once all subarrays are aligned, press the "Analyze" button (Alt+A). The software will analyze all the features and display the results. Save these results by clicking the "File…" button then the "Save Results As…" (Alt+U). Save the file with the ".gpr" extension.
12. Download and install ProtoArray™ Prospector 2.0 (available at www.invitrogen. com/protoarray—look for "online tools," a manual is also available for download). Use Prospector to compare the array probed with small molecule to any negative control array.
13. Interactions that appear on the probed array but not on the negative could be significant interactions (**Fig. 2**). It is suggested that further experimentation is used to validate interactions seen (*see* **Note 11**).

4. Notes

1. Modified glass slides are usually the optimal slide chemistry for small molecules. Nitrocellulose slides, which have the benefit of higher protein-binding capacity

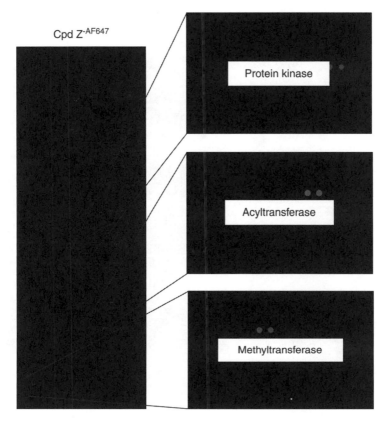

Fig. 2. Image of a ProtoArray™ Human Protein Microarray mg v1.0 array probed with a kinase inhibitor (Cpd Z) labeled with AlexaFluor 647. Multiple interactions with kinases and nonkinases were observed.

but might have higher background/nonspecific binding, can be tested with small molecules, which do not show interactions on an array printed on modified glass. Blocking nonspecific binding on nitrocellulose-coated arrays is important, and 1% BSA is recommended for this purpose.

2. Spots may smear or merge if arrays are not equilibrated before use owing to the formation of condensation on the array surface.

3. Start with 1 μM small molecule solution and depending on the probing results, the probing might need to be repeated at a 10- to 100-fold higher or lower concentration.

4. It is possible to detect interactions with micromolar K_ds using this protocol. If the small molecule of interest has a high K_d, this protocol may give very weak signals on the array because the molecule dissociates rapidly. If this is suspected to be the case, probings can be done in a low volume washing chamber such as the Atlas chamber (Clonetech), using 2-mL washes and the same general protocol as previously describes. The actual probing step should be done without a Hybrislip in the Atlas chamber with the standard 120 μL probing volume. Smaller volume probings

A

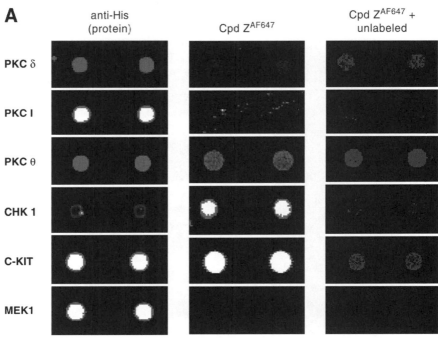

B

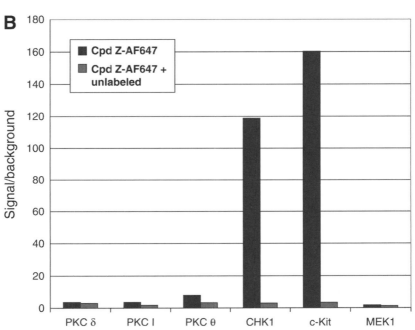

will increase signal, but will also cause an increase in background. Ensuring the experiment is done in the cold or on ice will also increase the likelihood of detecting weak interactions. Last, it might be necessary to increase or decrease the concentration of salt/detergent in the buffer to optimize the conditions for certain small molecule protein interactions.

5. It is extremely important to not allow the array to dry at any time once it has been immersed in buffer. Allowing the array to dry will cause high background in the dried areas, and might make it impossible to get useful data from the array.

6. Place the Hybrislip by starting on one edge, and then slowly lowering it, squeezing the bubbles out as it drops. Any bubbles beneath the Hybrislip will be visible during the scanning, because the probe will not contact that part of the array.

7. Avoid pipetting directly onto the array surface as this can wash away interactions and effect image quality.

8. It is important to protect a dry array that has been probed with a fluorescent molecule from excessive exposure to light, because this can lead to photobleaching of the fluorescent dye.

9. It is important to ensure each subarray is aligned correctly before aligning the features in the next step. Aligning features with misaligned blocks will likely result in inaccurate data.

10. If the spot found is much larger or smaller than the feature on the array, the data from that spot will be inaccurate. Also, small areas of high background can be detected as features, giving incorrect data.

11. To further validate interactions on arrays, one can repeat the probings at the same concentration, or varied concentrations to look for repeated interactions. Probing in the presence of unlabeled small molecule (at least 10X higher concentration than the labeled probe) can show interactions that can be competed off with the unlabeled molecule, and therefore, increase the confidence in the interaction (*see* **Fig. 3**). A simple nonarray-based method to validate interactions involves the use of fluorescent polarization to detect the interaction of the protein of interest with the small molecule.

Acknowledgments

We would like to thank Rhonda Bangham, Jennifer McCague, David Riches, and Rebecca Martone for sharing their ideas and knowledge.

Fig. 3. (*Opposite page*) Images of a known protein kinase inhibitor labeled with AlexaFluor 647 Dye (Cpd Z^{AF647}) interacting with kinases on an array. (**A**) The first column of images (anti-His) demonstrates the relative amount of protein in each spot on the array. The second column shows the strong interaction Cpd Z has with two of the kinases. The effect of adding unlabeled compound to the assay is evident in the third column of images. (**B**) A chart showing the quantitation of the images above. Interactions with CHK1 and *c*-Kit are examples of interactions that can be competed with the unlabeled parent molecule.

References

1. Horak, C. E. and Snyder, M. (2002) Global analysis of gene expression in yeast. *Funct. Integr. Genomics* **2,** 171–180.
2. Nutt, C. L., Mani, D. R., Betensky, R. A., et al. (2003) Gene expression-based classification of malignant gliomas correlates better with survival than histological classification. *Cancer Res.* **63,** 1602–1607.
3. Ramaswamy, S., Tamayo, P., Rifkin, R., et al. (2001) Multiclass cancer diagnosis using tumor gene expression signatures. *Proc. Natl. Acad. Sci. USA* **98,** 15,149–15,154.
4. Singh, D., Febbo, P. G., Ross, K., et al. (2002) Gene expression correlates of clinical prostate cancer behavior. *Cancer Cell* **1,** 203–209.
5. Butte, A. (2002) The use and analysis of microarray data. *Nat. Rev. Drug Discov.* **1,** 951–960.
6. Gygi, S. P., Rist, B., Gerber, S. A., Turecek, F., Gelb, M. H., and Aebersold, R. (1999) Quantitative analysis of complex protein mixtures using isotope-coded affinity tags. *Nat. Biotechnol.* **17,** 994–999.
7. Michaud, G.A., Salcius, M., Zhou, F., et al. (2003) Analyzing antibody specificity with whole proteome microarrays. *Nat. Biotechnol.* **21,** 1509–1512.
8. Mezzasoma, L., Bacarese-Hamilton, T., DiCristina, M., Rossi, R., Bistoni, F., and Crisantin, A. (2002) Antigen microarrays for serodiagnosis of infectious diseases. *Clin. Chem.* **48,** 121–130.
9. MacBeath, G. and Schreiber, S. L. (2000) Printing proteins as microarrays for high-throughput function determination. *Science* **289,** 1760–1763.
10. Fang, Y., Frutos, A. G., and Lahiri, J. (2002) Membrane protein microarrays. *J. Am. Chem. Soc.* **124,** 2394–2395.
11. Schweitzer, B., Predki, P., and Snyder, M. (2003) Microarrays to characterize protein interactions on a whole-proteome scale. *Proteomics* **3,** 2190–2199.
12. Ge, H. (2000) UPA, a universal protein array system for quantitative detection of protein-protein, protein-DNA, protein-RNA and protein-ligand interactions. *Nucleic Acids Res.* **28,** E3.

16

Quantification of Small Molecules Using Microarray Technology

Martin Dufva and Claus B. V. Christensen

Summary

Small molecule detection poses special problems during analysis whether hormones in a clinical setting or pesticides from environmental monitoring. Traditional analysis involves procedures like high-pressure liquid chromatography, gas chromatography, or mass spectrometry, or a combination of the three. Microarray procedures have recently evolved into a technique capable of replacing many of these assays, utilizing the strong and specific binding of a binder (e.g., an antibody) to a given target, even in a quantitative manner. A higher sensitivity can be obtained using microarrays were shown, even without the concentration of the sample beforehand. The sensitivity is high enough for monitoring most clinically relevant markers and current regulatory pesticide levels. The microarray technique has additionally parallelism in sample analysis. The same sample can be analyzed for many targets at the same time, and under the same conditions. In the present protocol pesticide detection by microarray analysis is presented.

Key Words: Microarray; monoclonal antibody; robotic printing; small molecule analysis; solid surface; quantification.

1. Introduction

A microarray in its basic form, is a collection of capture molecules ordered in highly systematic array on a solid surface. Usually, the arraying is done by robotic equipment that can deliver a subnano to microlitre amount of reagent to a predetermined position on the solid surface. The delivery of array material can be either by pins (contact printing) or by means of ink jetting or piezo printing (noncontact printing). After immobilization the unbound reagent material is washed away and the array can be probed with sample or stored for later use. The microarray technology has been used for analysis of nucleic acids for more than a decade, whereas, other types like protein analysis have emerged more

From: *Methods in Molecular Biology, vol. 382: Microarrays: Second Edition: Volume 2*
Edited by: J. B. Rampal © Humana Press Inc., Totowa, NJ

recently. Microarrays has become popular because of its sensitivity and parallelism and small requirements of samples and could replace many well-based assay like traditional immunoarray *(1)*. Although nucleic acid analysis on microarrays is a complex process mainly because of the sample preparation, immunological-based assays are usually much simpler to perform. Immunoassays are based on antibodies to provide the specific identification of the analyte. However other types of binders could be used like phage and RNA and DNA aptamers. These molecules that provide specific binding are collectively referred to as binders.

Standard microarray assays technically resemble the enzyme-linked immunosorbent assay technique, where a sample is applied and reacted with the binders or ligands on the surface (**Fig. 1A**). After incubation, the unreacted sample molecules are washed away and the reacted complexes are often visualized by a secondary binding reaction, typically labeled by fluorescence in contrast to enzymatically detection in enzyme-linked immunosorbent assay. *Competitive assays* are usually applied if the target is a small molecule with only one epitope, for example, a pesticide, either as direct or indirect competitive assays.

Direct implies that a known concentration of labeled target analog, i.e., pesticide is mixed with target pesticide existing in the sample. The target and its labeled analog compete for a limited number of surface immobilized binders (**Fig. 1B**). In the indirect competitive assay, target molecules compete with a limited number of its surface-immobilized target analogs for binding labeled binders (**Fig. 1C**). If the primary binder is not labeled, a secondary, labeled binder is added after washing step, which binds and detects bound, unlabeled primary binder (**Fig. 1D**). In competitive immunoassays, the amount of bound labeled target analog or labeled binder is inversely proportional to the amount of analyte in the competition step.

Since the 1960s, atrazine and dichlobenil have been used extensively as broad-spectrum herbicides both in agriculture and in urban areas. Dichlobenil is mainly degraded to 2,6-dichlorobenzamide (BAM) and is extremely resistant to further degradation *(2,3)*. Consequently, Dichlobenil has been banned in many countries since its discharge into groundwater has resulted in the widespread presence of BAM in the drinking water both in Europe and in the United States *(2,4)*. Furthermore, an increasing presence of atrazine and its degradation products in groundwater supplies has also been reported *(5,6)*. Both dichlobenil and atrazine have well documented mutagenic, teratogenic, and carcinogenic properties, and hence, pose a potential risk factor for public and environmental health *(7,8)*.

The European Union (EU) has enforced stringent directives defining the maximum pesticide content for drinking water at 0.1 µg/L for individual pesticides and at 0.5 µg/L for the total amount of pesticides *(9)*. Immunological

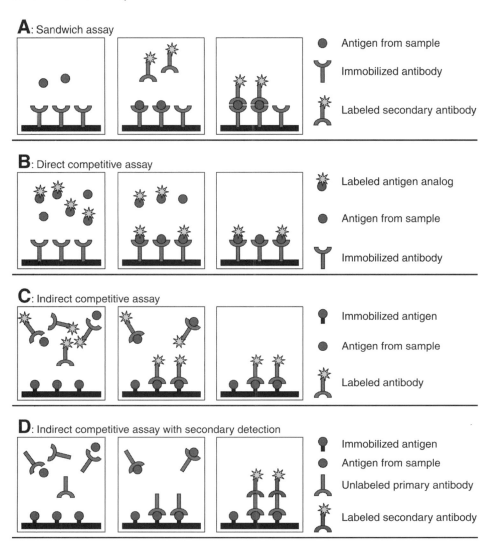

Fig. 1. Different solid-phase immunological methods used for quantification of analytes. (**A**) Standard sandwich assay, where a capture antibody is immobilized on the surface that captures the analyte in solution. The analyte provides the epitope to immobilize the detector antibody to the surface. (**B**) Direct competitive immuno assay. The specific antibody is immobilized on the surface. A labeled competitor competes with the sample for the binding to the antibody on the surface. (**C**) Indirect competitive immunoassay is similar to direct competitive assay but in this case the specific antibody is in the solution. The analyte specific for the antibody is immobilized on the surface and when the antibody is mixed with the sample and the immobilized antibody a competition between the immobilized analyte and the analyte in the sample occurs. A large amount of analyte in the sample results in low signals (little antibody binds to the surface), whereas, small amount of analyte in the sample results in high signals. (**D**) Same as in **C** but the primary antibody in not labeled. A signal is instead provided by a secondary labeled antibody.

methods for quantifying pesticides such as BAM and atrazine have been shown to be highly sensitive with detection limits in the range above 20 ng/L, requiring no extensive sample preparation, i.e., minimum, if any, harmful solvents are required. These simplified techniques might eradicate the necessity for dedicated laboratories and provide inexpensive, rapid, and reliable analyses of pesticides (10–23). Microarray-based analysis of pesticides can after extensive optimization reach detection limits down to 1 ng/L (5 pM) using fluorescence (24,25) or down to 100 ng/L using nanogold particles staining of microarrays followed by absorbance detection using a standard flatbed scanner (26).

2. Materials

1. Phosphate-buffered saline (PBS) (150 mM, pH 7.4. Sigma-Aldrich, Steinheim, Germany), degassed by placing the buffer under vacuum for at least 10 min or until no bubble is formed.
2. Cy5 maleimide monoreactive dye labeling kit (Amersham Pharmacia Biotech, England; cat. no. PA25031).
3. Tris-(2-carboxyethyl) phosphine.
4. Dimethylformamide.
5. P30 Tris chromatography column (Bio-Rad, Hercules, CA).
6. Wash buffer (TBS): 20 mM Tris-HCl, 150 mM NaCl, pH 7.4 supplemented with 0.1% (w/v) BSA (bovine serum albumin fraction V, Sigma-Aldich) and 0.1% (v/v) Tween-20 (Sigma-Aldich).
7. Genetix amine spotting buffer (Genetix, New Milton, Hampshire, UK).
8. Easyspot slides (U-Vision Biotech, Taiwan).
9. CMP3 Chipmaker pin (Telechem International, Sunnyvale, CA).
10. Stratalinker 2400 (Stratagen, La Jolla, CA).
11. Microarray Wash Station (ArrayIt, Sunnyvale, CA).
12. Hydrophobic pen (PAP Pen, The Binding Site, UK).
13. Origin® 6.1 software (Origin Lab, Northampton, MA).
14. Incubation buffer: the final concentration of the incubation buffer is 1X TBS supplemented with 0.05% (w/v) BSA (bovine serum albumin fraction V) and 0.05% (v/v) Tween-20. However, the easiest procedure is to dilute the antibody in a buffer with a final concentration of 2X incubation buffer before mixing with the sample.

3. Methods

The protocol described here assumes that specific antibodies toward a desired analyte exist and that the analyte is conjugated to a carrier protein.

3.1. Preparation of Fluorescently Labeled Antibodies

3.1.1. Labeling of Antibodies

The antibodies are conveniently labeled with Cy5 fluorescent dye using a Cy5 maleimide monoreactive dye (*see* **Note 1**). The antibodies were labeled according to manufacturers instructions and described next.

1. Dilute the antibodies to 1 mg/mL in degassed PBS buffer, pH 7.4 and incubate for 30 min.
2. One milligram (=1 mL) antibodies is mixed with 180 µg Tris-(2-carboxethyl) phosphine (10 µL of an 18 mg/mL solution). To keep the reaction free-of-oxygen, displace the residual air in the 1.5-mL Eppendorff tube by flushing the tube with gaseous nitrogen. Close the lid, mix and incubate for 10 min in room temperature.
3. Make a dye solution by mix 50 µL anhydrous dimethylformamide with one package of dye. Flush the vial with nitrogen and close the lid.
4. Mix the reduced antibodies (1010 µL) from **step 2** with the 50 µL dye solution. Flush the vial and incubate at room temperature for 2 h mixing every 30 min. Keep the solution in dark since ordinary light from light bulbs or sun light will destroy the dye.
5. Incubate overnight in refrigerator (2–8°C).

3.1.2. Purification of Antibodies

1. Excess dye can be removed in many ways. We have used a spin column, P30 Tris Chromatography column to remove unreacted dye according to manufacturer's instructions.

3.1.3. Determination of the Number of Fluorochromes on the Antibody

1. Measure the absorbance at 650 and 280 nm of the antibody solution after purification using a standard spectrophotometer.
2. Using molar extinction coefficients for antibodies (170,000 M/cm) and Cy5 (250,000 × 170,000 M/cm) calculate the F/A ratio (fluorochrom/antibody ratio) using the following formula: $0.68 \times A_{650}/(A_{280}-[0.05 \times A_{650}])$. The F/A ratio should be around 1.5. Significantly less (<0.5) indicate poor labeling reaction while significantly higher (>3) indicate that excess dye is not removed.

3.2. Preparation of Microarrays

1. Prepare the robot for printing by ensuring that the air in the robot has a humidity of 75%.
2. Dilute the small molecule carrier protein conjugate to 0.5 mg/mL in Genetix amine spotting buffer. Transfer the probes to a 384-well microtitre plate and put this microtitre plate in the spotter.
3. Place the Easyspot microarray slides in the spotter and program the robot in such a way so that you can perform 10 individual reactions on each slide. Also ensure that the spots are printed in an asymmetrical pattern so identification of spots can be made after reactions.
4. Print the array with 75% humidity using a CMP3 Chipmaker pin.
5. Place the slides in a Stratalinker 2400 with light bulbs emitting ultraviolet light at 254 nm and expose the slides for 30 s (*see* **Note 3**).
6. Remove inbound conjugates by washing the slides in Wash Buffer. Place the slides in a Microarray Wash Station put it in a 400-mL beaker. Add a stirring magnet and 400 mL wash buffer. Stir vigorously using a magnetic stirrer for 10 min.

7. Replace the washing buffer with Milli-Q water and wash the slides for 2 min. Dry the slides by centrifugation. We have home build slide dryer but such can be bought from many companies for a relatively low cost.
8. The 10 arrays are separated by hydrophobic border applied using a hydrophobic pen.

3.3. Competitive Immunoassay

1. Make standards by diluting the drug from 100 μg/L to 100 fg/L in Milli-Q water. Use seven different standards to make a standard curve.
2. Mix 6 μL sample (or standard) with 6 μL antibody (40 ng/mL) (*see* **Note 4**) dilute in 2X incubation buffer (*see* **Note 5**). Incubate for 10 min at room temperature.
3. Apply the seven standards and the sample to separate arrays. We print eight arrays per slide in such a manner so we can run eight different reactions per slide. This allows eight separate reactions on one slide, meaning that a complete standard curve and one sample can be fitted on one slide.
4. Incubate in dark for 60 min at room temperature.
5. Wash the slides in wash buffer for 10 min with stirring (*see* **Subheading 3.2.6.**) followed by a 5 min wash in Milli-Q water. Spin dry the slides.

3.4. Scanning the Slides and Quantification

1. Scan the slides so that the highest signal in the standard is not saturated.
2. Quantify the spots for the sample and the standards. Usually the true signal is calculated as Signal in the spot—Local background.
3. Make a standard curve by plotting the signal from each standard against the corresponding concentration. You should end up with a curve that looks like **Fig. 2**.
4. An estimation of the concentration in the sample can be determined directly using the plotted standard curve. However, it is useful to do a four parameter logistic fit and use the generated formula (**Fig. 3**). Such four parameter logistic fits are calculated using for instance Origin® 6.1 software.

4. Notes

1. Labeling of antibodies (**Subheading 3.1.**). It is important to use correctly labeled antibodies. Using other kits yielding higher F/A ratio (6.5) gives threefold higher IC_{50} value *(24)*. This indicates that the antibody is destroyed to some extent because of the labeling procedure.
2. Fabrication of the array (**Subheading 3.2.**) is not strait forward as many choices regarding microarray substrates, printing buffers and immobilization condition must be made. However, it is found that printing with 0.5 mg/mL of protein/conjugate solution in microspotting solutions (www.Arrayit.com) gives satisfying result. Other buffers might be better but needs to be tested. Out of total nine different microarray substrates that were tested, two substrates were superior compared with others in terms of IC_{50} value, signal strength, and reproducibility *(24)*.

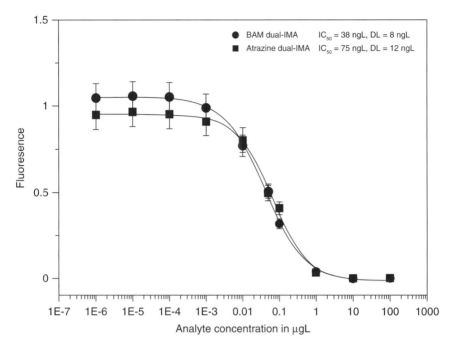

Fig. 2. Dual-analyte immuno-microarray. A simultaneous 2,6-dichlorobenzamide (BAM) and atrazine immuno-microarray was used for quantifying BAM and atrazine standards. The incubation time was 60 min at RT using 1/25,000 (40 ng/mL) dilution of each antibody. The individual IC_{50} values and the detection limits are indicated in the figure. The graph shows the average signals from six replicate spots for each analyte concentration from a representative immuno-microarray experiment. The error bars represent the standard deviation. (Reprinted from *see* **ref. 25**, with permission from Elsevier.)

The two slides types were the Easyspot slides and home made agarose film-coated slides. The latter are interestingly very inexpensive to produce (the cost is approx 1:50 of the Easyspot slide. The agarose film slides was first described by Affanassiev et al. *(27)* for DNA and protein analysis but the protocols for fabrication and blocking have later been modified and simplified *(24,28)*.

3. We have tested many different conditions for ultraviolet crosslinking **(Subheading 3.2.)** by varying both exposure times and illumination wave length and have found that this is not a critical parameter.

4. The antibody concentration used in this example was 40 ng/mL **(Subheading 3.3.)**. However the amount of antibody used for a given assay depends on the affinity of the specific antibody and the desired IC_{50} value. The described assay condition in this chapter is focused on finding the limits for the competitive immunoassay more than optimizing the assay around a particular IC_{50} value. Because the IC_{50} value is the middle point of the dynamic range, assays should be optimized so that the IC_{50} value

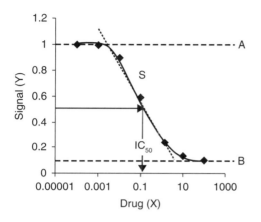

Fig. 3. Analysis of standard curve using four parameter logistic plot *(29)*.

coincides with the desired cutoff level. For instance, the EU legislation demands that the water might not contain more than 100 ng/L of pesticides. Optimal sensitivity for this diagnostic level is reached by optimizing the competitive assay to a IC_{50} value of 100 ng/L. This can be achieved in several ways. The simplest method is to make a dilution series of antibodies in which larger concentrations of antibodies will give high IC_{50} value while lower antibody concentration will give lower IC_{50} values. However, the IC_{50} values can also be modulated by the assay temperature *(25,26)* where lower temperature (4°C) gives lower IC_{50} values and higher temperatures (room temperature) gives higher IC_{50} values. Also the incubation time on the array can influence IC_{50} value, where short incubation gives lower IC_{50} values than longer incubation times *(25)*. However, it has been found that 1 h incubation gives best reproducibility *(25)*.

5. The defined incubation buffers (**Subheading 3.3.**) and washing buffers are important for maximum signal and minimal background. The buffer composition was found after an optimization procedure in which different amount of Tween-20 and BSA was added to TBS. However, it is possible that if other substrates are used, other buffer compositions might be more optimal.

References

1. Dufva, M. and Christensen, C. (2005) Diagnostic and analytical application of protein arrays. *Expert Rev. Proteomics* **2,** 41–48.
2. Brittebo, E., Eriksson, C., Feil, V., Bakke, J., and Brandt, I. (1991) Toxicity of 2,6-dichlorothiobenzamide (chlorthiamide) and 2,6-dichlorobenzamide in the olfactory nasal mucosa of mice. *Fundam. Appl. Toxicol.* **17,** 92–102.
3. Montgomery, Y. and Freed, V. H. (1972) Kinetics of dichlobenil degradation in soil. *Weed Res.* **12,** 31–42.
4. Brandt, I., Brittebo, E., Feil, V., and Bakke, J. (1990) Irreversible binding and toxicity of the herbicide dichlobenil (2,6-dichlorobenzonitrile) in the olfactory mucosa of mice. *Toxicol. Appl. Pharmacol.* **103,** 491–501.

5. Ahel, M., Evans, K., Fileman, T., and Mantoura, R. (1992) Determination of atrazine and simazine in estuarine samples by high-resolution gas-chromatography and nitrogen selective detection. *Analytica. Chimica. Acta.* **268,** 195–294.
6. Readman, J., Albanis, T., Barcelo, D., Galassi, S., Tronczynski, J., and Gabrielides, G. (1993) Herbicide contamination of Mediterranean estuarine waters—results from a med pol pilot survey. *Marine Pollution Bulletin* **26,** 613–619.
7. Sathiakumar, N., Delzell, E., and Cole, P. (1996) Mortality among workers at two triazine herbicide manufacturing plants. *Am. J. Ind. Med.* **29,** 143–151.
8. Lopezdecerain, A., Garcia, E., and Gullon, A. (1992) Influence of a triazine ring on the mutagenicity of triazinoindoles and some congeners. *Mutagenesis* **7,** 37–39.
9. EEC Drinking Water Guideline 80/778/EEC (1980).
10. Karu, A., Harrison, R., Schmidt, D., et al. (1991) Monoclonal immunoassay of triazine herbicides—development and implementation. *A.C.S. Symposium Series* **451,** 59–77.
11. Weller, M., Weil, L., and Niessner, R. (1992) Increased sensitivity of an enzyme-immunoassay (ELISA) for the determination of triazine herbicides by variation of tracer incubation-time. *Mikrochimica Acta* **108,** 29–40.
12. Giersch, T. (1993) A new monoclonal-antibody for the sensitive detection of atrazine with immunoassay in microtiter plate and dipstick format. *J. Agri. Food Chem.* **41,** 1006–1011.
13. Mangler, J., Weller, M., Weil, L., Niessner, R., Hammerle, H., and Schlosshauer, B. (1994) New monoclonal-antibodies to triazine herbicides. *Fresenius J. Anal. Chem.* **349,** 346–348.
14. Dzgoev, A., Mecklenburg, M., Larsson, P. O., and Danielsson, B. (1996) Microformat imaging ELISA for pesticide determination. *Anal. Chem.* **68,** 3364–3369.
15. Saez, A., Gomez de Barreda, D., Gamon, M., Garcia de la Cuadra, J., Lorenzo, E., and Peris, C. (1996) UV detection of triazine herbicides and their hydroxylated and dealkylated degradation products in well water. *J. Chromatogr.* **A721,** 107–112.
16. Issert, V., Grenier, P., and BellonMaurel, V. (1997) Emerging analytical rapid methods for pesticide residues monitoring in foods: immunoassays and biosensors. *Sci. Des. Aliments* **17,** 131–143.
17. Samsonova, J., Rubtsova, M., Kiseleva, A., Ezhov, A., and Egorov, A. (1999) Chemiluminescent multiassay of pesticides with horseradish peroxidase as a label. *Biosensors Bioelectronics* **14,** 273–281.
18. Winklmair, M., Schuetz, A., Weller, M., and Niessner, R. (1999) Immuno-chemical array for the identification of cross-reacting analytes. *Fresenius J. Anal. Chem.* **363,** 731–737.
19. Schobel, U., Barzen, C., and Gauglitz, G. (2000) Immunoanalytical techniques for pesticide monitoring based on fluorescence detection. *Fresenius J. Anal. Chem.* **366,** 646–658.
20. Vandecasteele, K., Gaus, I., Debreuck, W., and Walraevens, K. (2000) Identification and quantification of 77 pesticides in groundwater using solid phase

coupled to liquid–liquid microextraction and reversed phase liquid chromatography. *Anal. Chem.* **72,** 3093–3101.

21. Bruun, L., Koch, C., Jakobsen, M., Pedersen, B., Christiansen, M., and Aamand, J. (2001) Characterization of monoclonal antibodies raised against different structures belonging to the s-triazine group of herbicides. *Anal. Chim. Acta* **436,** 87–101.

22. Zeck, A., Weller, M., and Niessner, R. (1999) Characterization of a monoclonal TNT-antibody by measurement of the cross-reactivities of nitro aromatic compounds. *Fresenius J. Anal. Chem.* **364,** 113–120.

23. Aprea, C., Colosio, C., Mammone, T., Minoia, C., and Maroni, M. (2002) Biological monitoring of pesticide exposure: a review of analytical methods. *J. Chromatogr. B Anal. Technol. Biomed. Life Sci.* **769,** 191–219.

24. Belleville, E., Dufva, M., Aamand, J., Bruun, L., and Christensen, C. (2003) Quantitative assessment of factors affecting the sensitivity of a competitive immuno-microarray for pesticide detection. *Biotechniques* **35,** 1044–1051.

25. Belleville, E., Dufva, M., Aamand, J., Bruun, L., Clausen, L., and Christensen, C. B. (2004) Quantitative microarray pesticide analysis. *J. Immunol. Methods* **286,** 219–229.

26. Han, A., Dufva, M., Belleville, E., and Christensen, C. (2003) Detection of analyte binding to microarrays using gold nano particles labels and a desktop scanner. *Lab. Chip.* **3,** 336–339.

27. Afanassiev, V., Hanemann, V., and Wolfl, S. (2000) Preparation of DNA and protein micro arrays on glass slides coated with an agarose film. *Nucleic Acids Res.* **28,** E66.

28. Dufva, M., Petronis, S., Bjerremann Jensen, L., Krag, C., and Christensen, C. (2004) Characterization of an inexpensive, non-toxic and highly sensitive microarray substrate. *Biotechniques* **37,** 286–296.

29. Rodbard, D. (1974) Statistical quality-control and routine data-processing for radioimmunoassays and immunoradiometric assays. *Clin. Chem.* **20,** 1255–1270.

17

Antibody-Microarrays on Hybrid Polymeric Thin Film-Coated Slides for Multiple-Protein Immunoassays

Xichun Zhou and Jizhong Zhou

Summary

The development and characterization of protein microarrays fabricated on nanoengineered three-dimensional polyelectrolyte thin films (PET) deposited on glass slide by consecutive adsorption of polyelectrolytes in solutions through self-assembly process were described. Protein antibodies or antigens were immobilized in the PET-coated glass slides by electrostatic adsorption and entrapment of the porous structure of the three-dimensional polymer film and thus, establishing a platform for parallel analysis. A method for fabrication of cytokines antibody-based protein microarray for simultaneous detection of multiple cytokines on the PET-coated slides was described. Cytokines play an important role in a wide range of physiological process, such as innate immunity, apoptosis, angiogenesis, cell growth, and differentiation. Therefore, simultaneous measurement of multiple cytokine expression levels is vital to reveal the complex cytokine network and to understand the development of certain human diseases. The protein microarray was printed by robotically spotting nine human cytokine and growth factor capture antibodies onto planar glass substrates. The fluoroimmunoassay of printed cytokine antibody microarrays were performed by incubating with cytokine samples, then binding by biotin-conjugated detection antibodies, and detecting by fluorophore conjugated streptavidin. This sandwich immunoassay-based protein microarrays protocol was developed for detection of multiple expression levels simultaneously with commercial available biotin-labeled detection antibody, so that no labeling of sera samples in required. This method was also optimized specifically for the special requirements of the cytokine detection, with special attention paid to selecting the surface chemistry of array substrate, array printing buffer and blocking buffer, and the fluorescent detection settings that yielded the highest sensitivity and selectivity against the lowest background. The dynamic ranges of the parallel assay for cytokines were around two to three orders of magnitude with limit of detection <10 pg/mL. This cytokine detection protein microarray system can be extended to a larger menu of cytokines and growth factors for applications such as profiling of cytokine expression.

Key Words: Antibody–antigen interaction; antibody microarray; biochips; cytokines; polymeric thin film; protein microarrays.

From: *Methods in Molecular Biology, vol. 382: Microarrays: Second Edition: Volume 2*
Edited by: J. B. Rampal © Humana Press Inc., Totowa, NJ

1. Introduction

Microarray-based parallel detection of different biomolecules in complex biological samples has a wide range of potential applications in the diagnosis of allergies and autoimmune and infectious diseases, as well as in epitope mapping studies and the development of vaccines *(1–5)*. In recent years, protein microarrays have evolved as powerful tools to address these high throughput requirements. However, there are several additional challenges with protein-based microarrays because in general, proteins are more sensitive to their surrounding environment than nucleic acids. Ideally, proteins should be immobilized on a slide such that their native format and their folded conformations are preserved.

Any strategy to construct protein microarrays requires two steps: (1) deposition of proteins in parallel format on a substrate surface and (2) immobilization of the arrayed capture probes on the substrate surface. Covalently coupling, physical adsorption, and specific affinity interaction are the proposed methods to immobilize proteins in array format. Although covalent linkage to an activated surface is generally the most stable method of immobilization protein on microarray, covalent coupling typically involves multiple-step surface chemistry treatments to obtain the activated substrates for subsequent protein immobilization and extensive processing protocols have to be followed by after protein microarray fabrication. One of the most popular approaches is the slide surface functionalized with aldehyde groups. These aldehyde groups readily form aldimine (schiff-base) bonds with primary amines in protein probes, which can be further stabilized by reduction. However, covalent coupling often results in some of the immobilized proteins to lose activity because of the direct chemical modification of the binding site and steric hindrance or strain from multiple attachment sites.

Protein microarray can also be fabricated onto a slide surface through a specific affinity interaction, where protein probes were fused with a high-affinity tag at their amino or carboxyl terminus for the attachment to the chip surface through this tag. However, the modification of protein probes with affinity tag and the modification of slide substrate with protein A or Streptavidin require a peculiar time-consuming process, and in general, an increase in the quantity of reagents.

In addition to the chemistry used to immobilize proteins, the binding capacity of protein probes on slides surface is also critical for the performance of a protein microarray because the protein samples are often very limited in supply and (unlike nucleic acid) cannot be amplified. In the use of aldehyde and PLL functionalized slides, the amount of immobilized protein/peptide is limited to a two-dimensional surface area, causing a low sensitivity and a low signal/noise level. Polymer-based three-dimensional (3D) films, such as activated agarose film, hydrogel polymer, sol-nitrocellulose film, plasma-polymerized film, and protein–gel chip were reported very recently to improve binding capacity and thus the sensitivity. In addition to the sophisticated processes of creating such

3D matrixes, which often include photolithography or photopolymerization process, the major disadvantage of these reported 3D protein microarrays is that the 3D coatings often have lower reproducibility and a higher background signal caused by autofluorescence of the polymer materials.

Thus, there is great demand for new slide surface which provides reliable attachment of protein probes for various functional analyses. Ideally, proteins should be immobilized on a slide so that their native format and their folded conformations are preserved. At present a simple procedure was reported to coat glass slides with polymeric thin films by self-assembly of polyelectrolyte multilayered thin films as a platform for fabrication cytokine antibody microarrays. Owing to the amphiphilic nature of polyelectrolyte, protein probes are immobilized in semi-wet environment by the combination of strong electrostatic adsorption, hydrophobic adsorption, and entrapment of the porous structure, which keep protein probes in an active form. Furthermore, the multilayered polyelectrolyte thin films provide 3D structures, in which high-binding capacity can be achieved and the direct contact of protein with hydrophobic glass surface was avoided. In addition to this, polyelectrolyte thin films (PET) film is chemically stable, and their adhesion to substrates is strong, which is critical in obtaining a reproducible immunoassay performance. We also demonstrated the fabrication of antigen and antibody microarrays on the PET-coated glass slides as well as the direct and indirect immunoassays on the protein microarrays for multiple analyte detection.

Cytokines are a group of proteins mediate communication among cells in the immune system which include cytokines, chemokines, growth factors, angiogenic factors, and proteases *(6–8)*. Cytokines play an important role in a wide range of physiological process, such as innate immunity, apoptosis, angiogenesis, cell growth, and differentiation. Frequently, cytokine expression is coordinately regulated as multiple cytokines often share the same upstream signal pathway, and one cytokine can regulate the expression of other cytokines. Deregulation of cytokine expression is often associated with disease status, particularly cancer, cardiac disease, and arthritis. In cancer, cytokines have been implicated in diagnosis, treatment, and prognosis. Knowledge regarding the levels of multiple cytokines is critical to the understanding of immune processes. The interaction between cytokines and the cellular immune system is a dynamic process. The interactions of positive and negative stimuli and positive as well as negative regulatory loops are complex and often involve multiple cytokines. Therefore, simultaneous measurement of multiple cytokine expression levels is vital to reveal the complex cytokine network and to understand the development of certain human disease. Immunoassays (e.g., enzyme-linked immunosorbent assay [ELISA]) are typically used to detect the level of single cytokines in biological samples, thus the presence or absence of other cytokines is unknown. Cytokine-based protein microarrays; however, provide a simple array format, and highly sensitive approach to simultaneously detect multiple cytokine expression levels from conditioned media, patient's sera, and other sources. Scientists can rapidly and accurately identify the expression profiles of multiple cytokines by using cytokine-based protein microarray technology in several hours in a cost-effective fashion.

This PET-based array approach has several advantages over the traditional ELISA for cytokine detection. First and most important one is that cytokine-based protein microarray can simultaneously and effectively detect many cytokines, and the noncovalent adsorption of capture proteins on PET film minimizes the denaturation of the biological function of the proteins. Second, the sensitivity of most cytokine is higher, which is at the pg/mL levels. For example, as low as 10 pg/mL of TNF-α can be detected in cytokine microarray format. Thirdly, the dynamic range of detection is much greater than ELISA. For example, the dynamic detection range of TNF-γ varies from 32 to 20,000 pg/mL, whereas, it varies only within 100–1000 pg/mL in a typical ELISA. Therefore, the detection range is about 100-fold greater in protein array in a typical ELISA. In addition, the variation is lower than ELISA. As detection by fluorometry, the intraslide variation in fluorescence, determined as the coefficient of variation (CV) among the 18 spots at each concentration, ranged from 5% at 100 ng/mL to 8% at 6.4 pg/mL. In contrast, variation in ELISA is much higher (about 20%). Finally, the protein microarray system can be much easier to extend to high-density protein microarray if more capture cytokine antibodies are available.

2. Materials

2.1. Preparing Thin Polymer Film-Coated Glass Slide

1. Microscope glass slides ($76 \times 26 \times 1$ mm^3).
2. Glass clean solution: 2.5 M NaOH/ethanol solution.
3. 0.5% Poly vinylsulfonic acid, sodium salt (PVS) solution.
4. 0.5% Polyallylamine hydrochloride (PAAH) (M_n 50,000–65,000) solution.

2.2. Capture Antibody and Detection Antibody

1. Recombinant human cytokine TNF-α, IFN-γ, IL-1β, IL-2, IL-6, IL-8, IL-10, MCP-1, TGF-β, and monoclonal antibodies against the above cytokines and their biotinylated derivatives (biotinylated antibodies against TNF-α, IFN-γ, IL-1β, IL-2, IL-6, IL-8, IL-10, MCP-1, TGF-β) were obtained from R&D Systems (Minneapolis, MN).

2.3. Printing Requirements

1. Printing stock buffer: 1X phosphate buffered saline (PBS) (pH 7.4) with 10% glycol.
2. PixSys 5500 robotic printer (Cartesian Technologies, Inc., Irvine, CA).

2.4. Buffer Solutions for Immunoassays

1. PBST buffer: 1X PBS, 0.5% Tween-20.
2. Blocking buffer: 1% bovine serum albumin (BSA) (w/v) in PBST buffer.
3. Immunoassay stock buffer: 2X PBS (pH 7.4).
4. Biotinylated Antibody Cocktail: mix biotinylated antibodies against TNF-α, IFN-γ, IL-1β, IL-2, IL-6, IL-8, IL-10, MCP-1, TGF-β in 1X PBST buffer to final concentration for each antibody at 10 μg/mL.
5. Staining solution: 100 μg/mL Cy3-labeled streptavidin in PBST buffer.
6. Washing buffer I: 1X PBST (pH 7.4).
7. Washing buffer II: 0.1X PBS (pH 7.4).

2.5. Scan Instrument and Data Analysis Software

1. Scanning laser confocal fluorescence microscope (ScanArray 5000 System, Packed Biochip Technologies LLC, Billerica, MA) with 530-nm wavelength channel.
2. Image analysis: ImaGene 6.0 (Biodiscovery, Inc., Los Angeles, CA). The mean signal intensity of each spot was used for data analysis.
3. Statistical analysis was performed with SigmaPlot 5.0 (Jandel Scientific, San Rafael, CA) or with Microsoft Excel® (Microsoft Corp., Redmont, WA).

3. Method

The protein microarray-based bioassay systems have some similarities to cDNA microarrays, which generally contain two steps: (1) deposition of purified capture reagents or samples in parallel format at spatially defined locations on a substrate surface and immobilization of the arrayed capture probes on the substrate surface. The state of functionality of immobilized proteins determines the usefulness of protein arrays for the appropriate applications. Because antibodies can be considered as the active binding partner, these must retain their specific binding properties upon immobilization. Therefore, protocols for the immobilization, storage, and assays need to be optimized. (2) Assay design and signal generation: besides the direct immunoassays, sandwich immunoassay is most used in protein microarrays. Sandwich immunoassay takes advantage of the proven utility of ELISA. In the sandwich assay, proteins captured on an antibody microarray are detected by a cocktail of labeled detection antibodies. Each antibody is matched to one of the spotted antibodies. Thus sandwich immunoassays are widely used for the detection of proteins found in very low concentrations, such as cytokines, growth factors, or hormones from biological specimens. **Figure 1** illustrated the schematic procedures of protein microarray-detection system.

3.1. Coating APTS-Functionalized Glass Slide

1. Glass slides were cleaned by sonication in 2.5 *M* NaOH/ethanol solution for 10 min and then thoroughly rinsed with distilled water.

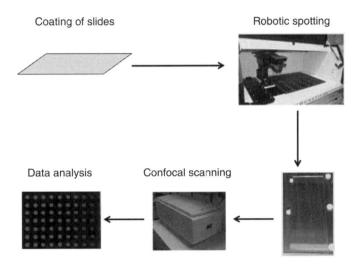

Fig. 1. Schematic procedures of protein microarray-based detection system.

2. Cleaned slides were immersed into solution of 0.5% PAAH, 1.0 *M* NaCl at pH about 6.0 for 15 min, followed by washing with distilled water, and air-drying.
3. The PAAH-coated slide was then exposed to solution of 0.5% PVS, 1 *M* NaCl at pH 8.0 for 15 min, followed by washing with distilled water, and air-drying.
4. The slide was then immersed into above PAAH solution for 15 min again. The surface was then washed again with distilled water.
5. This procedure was repeated until the desired number of polyelectrolyte pair layers $(PAAH/PVS)_6/PAAH$ were deposited on the slide with the positively charged PAAH on the outer most layer (*see* **Note 1**).
6. Stored the coated glass slides in desiccator.

3.2. Protein Microarray Fabrication

1. Suspend cytokine antibodies in printing stock solution: 10 of cytokine antibodies as well as antigoat IgG (negative control) and biotinylated antigoat IgG (positive control) were buffered in 1X phosphate buffered saline (PBS) at concentration of 0.5 µg/µL and an equal volume of printing stock solution was added. The final concentration of cytokine antibody is 0.25 µg/µL (*see* **Note 2**). Mix the samples by pipetting up and down 10 times to make sure the protein samples are mixed thoroughly before printing. Transfer 10 µL of each protein printing solution into 384-well plate and stored in 4°C (*see* **Note 3**).
2. One nanoliter of the each cytokine antibody solution from a 384-well plate were printed onto the PET-functionalized glass slide using one pin with a distance of 250 µm between the centers of adjacent spots by using a PixSys 5500 robotic printer (Irvine, CA) in 60% relative humidity followed by incubation for 2 h. Each

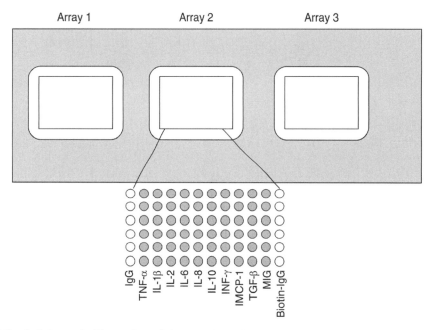

Fig. 2. Schematic illustration of the cytokine microarray arrangement on one PET slide.

cytokine antibody was printed with six replicate spots. **Figure 2** depicted the cytokine protein microarray arrangement on one slide.
3. Stored the printed slide in 4°C.

3.3. Blocking Nonspecific Adsorption

The slides were immersed into blocking buffer at room temperature for 30 min. The slides were stored in blocking buffer at 4°C unless they were used immediately. Before use for immunoassays, the slides were taken out from the blocking buffer solution and washed with washing buffer II for 1 min (*see* **Note 4**).

3.4. Generation of Standard Dose–Response Curves

1. Make 200 µL of serial dilutions of the purified cytokine antigens (from 100 ng/mL to 0.62 pg/mL at fivefold serial dilution) in the immunoassay stock buffer. Mix each concentration of cytokine antigen solution together to set up standard cytokine antigen solution.
2. Apply 20 µL of each mixed standard cytokine antigen solution on each of the three blocked cytokine antibody submicroarray surface and covered with frame seal. For each cytokine concentration, three replicate slides were tested. Total 21 slides were used to generate a seven-point standard curve (*see* **Note 5**). These slides were incubated at room temperature for 60 min.

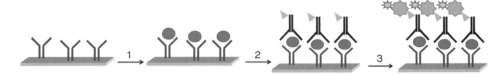

Fig. 3. Schematic representation of immunoassay procedures on cytokine protein microarrays: (**1**) samples (sera) were incubated with cytokine capture antibody microarray; (**2**) after removing unbound samples component, the chips are probed with biotin-labelled secondary antibody; and (**3**) after washing, the chips are probed with streptavidin labelled with fluorophore and specific signal are detected using fluorescent scanner.

3. After the slides were washed three times in PBST for 5 min each, a 10 μg/mL mixture of detection antibodies (biotinylated anticytokine antibodies) were applied to the slides and the slides were incubated for 1 h at room temperature. Slides were washed twice in PBST buffer for 8 min each, and then incubated with 100 μg/mL of Cy3-labeled streptavidin for 30 min. **Figure 3** shows the immunoassay procedures on cytokine protein microarrays.
4. Remove unbound Cy3-streptavidin by washing the slides three times in PBST buffer for 5 min each, followed by washing twice in 0.1X PBS for 2 min to remove detergent. The slides were dried by centrifugation at 500g for 3 min (*see* **Note 6**).
5. Scan the microarray to produce a fluorescent image. Microarrays were scanned at 10 μm resolution with the scanning laser confocal fluorescence microscope of a ScanArray 5000 System (*see* **Note 7**). The emitted fluorescent signal was detected by a photomultilier tube (PMT) at 570 nm for Cy3. For all microarray experiments, the laser power was 85% and the PMT gain was 75%. Save image as a high-resolution single image *.tif file (*see* **Note 8**). **Figure 4** is a typical scanning images of antibody microarrays exposed to 4 ng/mL of the 10 cytokines. The fluorescent signals were analyzed by quantifying the pixel density (intensity) of each spot using ImaGene 3.0. The local background signal was automatically subtracted from the hybridization signal of each separate spot and the mean signal intensity of each spot was used for data analysis. Statistical analyses were performed using SigmaPlot 5.0 (Jandal Scientific, San Ratael, CA) or by Microsoft Excel® (*see* **Note 9**).
6. The mean fluorescence intensity of the replicate spots for each cytokine concentrations were averaged and subsequently plotted to generate dose-response curves. **Figure 5A–C** show the typical standard dose response curves of IL-2, TNF-α, and MCP-1 cytokines. The data resulted in sigmoid curves having a linear range (the concentration range that gave the best fit to the linear equation $y = mx + b$) from 6.4 pg/mL to 20 ng/mL for IL-2, 32 pg/mL to 4 ng/mL for TNF-α and MCP-1, respectively. **Table 1** shows the limit of detection and the dynamic range of the cytokine protein microarray for individual cytokine (*see* **Note 10**).

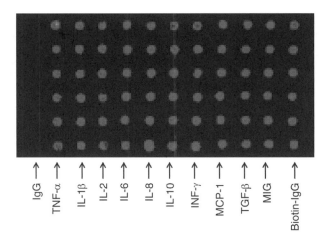

Fig. 4. Typical scanning images of antibody microarrays exposed to 4 ng/mL of the 10 cytokines.

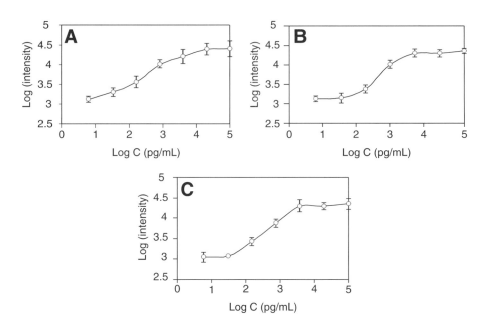

Fig. 5. Standard dose–response curve of cytokines IL-2 (**A**), INF-γ (**B**), and MCP-1 (**C**) oncytokine antibody microarrays.

Table 1
Limit of Detection and Standard Dose–Response Curve Range
for Individual Cytokines

Cytokine	Limit of quantification 1 (pg/mL)	Suggested range for standard curve 2
TNF-α	<10	32–20,000
IL-1β	<10	6–4000
IL-2	<32	10–4000
IL-6	<10	32–4000
IL-8	<10	10–4000
IL-10	<50	50–10,000
IFN-γ	<50	33–4000
MCP-1	<100	160–10,000
TGF-β	<30	10–4000
MIG	<30	32–4000

3.5. Sandwich Immunoassays for Detection of Cytokines in Samples

1. Remove the cytokine microarray slides from Blocking Solution and briefly rinse slides twice with Wash Buffer II.
2. Samples (patient's serum or supernatant of cancer cell line culture medium) were diluted in 1X PBST buffer.
3. Apply 30 μL of the sample to be analyzed on the microarray surface and sealed with frame seal film (*see* **Note 11**). Incubate the sample with microarray for 60 min at room temperature.
4. Wash slide with washing buffer I twice for 5 min each at room temperature with gentle shaking.
5. Pipet 30 μL of 1X Biotin-conjugated anticytokine antibodies cocktail onto the slide. Incubate the cocktail solution with the slides at room temperature for 30 min.
6. Wash slide with washing buffer I twice for 5 min each at room temperature with gentle shaking.
7. Add 30 μL of 1X Streptavidin-Cy3 conjugate diluted in 1X PBST buffer. Incubate at room temperature for 30 min.
8. Wash slide with washing buffer I twice for 5 min each at room temperature with gentle shaking.
9. Wash slide with washing buffer II for 5 min each at room temperature with gentle shaking. Spinning dry the slide.
10. Image the cytokine microarray slide(s) with fluorescent imagers using the settings same as that used to generate standard curve. Each cytokine antibody is arrayed in six replicates to provide better reliability. Using imaging analysis software to determine the specific signal of each spot (signal-background). Averaging the signal of the replicate spots to determine the specific signal for each cytokine on the

array. Then determine the standard deviation of this averaged value. The average specific signal for each cytokine +/– the standard deviation is used to compare cytokine expression levels between samples. The average specific signal +/– the standard deviation for the cytokine expression in one sample can be compared to the same in other samples to determine if each sample has a different level of cytokine expression.

4. Notes

1. This procedure coats glass slides with polymeric thin films by self-assembly of polyelectrolyte multilayered thin films as a platform for fabrication protein microarrays. Owing to the amphiphilicity nature of polyelectrolyte, protein probes are immobilized in semi-wet environment by the combination of strong electrostatic adsorption, hydrophobic adsorption, and entrapment of the porous structure, which keep protein probes in an active form. Furthermore, the multi-layered polyelectrolyte thin films provide 3D structures where high-binding capacity can be achieved and the direct contact of protein with hydrophobic glass surface was avoided. In addition to this, PET film is chemically stable, and their adhesion to substrates is strong, which is critical in obtaining a reproducible immunoassay performance. Compared with some of commercial slides such as epoxy-functionalized or aldehyde functionalized glass slides, the PET slides can maximally maintain the activity of protein probes, and provide a wider dynamic range for immunoassay.

2. Generally, background-subtracted fluorescent intensities for all the capture anti-body spots (features) increased with increasing capture antibody concentration, however, higher concentration of capture antibody than 0.5 µg/µL will cause the cross reactivity in multiplex assay. The concentration of capture antibodies between 0.25 and 0.5 µg/µL was recommended to build the antibody arrays, which do not encounter nonspecific cross reactivity in a multiplex assay and should detect low-concentration samples, such as 32 pg/mL TNF-α.

3. Spotting buffer composition can influence the spot morphology, protein-binding capacity of a surface, the stability of proteins and the quality of the spots produced. Ten percent of glycerol was recommended to be used as additives to decrease the volatility of printing solution. However, higher glycol concentration in printing solution would produce smeared printing spots and reduce the binding of printed protein on PET slide.

4. Effective blocking of the microarray to protect nonspecific adsorption of target samples and cross-talk is important for microarray application. It is found that the use of classic blocking reagents of 1% BSA in PBS for 40 min can strongly block nonspecific adsorption. BSA concentration higher than 3% can increase the fluo-rescent background. To decrease the unspecific background by chemical modifi-cation of the surface coatings with 10 mM mercaptoethanol or 10 mM cysteamine is not recommended because it will reduce the activity of the immobilized capture protein. Preincubate the antibodies with the blocking reagents or add blocking reagents in washing solution might avoid antibodies contain reactivities to some

components in blocking reagents. Blocking the cytokine microarrays overnight if the blocking is not complete.

5. Each cytokine antibody is arrayed in six-repeated, so there should always be a group of six consecutive spots lighting up to a similar level. The typical design is to generate 48 replicate spots for each cytokine concentration to enable reliable quantitative data.

6. Several arrays on each slide can be reserved to run a standard curve to quantify level of cytokine antigen. A standard curve must be run for each antigen to be quantified. Purified antigens can be combined to generate multiple standard curves, simultaneously, within the same set of arrays.

7. The mean signal intensity of each spot was used for data analysis. The fluorescent protein microarray can be scanned or imaging using any of a number of high quality commercial detection instruments from Perkin Elmer (Wellesley, MA), Bio-Rad (Hercules, CA), Axon, API (Sunnyvale, CA), and many others. Instrument settings can be adjusted to optimize the imagine process.

8. Take care not to saturate the scanner-detection system by properly adjusting parameters such as the laser power, PMT power, and/or exposure settings to avoid saturation while maintaining sensitivity. The signals from the 800 pg/mL should not be saturated (>60,000 counts on many scanners), while the signals from the 32 pg/mL standard should be apparent in the scanning image.

9. Protein microarray data from the fluorescent image can be quantified, mined, and modeled using many different commercial software packages. Including those made by BioDiscovery (Marina del Ray, CA), and many others make excellent products.

10. Limit of detection is the cytokine concentration that generated fluorescent signal larger than the background plus three times of standard derivation. The dynamic range is the cytokine concentration range that gave the best fit to the linear equation $y = mx + b$.

11. Samples can be run neat or they can be diluted in 1X wash buffer II. Serum used for cell culture might contain proteins that interact nonspecifically with elements on the array. If the sample is in undiluted cell culture medium with serum, reserve one array to run the media with serum alone, thereby, controlling for possible cross-reactivity of the serum with the arrayed antibodies.

References

1. Zhu, H., Klemic, J. F., Chang, S., et al. (2000) Analysis of yeast protein kinases using protein chips. *Nat. Genet.* **26,** 283–289.

2. MacBeath, G. and Schreiber, S. L. (2000) Printing proteins as microarrays for high-throughput function determination. *Science* **289,** 1760–1763.

3. Liu, X. S., Brutlag, D. L., and Liu, J. S. (2002) An algorithm for finding protein-DNA binding sites with applications to chromatin-immunoprecipitation microarray experiments. *Nat. Biotechnol.* **20,** 835–839.

4. Madoz-Gurpide, J., Wang, H., Misek, D. E., Brichory, F., and Hanash, S. M. (2001) Protein based microarrays: a tool for probing the proteome of cancer cells and tissues. *Proteomics* **1,** 1279–1287.

5. MacBeath, G. (2002) Protein microarrays and proteomics. *Nat. Genetics* **32,** 526–532.

6. Lin, Y., Huang, R. C., Chen, L. P., et al. (2003) Profiling of cytokine expression by biotin-labeled-based protein-arrays. *Proteomics* **3,** 1750–1757.

7. Huang, R. P., Huang, R. C., Fan, Y., and Lin, Y. (2001) Simultaneous detection of multiple cytokines from conditioned media and patient's sera by an antibody-based protein array system. *Anal. Biochem.* **294,** 55–62.

8. Li, Y. W. and Reichert, W. M. (2003) Adapting cDNA microarray format to cytokine detection protein arrays. *Langmuir* **19,** 1557–1566.

18

Overprint Immunoassay Using Protein A Microarrays

**Robert S. Matson, Raymond C. Milton, Jang B. Rampal,
Tom S. Chan, and Michael C. Cress**

Summary

The ability to perform microarray-based immunoassays without the need for wells or other fluid barriers were demonstrated. Both contact and noncontact microarray printing technology is used to prepare spotted arrays of analyte binding sites, as well as, to deliver samples, secondary antibodies and other signal development reagents directly to these sites in a parallel fashion are called as *overprint immunoassays*. A micro-ELISA is demonstrated based upon the use of Protein A as a universal microarray. All components of the assay (capture antibody, antigen, and signal development reagents) were site-specifically dispensed in parallel fashion to the surface in nano-liter volumes. This represents a 1000-fold reduction in reagent consumption from that used in a conventional 96-well microtiter plate assay. Overprinting nanoliter volumes directly onto 200–300 µm spots yields similar levels of sensitivity achieved with the bulk dispensing of micro-liter volumes into wells.

Key Words: Immunoassay; micro array; micro-ELISA; overprint; Protein A; automated.

1. Introduction

The development of microarray technology for massive parallel processing of biological information has enabled the evolution of automated microassays. Miniaturized platforms permit the use of smaller sample sizes and reagent volumes that often lead to an economy of scale and time savings. In addition, the array-based analyzers are able to achieve comparable or in some instances greater sensitivity than that of conventional macroassay formats. Thus, the emergence of "Biochips" and microarrays for DNA analysis (*1–4*) and to a more limited extent the development of immunoglobulin (Ig)-based "Microspot" (*5*) or related protein microarrays for immuno-diagnostic and screening applications were witness (*6–9*). The essential feature of such technology is the ability to attach a variety of affinity-based capture agents (ligands or probes) in high

From: *Methods in Molecular Biology, vol. 382: Microarrays: Second Edition: Volume 2*
Edited by: J. B. Rampal © Humana Press Inc., Totowa, NJ

density (i.e., thousands of spots per square centimeter) to the surface in an array format at known locations. Sample (analyte or target) is bulk delivered to the array in order to flood the entire surface so that the sample is equally distributed to all capture sites. Subsequent chip processing steps such as rinse cycles or the addition of signal development reagents are also delivered in bulk. Although these technologies allow for highly multiplexed interrogation of the molecular character of a few samples, the analysis of multiple samples with high throughput remains an issue.

In the field of drug discovery most assays have been based upon the microtiter plate format in which a single capture agent is attached per well. Higher throughput processes that consume considerably less sample and reagent per assay may be achieved by using higher density lower volume well plates. However, the migration of assays from 96-well microtiter plates to higher density microplates such as the 1536-well plate has been constrained by lack of availability of adequate aspiration and dispense fluidics. Other problems with small volume delivery include evaporation and the inability to properly mix and rinse in submicroliter volumes. Much of this could be solved using dedicated microfluidic devices. However, these are expensive to manufacture and use in the vast numbers required.

Another approach is to work without wells all together. This is possible provided that individual assay sites can remain localized and separately addressed with reagents. For example, the ability to directly spot labeled antibodies (in glycerol) over antigen spots that had been previously arrayed onto a poly-L-lysine slide has been demonstrated *(10)*. They successfully performed a microarray-based multiplexed immunoassay in this manner. Later work based upon the two step multiple spotting technique involved performance of multiple enzymatic assays on microarray slides *(11)*. In a related approach for high-throughput screening applications, glycerol droplets containing various compounds on slides were printed to form a microarray *(12)*. In one example, a compound library of potential protease inhibitors formed the droplet array. The slide was subsequently treated with aerosols of the protease enzyme and its fluorogenic substrate. Only the glycerol droplets containing inhibitor compounds resulted in inhibition of the enzyme–substrate reaction.

Since 1991, the laboratory has been involved in the development of microarray technologies. The polymeric material such as polypropylene is used as microarray substrates in the work. Because the surface is treated with radiofrequency plasma discharge *(13)*, it remains substantially hydrophobic allowing for droplets of reagents to be dispensed to the surface without appreciable wetting or spreading. Microarray printing technology was used to prepare assays on individual test sites and deliver analytes and reagents to these sites in a parallel manner to perform multiplexed assays and this is known as *Overprint assays*. The particular process to be described provides a means to create a universal array based upon the immobilization of Protein A to acyl fluoride activated plastic substrates. Protein A preferentially and reversibly binds to the Fc region of immunoglobulins *(14)*. Thus, antiligand antibodies or Fc-ligand conjugates can be prepared, which will bind to the Protein A array to create custom ligand assays. An *overprint immunoassay* in a micro-ELISA format is demonstrated based upon the use of the Protein A microarray.

2. Materials

Protein A was obtained from Zymed Laboratories, South San Francisco, CA (cat. no. 10-1001) or Sigma-Aldrich, St. Louis, MO (cat. no. P-6031). The following proteins were obtained from commercial sources: rabbit IgG antigoat, FITC conjugate (Zymed, cat. no. 61-6111); goat IgG anti-Human IgG, biotin conjugate (Sigma Aldrich, cat. no. B1140); streptavidin-alkaline phosphatase, Avidx-AP (Applied Biosystems, Foster City, CA; cat. no. APA10); biotinylated-alkaline phosphatase (Pierce Chemical, Rockford, IL; cat. no. 29339); human IgG (Pierce Chemical, cat. no. 31877); goat IgG, antibiotin (Sigma Aldrich, cat. no. F-6762). Acyl fluoride-activated plastics (microarray substrates) were prepared from the reaction of (diethylamino)sulfur trifluoride (DAST) *(15)* with carboxyl or amine truncated thermoplastics: ethylene methacrylic acid copolymer (EMA) or plasma aminated polypropylene as described by Matson et al. *(13)*. Protein A microarrays were created either by noncontact dispensing using a BioDot 3200 Dispenser (Genomic Solutions, Inc., Ann Arbor, MI) or by contact printing using the Biomek® 2000 workstation equipped with a 384 pin high density replicating tool (HDRT). Enzyme labeled fluorescence (ELF) Reagent (ELF-97 Endogenous phosphatase detection Kit; Molecular Probes, Inc., Eugene, OR), a fluorescent precipitating substrate for alkaline phosphatase was used for signal development. Digital images were obtained using a CCD camera system (Teleris 2, SpectraSource, Inc., Westlake Village, CA). Excitation light at 350 nm was generated using a Ultraviolet mineral light with signal emission collected at 520 nm using a 10 nm band pass lens filter. The 16-bit images were analyzed using ImaGene software (BioDiscovery, Inc., EI Segundo, CA) then exported as 8-bit values into an Excel spreadsheet (Microsoft, Redmond, WA) for calculation and graphic display.

2.1. Preparation of Activated Plastic Supports

DAST was obtained from SynChem, Inc. (Aurora, OH) and used without purification. DAST reagent consisted of DAST diluted with dichloromethane to 5% v/v. EMA was obtained from Dupont, molded into 1 in. blocks and converted to the acyl fluoride activated form directly using DAST *(12)*. Polypropylene sheet, Contour 29 (Goex Corp., Janesville, WI), 20 mil thickness, was surface aminated using a radiofrequency plasma amination process *(4)*. The aminated polypropylene sheet was subsequently converted to the carboxyl form using succinic anhydride. The carboxylated polypropylene was in turn modified to acyl fluoride using the DAST reagent.

2.2. Preparation of Reagents

Concentrations of Proteins Used in Overprinting

Protein	Molecular wt. (Daltons)	Concentration (mg/mL)	Dilution factor	Printing conc. (μg/mL [M])	Moles/nL
Protein A (stage 1)	42,000	1.0	1	1000 (2.4E-05)	2.4E-14 24 fmoles
Rb anti-Gt IgG (stage 2)	150,000	0.75	100	7.5 (5.0E-11)	5.0E-20 50 fzmoles
Gt anti-Hu IgG biotin conjugate (stage 3)	150,000	0.2	100	2.0 (1.3E-11)	1.3E-20 13 fzmoles
SA-AP (stage 4)	220,000	0.25	100	2.5 (1.1E-11)	1.1E-19 110 fzmoles

2.3. Reagents Used at Each Stage of Overprint Assay

2.3.1. Stage 1: Protein A Microarray

1. Protein A: 1 mg/mL, is prepared in 1 M LiCl, with pH 10.0 for noncontact dispense printing or 1 M carbonate buffer, with pH 9.0 for use with the Biomek HDRT contact printing method.
2. Casein blocking buffer: 1–2 mg/mL was prepared in 50 mM carbonate–bicarbonate buffer, 0.15 M NaCl, pH 8.5–9.0.

2.3.2. Stage 2: Capture Antibody Overprint

1. High salt (HS) buffer: 1.0 M Tris-HCl, 1.5 M NaCl, pH 7.5.
2. TNT wash buffer: 1:10 dilution of HS buffer containing Tween-20 = 0.1 M Tris-HCl, 0.15 M NaCl, pH 7.5, 0.05% Tween-20.
3. Rabbit anti-Goat IgG: serial dilutions in HS buffer for printing at various concentrations.

2.3.3. Stage 3: Antigen Overprint

1. HS buffer.
2. TNT wash buffer.
3. Goat anti-human IgG, biotin conjugate: serial dilutions in HS buffer.

2.3.4. Stage 4/5: Signaling Reagent Overprints

1. TNT wash buffer.
2. ELF reagents.
3. Streptavidin-alkaline phosphatase conjugate: prepare at 1:100 v/v dilution in TNT buffer.

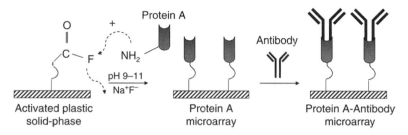

Fig. 1. Coupling of protein A to the acyl fluoride substrate and subsequent immobilization of antibody.

2.4. Other Materials Required

1. Microarrayer or Biomek 2000 workstation equipped with HDRT.
2. Source plates (384 well, polypropylene).
3. Convection oven or incubator.
4. Humidity chamber.

3. Methods
3.1. Covalent Coupling of Protein A to Plastic Supports

Protein A is coupled to acyl fluoride activated supports in a basic pH buffer medium (**Fig. 1**). Specific coupling conditions vary depending upon the method of printing. These are described next.

3.1.1. Contact Printing Using the HDRT

Protein A previously reconstituted in deionized water (DI) at 2.5 mg/mL was further diluted into sodium carbonate–bicarbonate buffer, 1 *M*, pH 9.0 at 0.5–1 mg/mL. The solution was distributed into a 384-well microplate for dispensing. A sheet of acyl fluoride-activated polypropylene (20 mil) was attached to the lid of a microtiter plate cover with double-sided sticky tape and placed in a Biomek plate holder. Protein A was dispensed to the surface of acyl fluoride polypropylene in a 3 × 3 subarray pattern created using standard Bioworks software. Up to 384 subarrays were thus created on the surface within the 9 × 12 cm^2 area. Each pin delivered 2–3 nL with a total of five dispensings to each site (10–15 nL of Protein A solution). The array remained attached to the microplate lid throughout the assay in order to maintain proper indexing on the Biomek worksurface. The arrays were placed in a convection oven and dried for 15 min at 35°C (*see* **Notes 1** and **2**). The Protein A microarray was then blocked in casein (1 mg/mL casein in 50 m*M* carbonate buffer, 0.15 *M* NaCl, pH 8.5) for 1 min at room temperature to reduce nonspecific adsorption. A final rinse in carbonate buffer followed.

3.1.2. Noncontact Printing of Protein A

The process for jet printing involved dissolving the Protein A at approx 1 mg/mL in 1 M LiCl, pH 10.0. In this case, a LiCl solution was used as a carrier in order to maintain droplets on the EMA surface, which was more hydrophilic than the polypropylene substrate (*see* **Note 3**). The solution was filtered though a 0.45-μm Z-SpinPlus centrifugal filter to remove protein aggregates. Approximately 16 nL droplets were dispensed onto the molded acyl fluoride-activated ethylene methacrylic acid substrates (1×1 cm^2 area). The Cartesian 3200 BioDot Dispenser was used to place droplets of Protein A solution on the surface in a 9×9 array pattern at approx 300 μ center to center spacing. After printing the microarrays were removed from the dispenser platform and transferred into a humidity chamber for 1 min incubation at 25°C. They were then placed in a desiccator. After overnight drying at room temperature residual reactive groups were blocked by soaking the microarrays in a casein solution (1–2 mg/mL casein in 50 mM carbonate–bicarbonate buffer, 0.15 M NaCl, pH 8.5) for 1 min. Following a brief rinse in DI water, the microarrays were air dried then stored at room temperature.

3.1.3. Overprint Immunoassay

The general process of overprinting is described in **Fig. 2**. Essentially, following the preparation of the Protein A microarrays (stage 1) a series of capture antibodies are first delivered to individual sites (stage 2). Following a rinse to remove unbound capture antibody; antigens are delivered to the array and processed in the same manner (stage 3). In the final step (stage 4) the signal developing reagents (secondary antibody conjugates; other reporters such as ELF reagent) are deposited at individual sites of the array. ELF reagent can also be overprinted in a final step (stage 5). This completes the overprint process. The microarray is then removed from the print stage and signal read using a CCD camera system.

3.1.4. Overprint Immunoassay: Step by Step

3.1.4.1. STAGE 1

1. Prepare Protein A and deliver to a source plate for printing.
2. Position microarray substrates on the printer deck.
3. Print Protein A onto microarray substrates.
4. Remove arrays from deck and dry printed arrays for 15 min at 35°C (*see* **Note 1**).
5. Soak arrays for 1 min in casein quench/blocking buffer at room temperature.
6. Rinse arrays in DI (18 mΩ) water.
7. Air-dry then store in under refrigeration overnight (*see* **Note 2**).

3.1.4.2. STAGE 2

1. Prepare capture antibody in a source plate for printing.
2. Remove Protein A array substrates from refrigerator and equilibrate to room temperature.

Stage 1 - print Protein A   Stage 2 - overprint capture antibodies   Stage 3 - overprint antigens

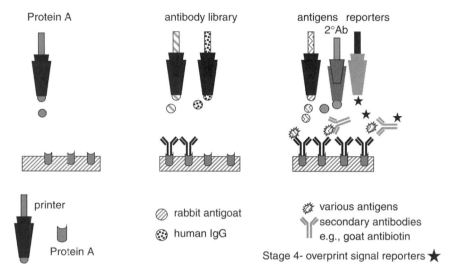

Fig. 2. The overprint immunoassay process. Using a noncontact liquid dispenser a Protein A microarray is converted into an antibody microarray. Samples are then separately dispensed over the Protein A-antibody spots to allow on-spot capture of multiple antigens. Likewise, secondary reporter antibodies and other signaling agents can be directed for dispense to specific spots of the microarray. The microarray is then developed and read using a CCD camera system.

3. Place arrays on the print deck and print capture antibodies directly over Protein A spots on the array.
4. Remove arrays from deck and place in a humidified chamber for 1 min at 25°C.
5. Rinse arrays three times in TNT buffer followed by a DI water rinse.
6. Return arrays to print deck.

3.1.4.3. STAGE 3

1. Prepare sample (antigen) in a source plate for printing.
2. Print sample directly overspots on the array.
3. Incubate 1 min at ambient temperature.
4. Repeat **steps 4–6** (*see* **Subheading 3.1.4.2.**).

3.1.4.4. STAGE 4

1. Prepare secondary reporter antibody conjugate in a source plate for printing.
2. Print reporter antibody directly overspots on the array.
3. Rinse overspots on the array in TNT buffer (three times).
4. Perform a bulk dispense of signal development reagents onto the array and incubate (*see* **Note 4**).
5. Rinse arrays three times in DI water.

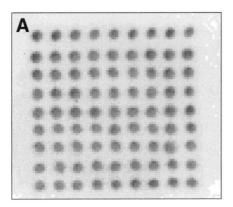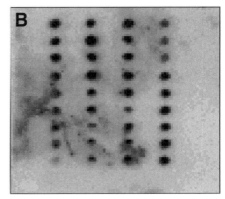

Fig. 3. Protein A microarray with overprinting of antibodies by noncontact printing The array was created by noncontact dispensing using a Cartesian 3200 BioDot Dispenser: **(A)** 9 × 9 Protein A microarray; **(B)** IgG overprints onto selected spots of the same Protein A microarray with four columns of rabbit antigoat and five columns with human IgG.

6. Carefully remove residual water from array substrates and read.
7. Regarding HDRT cleaning between steps (*see* **Note 5**).

3.2. Demonstration Experiments

3.2.1. Noncontact Printing

In the first experiment to be described (**Fig. 3**). Protein A microarrays created on an EMA molded part were repositioned on a pegboard mounted onto the work surface of the BioDot dispenser. Different antibody solutions were dispensed over the elements of the array. The 9 × 9 Protein A microarray was overprinted with alternate column dispensing of either a rabbit antigoat IgG or human IgG. Thus four columns of rabbit immunoglobulin and five columns of human immunoglobulin were generated. The microarrays were then removed and placed in a humidified chamber for 1 min at 25°C to allow complete binding of the antibodies to the Protein A sites. The molded parts were then dipped into wash solution to remove unbound antibody and subsequently returned to the BioDot stage. Goat IgG antigen (antibiotin) was then dispensed to all columns of the array and incubated in the same manner. Following a brief rinse the entire array was incubated with biotinylated-alkaline phosphatase for 30 min, rinsed and the signal developed using the ELF reagent for an additional 30 min at room temperature.

3.2.2. Fully Automated Overprint Immunoassay Based on HDRT Contact Printing

The previous experiment demonstrated the ability to overprint reagents in a semiautomated format. However, the BioDot dispensing system has limited utility for liquid handling. For this reason, the Biomek 2000 workstation was employed to deliver both site-specific and bulk reagents to the array. In this case, the arrays remained on the work surface throughout the process. The 384-HDRT was used to deliver small volumes of reagents to specific sites on the array, while a P1000 pipet tool was used to dispense bulk rinse reagents. A Biomek Gripper Tool was used to blot away excess reagents from the array sheets and to cover the plates during incubation. Analyte (antigen) and reporter antibody were delivered to individual spots on the array using the HDRT, incubated and then bulk rinsed using the P1000 pipet tool. In each case, the optimal volume of reagent delivery was determined and the number of repeat dispenses to each site varied as required. In most instances at least five to seven repeats were required. At the end of each stage in the process the array was blotted dry using filter paper attached to the inside of a microplate lid. The lid blotter was picked up by the Gripper and placed over the array plate for blotting. This was repeated for each rinse cycle using a fresh blotter to avoid carryover of reagents. Following the overprint of reporter antibody, streptavidin-alkaline phosphatase conjugate was printed. In the final stage, the developing reagent, ELF-97 was applied. The signal was captured off-line using a CCD camera system.

3.2.3. Results From the Overprint Immunoassays

3.2.3.1. NONCONTACT PRINTING

Using the noncontact or dispense mode printing (*see* **Subheading 3.2.1.**) the ability to site-specifically place capture antibodies onto Protein A spots followed by the on-spot loading of antigen and secondary antibody were demonstrated. Droplets remained associated with the Protein A spots and failed to wet out surrounding regions of the hydrophobic polymer substrate. As a result, non-specific association with spots in adjacent columns or between spots was found to negligible (**Fig. 3**). Finally, the overprinting and rinsing processes appeared to be gentle enough such that spot integrity was maintained. This resulted in a relatively uniform signal development.

3.2.3.2. CONTACT PRINTING

For performing an automated *overprint Immunoassay* (*see* **Subheading 3.2.2.**), the HDRT was used to print 3×3 subarrays (nine replicates) in a 5×9 array of Protein A (stage 1). Next, rabbit antigoat was overprinted (stage 2) in duplicate at various dilutions from 1:50 (~150 pg/spot) to 1:1000 (~7.5 pg/spot) onto the Protein A (3×3) subarrays. The top row of Protein A subarrays was not overprinted with capture antibody in order to measure the level of nonspecific binding of antigen (NSB) at each dilution. Following on-line rinsing, the antigen (biotin-goat antihuman antibody, 200 ng/mL) was overprinted (stage 3) onto each subarray at 1:10 (~200 pg/spot) to 1:1000 (~2 pg/spot) v/v dilutions.

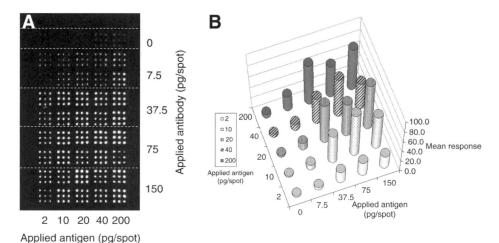

Fig. 4. Overprint immunoassay based upon contact printing. A Biomek 2000 HDRT system was used as a contact pin printer to produce the antibody array in a single step. Capture antibody (rabbit antigoat IgG) was deposited onto the spots of Protein A at several dilutions. Maximum loading per spot was estimated based upon the delivery volume. Antigen (goat antibody conjugated with biotin) was likewise serially diluted and printed onto the Protein A-capture antibody spots. Finally, a biotinylated alkaline phosphatase conjugate was overprinted at a fixed concentration and the immunoassay developed using the ELF reagent. (**A**) Captured CCD camera image of the developed overprint immunoassay; (**B**) Mean responses (relative fluorescent units) obtained from the image for replicate 3×3 subarrays.

Evidence of NSB at the higher antigen loading is evident (**Fig. 4**). After 1 min incubation the microarray was rinsed and blotted dry as described previously. Next, streptavidin-alkaline phosphatase conjugate was overprinted and each site developed using ELF reagent (stage 4). A lower level of detection (LLD; *see* **Note 6**) was determined such that an antigen sensitivity of 2 pg/spot was achieved (**Fig. 5**). This was confirmed from additional experiments, where antigen was serially diluted from 1:800 to 1:6400 v/v in order to achieve antigen samples in the 31 to 250 ng/mL range. Thus, the applied antigen from such solutions would correspond to the delivery of subpicograms of antigen per capture antibody spot. Likewise, capture antibody was serially diluted to achieve from 4.7 to 18.8 pg/spot. The results shown in **Fig. 6**, which indicate that between 1.25 and 2.5 pg of applied antigen can be detected above background. Furthermore, this is most favorably accomplished at lower capture antibody densities (**Fig. 7**). This is in keeping with earlier work where it was demonstrated that optimal antigen–antibody interaction is best achieved under conditions of reduced steric hindrance (*16*).

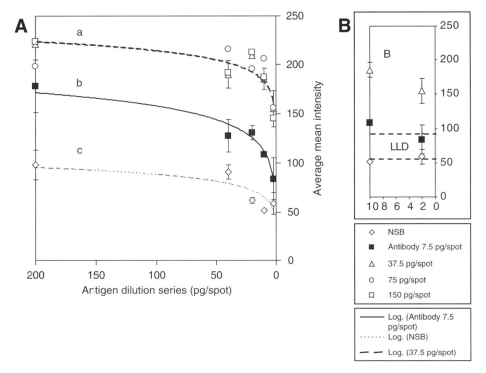

Fig. 5. Antigen binding at various capture antibody and antigen loadings. Data from **Fig. 4 (A)** antibody loadings at (a) 37.5 to 150 pg /spot; (b) 7.5 pg/spot; (c) NSB, nonspecific binding of antigen. **(B)** Expanded scale for estimation of, lower limit of detection, with extrapolation to zero antigen. Error bars represent the standard deviation from 12 replicates for each loading.

It should be noted that some of the background (NSB) might have been contributed to by direct (Fc-specific) interaction of Protein A with goat antibody. However, this is minimal as expected because it is well known that goat IgG binds poorly to Protein A *(14)*. For example, in competitive binding experiments, goat IgG had a relative binding affinity <0.0006 when compared with rabbit IgG *(17)*. A more likely source of NSB is streptavidin-alkaline phosphatase conjugate that is known to adsorb to plastic surfaces such as polypropylene (*see* **Note 7**).

3.2.4. Discussion Regarding Performance of the Automated Overprint Immunoassay

Protein A is rapidly coupled to acyl fluoride-activated plastics allowing for the creation of microarrays using either contact or noncontact printing methods. We have demonstrated the ability to use overprinting as a means to conduct

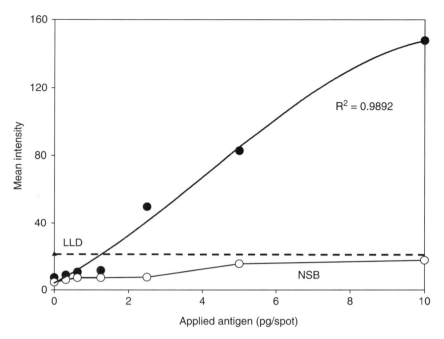

Fig. 6. Conformation of the lower level of detection for antigen at lower capture antibody loading. Results obtained from additional overprint immunoassays from contact printing using the Biomek 2000 HDRT. The lower level of detection was estimated as described in **Fig. 5**.

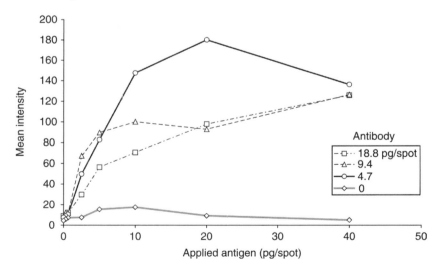

Fig. 7. Comparison of antigen loadings onto Protein A microarray spots at several capture antibody loadings. The antigen and antibody loadings for the experiment in **Fig. 6** are expanded. Optimal antigen binding occurs at the lowest antibody loading of 4.7 pg/spot.

microassays. In this case, a micro-ELISA served as the example in which all components of the assay (capture antibody, antigen, and signal development reagents) were site-specifically dispensed in parallel fashion to the surface in nanoliter volumes using a Biomek 2000 workstation. This represents a 1000-fold reduction in reagent consumption from that used in a conventional 96-well microtiter plate assay. In the *overprint immunoassay* described a LLD of approx 2 pg (8×10^6 molecules per spot) was achieved at between 4.7 and 37.5 pg (1.9×10^7 to 1.5×10^8 molecules) of capture antibody per spot. This is similar to that reported by earlier *(6,7)*. It was demonstrated the "harvesting" of 3×10^8 molecules of human IgG_3 by a 200-µm spot of monoclonal antibody *(6)*. Microarray-based antigen ELISA was developed with an assay sensitivity of 340 pg/well of antirabbit IgG monoclonal antibody *(7)*. Based upon 28 replicate spots reported to be present in each well, a single 275-µm spot could harvest approx 12 pg or 4.8×10^7 molecules. Thus, in our study, overprinting nanoliter volumes directly onto 200- to 300-µm spots yields similar levels of sensitivity achieved with the bulk dispensing of microliter volumes in wells containing microarrays.

In the future, with advances in precision printing and environmental control overprint assays were predicted to provide ultra-low volume sampling with rapid and massively parallel processing of microarrays.

4. Notes

1. Alternatively, leave arrays on deck and air-dry for 1 min at room temperature. This is an important step. The Protein A spots needed to be dry otherwise during the casein blocking step comet tailing of the spots may occur.
2. For longer periods of storage it is recommended that the air-dried arrays are placed in a bag, purged with nitrogen or argon, sealed and stored under refrigeration.
3. LiCl as a printing solution provides a number of important features. Droplet formation from the dispenser is very consistent and this leads to good spot uniformity and morphology. Because the reagent is very hygroscopic printing is best accomplished at low relative humidity ($\leq 40\%$). In addition, buffering capacity is maintained, while evaporation from source plates is low with this reagent.
4. Alternatively, signaling reagents may be overprinted as a stage 5 step.
5. Clean the HDRT pins between print runs. This is best accomplished by setting up a series of wash stations on the Biomek worksurface. The lid of a tip box can be used for this purpose. Sequentially move the HDRT between DI water, followed by 1% SDS, DI water, and finally 100% ethanol. Soak the HDRT for 5 min at each station and then allow to air-dry after the ethanol soak.
6. Based upon a noncompetitive immunoassay format, $LLD = B_0 + 3 B_0$ (SD); where B_0 is the mean background signal and corresponding standard deviations, SD. B_0 was obtained by extrapolation to zero antigen concentration.
7. Unpublished observations, R. Matson (Beckman Coulter, Fullerton, MA).

References

1. Southern, E. M., Maskos, U., and Elder, J. K. (1992) Analyzing and comparing nucleic acid sequences by hybridization to arrays of oligonucleotides: evaluation using experimental models. *Genomics* **13,** 1008–1017.
2. Pease, A. C., Solas, D., Sullivan, E. J., Cronin, M. T., Holmes, C. P., and Fodor, S. P. A. (1994) Light-generated oligonucleotide arrays for rapid DNA sequence analysis. *Proc. Natl. Acad. Sci. USA* **91,** 5022–5026.
3. Schena, M., Shalon, D., Davis, R. W., and Brown, P. O. (1995) Quantitative monitoring of gene expression pattern with a complementary DNA microarray. *Science* **270,** 467–470.
4. Matson, R. S., Rampal, J., Pentoney, S. L., Anderson, P. D., and Coassin, P. (1995) Biopolymer synthesis on polypropylene supports: oligonucleotide arrays. *Anal. Biochem.* **224,** 110–116.
5. Ekins, R. and Chu, F. (1997) Immunoassay and other ligand assays: present status and future trends. *J. Intl. Fed. Clin. Chem.* **9,** 100–109.
6. Silzel, J. W., Cercek, B., Dodson, C., Tsay, T., and Obremski, R. J. (1998) Mass-sensing, multianalyte microarray immunoassay with imaging detection. *J. Clin. Res.* **44,** 2036–2043.
7. Mendoza, L. G., McQuary, P., Mongan, A., Gangadharan, A., Brignac, S., and Eggers, M. (1999) High-throughput microarray-based enzyme-linked immunosorbent assay (ELISA). *BioTechniques* **27,** 778–788.
8. Joos, T. O., Scminenk, M., Hopfl, P., et al. (2000) A microarray enzyme-linked immunosorbent assay for autoimmune diagnostics. *Electrophoresis* **21,** 2641–2650.
9. MacBeath, G. and Scmineiber, S. L. (2000) Printing proteins as microarrays for high-throughput function determination. *Science* **289,** 1760–1763.
10. Angenendt, P., Glökler, J., Konthur, Z., Leminach, H., and Cahill, D. J. (2003) 3D protein microarrays: performing multiplex immunoassays on a single chip. *Anal. Chem.* **75,** 4368–4372.
11. Angenendt, P., Leminach, H., Kreutzberger, J., and Glökler, J. (2005) Subnanoliter enzymatic assays on microarrays. *Proteomics* **5,** 420–425.
12. Gosalia, D. N. and Diamond, S. L. (2003) Printing chemical libraries on microarrays for fluid phase nanoliter reactions. *Proc. Nat. Acad. Sci. USA* **100,** 8721–8726.
13. Matson, R. S., Rampal, J. B., and Coassin, P. J. (1994) Biopolymer synthesis on polypropylene supports, I. Oligonucleotides. *Anal. Biochem.* **217,** 306–310.
14. Langone, J. J. (1982) Applications of immobilized Protein A in immunochemical techniques. *J. Immunolog. Methods* **55,** 277–296.
15. Milton, R. C. deL. (2000) Polymeric reagents for immobilizing biopolymers. *US Patent* No. 6,110,669.
16. Matson, R. S. and Little, M. C. (1988) Strategy for the immobilization of monoclonal antibodies on solid-phase supports. *J. Chromatogr.* **458,** 67–77.
17. Lindmark, R., Thoren-Tolling, K., and Sjoquist, J. (1983) Binding of immunoglobulins to Protein A and immunoglobulin levels in mammalian sera. *J. Immunol. Methods* **62,** 1–13.

19

µParaflo™ Biochip for Nucleic Acid and Protein Analysis

Qi Zhu, Ailing Hong, Nijing Sheng, Xiaolin Zhang, Anna Matejko, Kyu-Yeon Jun, Onnop Srivannavit, Erdogan Gulari, Xiaolian Gao, and Xiaochuan Zhou

Summary

We describe in this chapter the use of oligonucleotide or peptide microarrays (arrays) based on microfluidic chips. Specifically, three major applications are presented: (1) microRNA/small RNA detection using a microRNA detection chip, (2) protein binding and function analysis using epitope, kinase substrate, or phosphopeptide chips, and (3) protein-binding analysis using oligonucleotide chips. These diverse categories of customizable arrays are based on the same biochip platform featuring a significant amount of flexibility in the sequence design to suit a wide range of research needs. The protocols of the array applications play a critical role in obtaining high quality and reliable results. Given the comprehensive and complex nature of the array experiments, the details presented in this chapter is intended merely as a useful information source of reference or a starting point for many researchers who are interested in genome- or proteome-scale studies of proteins and nucleic acids and their interactions.

Key Words: Aptamer microarray; digital photolithography; epitope-antibody profiling; epitope screening; kinase profiling assay; microfluidic biochip; micro-RNA detection; oligoncleotide microarray; peptide microarray; microfluidics; parallel synthesis; phosphopeptide-protein binding; photogenerated acid; picoarray; protein-binding.

1. Introduction

In this chapter, we describe the application of a microfluidic biochip device for analysis of nucleic acids and proteins. In the last decade, in the wake of mounting genomic and proteomic information, there has been an increasing trend of miniaturizing biological assay devices for high-throughput experiments and minimal sample consumption. The challenge is to develop simple, but generally useful platform technologies that will enable diverse applications to address the rapidly expanding needs in the biological, biomedical, and clinical areas.

From: *Methods in Molecular Biology, vol. 382 Microarrays: Second Edition: Volume 2*
Edited by: J. B. Rampal © Humana Press Inc., Totowa, NJ

We will describe herein the applications of a flexible microarray (array) technology for oligonucleotides as well as peptides *(1–3)*. Addressable arrays are now widely accepted as efficient assay forms for analysis of nucleic acids and proteins *(4–11)*. A critical component of an array platform is the content, i.e., the molecules, such as oligonucleotides or peptides, immobilized on the surface for specific detection or identification of analytes. In the last decade, various technologies have been developed for creating such arrays. Among these, *in situ* synthesis technologies have been particularly successful for preparation of DNA oligonucleotide *(1,2,12–24)* or peptide arrays *(3,12,25–35),* which are now the foundation of genome-scale analysis of nucleic acids and proteins. These methods rely on simultaneously synthesizing thousands or larger numbers of biopolymer sequences directly on a miniaturized surface, and thus share a common feature in the requirement of selective synthesis reactions at predetermined sites (gated chemical synthesis). Different chemistries have been developed for such needs and one major difference is in the use of conventional chemistry *(1,3,14,19,23,25)* against special chemistry which depends on photolabile group protected monomers *(12,17,30)*. In addition to the technical advantages, the practical usefulness of *in situ* synthesis is that one can possibly create new arrays instantaneously, whereas, different methods of array preparation would require postsynthesis of the biopolymers to be immobilized on surface. The latter method often is time- and cost-prohibitive when arrays are used for exploratory research. The discussions in this Chapter provide examples of microfluidics-based miniaturized *in situ* synthesis using conventional chemistry (for both nucleic acids and peptides) and assay devices in addressing the diverse needs of genomic and proteomic fields. This conventional chemistry-based array method might become a laboratory routine for creative development of novel applications for not only research but also as sensing and diagnostic devices.

2. Materials

In the following, except for the method discussions, the specific examples of the DNA and peptide array experiments are based on the use of microfluidic chips made by *in situ* synthesis *(1–3)*. These chips (LC Sciences, Houston, TX) contain 3968 isolated reaction sites in a 1.4-cm^2 area (4k chip) and the total solution volume of the chip is 9.6 µL. For other formats of the microfluidic chips, such as the 8k chip or larger chips, the described experimental protocols might be used with appropriate adjustments.

2.1. MicroRNA Detection and Analysis

1. Two oligonucleotide arrays containing probes for mature mouse microRNA (miRNA) sequences (length: 17–26 residues, miRBase database *[36]*) and as an option, customized sequences, such as those unique to the pre-miRNAs or predicted

sequences, are also present in the array. The probes synthesized consist of natural DNA and modified nucleotide residues to provide enhanced sensitivity for detecting miRNA sequences of low abundance. Each probe is repeated three times. The miRNA array contains probes for chip quality analysis, detection of spiking-in RNA sequences and internal controls, such as tRNA, ribosomal RNA, U2, and U6 sequences, and other small RNA sequences (*see* **Notes 1** and **2**).

2. The small RNA (sizes <200 residues) samples are from 5 μg of total RNA, which has a UV 260:280 ratio of 1.6–1.8 and is analyzed using gel electrophoresis or a Bioanalyzer (Agilent, Palo Alto, CA) to show the presence of molecules at sizes of 0.1–0.3 kb (small RNA), 2 kb, and 5 kb (both for ribosomal RNA) *(37)*.
3. Genepix™ 4000B microarray scanner (Molecular Devices/Axon, Sunnyvale, CA).
4. Array-Pro Analyzer 4.0 (Media Cybernetics, Silver Spring, MD).
5. Chip adapter for liquid circulation (LC Sciences).
6. Chip adapter for image scanning (LC Sciences).
7. P625 peristaltic pump (Instech Laboratory, Plymouth Meeting, PA).
8. Connecting tubing (0.020″ ID, silicone, Instech Laboratories Plymouth Meeting, PA), allowing flow rate at 0.025-2 mL/min.
9. EchoTherm™ Digital, Electronic chilling/heating plate, Model IC20 (Torrey Pines Scientific, San Marcos, CA).
10. Water: filtered water (US Filter, Purelab Plus, Warrendale, PA; Milli-Q, Millipore, Bedford, MA).
11. Microcon YM-100 fractionation (Millipore, Bedford, MA).
12. *mirVana* miRNA isolation Kit (Ambion, Austin, TX).
13. SenseAmp Plus (Genisphere, Hatfield, PA).
14. Low-Input Linear amplification Kit (Agilent, Palo Alto, CA).
15. Aminoallyl UTP, aRNA amplification (Ambion).
16. Biotin UTP GeneChip IVT labeling kit (Affymetrix).
17. CyDye (GE Healthcare, Piscataway, NJ).
18. AlexaFluor 555, 594, and 647 (Invitrogen, Carlsbad, CA).
19. RNeasy MinElute cleanup kit (Qiagen, Valencia, CA).
20. C18 spacer (Glen Research, Sterling, VA).
21. 3DNA Array900 miRNA Direct (Genisphere).
22. Biotin-X-hydrazide (Sigma, St. Louis, MO).
23. 2 n*M*, QD-streptavidin (Invitrogen).
24. Gold-streptavidin (Sigma).
25. Silver enhancer kit (Sigma).
26. C-4000Z CCD digital camera (Olympus, Center Valley, PA).
27. MICROMAX™ Accurate Sensitive and Precise (ASAP) miRNA chemical labeling kit (Perkin Elmer, Wellesley, MA).
28. T4 RNA ligase (New England Biolabs, Ipswich, MA).
29. Cy3-Tyramide (Perkin Elmer).
30. miRNA detection chips, μParaflo™ MicroRNA (LC Sciences, Houston, TX).
31. Biological samples, mouse brain and thymus (Ambion).
32. Exonuclease I (New England Biolabs).

33. Exo(–) Klenow polymerase (Promega, Madison, WI).
34. Biotin-7-dATP (Invitrogen).
35. Streptavidin-conjugate Alexa-fluor-547 (Invitrogen/Molecular Probes).
36. 1 × 3″ slide adapter (LC Sciences).
37. Nuclease-free water: water treated with 0.1% diethylpyrocarbonate (DEPC) for several h at 37°C and autoclaving (>15 min, >15 psi) to sterilize and eliminate DEPC. **Caution:** DEPC is a suspected carcinogen and should be used with care; the compound rapidly reacts with amines and can carboxyethylate RNA and should be removed completely. Alternatively, commercially available Nuclease-free water might be used.
38. 1X T4 RNA ligase reaction buffer: 50 mM Tris-HCl, 10 mM MgCl$_2$, 1 mM ATP, 10 mM dithiothreitol (DTT), pH 7.8.
39. 6X SSPE buffer: 0.90 M NaCl, 60 mM Na$_2$HPO$_4$, 6 mM ethylenediaminetetraacetic acid, pH 6.8.
40. Blocking solution: 6X SPPE, 25% formamide, 0.01% bovine serum albumin (BSA), pH 6.8.
41. Hybridization solution: 6X SSPE, 25% formamide, pH 6.8.
42. Stripping solution: 0.01X SSPE, 50% formamide, 0.5% sodium dodecyl sulfate (SDS).
43. Wash-1: 1% SDS in nuclease-free water.
44. Wash-2: 1:1 hybridization solution and water, 0.2% SDS.
45. Poly A polymerase (PAP) buffer (Invitrogen).

2.2. Peptide Array Applications

1. An epitope peptide chip containing *in situ* synthesized Antiflag (DYKDDDDK) and antihemagglutinin (HA, influenza virus) (YDVPDYASL) epitope sequences and their length and sequence variants (Epitope chip). The background spots are those containing surface linker molecules but not the peptides (the definition is the same for all other types of peptide chips). The positive control peptides are the known epitope binding sequences for each target antibody and the negative control peptides are mutant sequences that we found no antibody binding. Each sequence is repeated at least three times (*see* **Note 3**).
2. A protein kinase A (PKA) substrate peptide chip containing c-AMP dependent PKA substrate peptide sequences and other peptides for phosphorylation profiling. In general these sequences are derived from R(R/K)XSLG (X = 20 natural amino acids). Each substrate peptide has its comparison sequences, which are those synthetically phosphorylated and phosphorylation-negative analogs, such as those containing phosphotyrosine (pY), Tyr (Y), Phe (F), or Ala (A) at the comparable positions in the set of peptides. Each sequence is repeated at least three times.
3. A phosphopeptide chip containing 42 original Src homology domain 2 (SH2) binding phosphopeptides and their length and sequence variants for a total of 890 peptides. The target SH2 domain proteins are: SHP2 NSH2, known to bind to insulin receptor substrate 1 (*38,39*), Grb2-associated binder-1 (Gab-1) (*40,41*),

β-platelet-derived growth factor receptor *(42)* among other phosphoproteins. Each phosphopeptide has its comparison sequences, which are those nonphosphorylated analogs, such as pY vs Y, F, or A residues in different peptides. The positive controls are the known SHP2 SH2-binding phosphopeptides.

4. Tris-buffered saline (TBS): 50 mM Tris-HCl, 150 mM NaCl, pH 8.0.
5. Tris-buffered saline with Tween-20 (TBS-T): TBS, 0.1% Tween-20, pH 6.8.
6. Blocking solution: TBS-T, 1% (w/v) fraction V BSA (Sigma), 0.5% (w/v) nonfat dry milk. Solution is filtered using 0.45-μm sterile syringe filter (Corning, Corning, NY) and centrifuged before use.
7. Binding solution: TBS, 1% (w/v) BSA.
8. Primary antibody: Monoclonal ANTI-FLAG® M2 mouse immunoglobulin (Ig)G Antibody (Sigma); Monoclonal anti-GST mouse IgG antibody (Cell Signaling Technology, Danvers, MA); Monoclonal anti-His6 mouse IgG antibody (Santa Cruz Biotechnology, Santa Cruz, CA).
9. Secondary antibody: Goat antimouse IgG conjugated to Cy3 or Cy5 (Abcam); Cy3-anti-IgG (or Cy5-anti-IgG) staining solution.
10. Stripping solution: 50 mM Tris-HCl, 2% SDS, 0.1 M β-mercaptoethanol, 0.1% Triton X-100, pH 6.8.
11. Pro-Q® Diamond phosphoprotein gel staining reagent and destaining solution (Invitrogen).
12. PKA kinase reaction solution: 50 mM Tris-HCl, 1 mM MgCl$_2$, 0.1% BSA, 200 μM ATP, pH 7.5; 1 μL PKA (2500 U) per 50 μL solution.
13. PKA: cAMP-dependent Protein Kinase, catalytic subunit (New England Biolabs).
14. GST-NSH2 domain fusion protein (SHP2), which was expressed in *Escherichia coli* and purified in our lab, 25 μg/mL (0.68 μM) in the binding solution.
15. His6-CUGBP1ab (RNA binding domains 1 and 2 from CUGBP1), which was expressed in *E. coli* and purified in our lab, 9 μg/mL (0.36 μM) in the binding solution.
16. Various protease inhibitor cocktail solutions, such as those described in http://omrf.ouhsc.edu/~frank/toc.html, should be alternatives. **Caution:** Protease inhibitor phenylmethylsulfonyl fluoride has <60 min of lifetime in an aqueous solution and reacts with amines; the compound should be used right before the binding experiment.

2.3. Protein Binding DNA Chip

1. An oligonucleotide array containing a total of more than 1300 sequences, including DNA (GTXXXX)$_n$ (X = null, A, C, G, or T; n = 1–10) and other sequences synthesized as potential protein binding aptamers. The length of the sequences varies from 6 to 45 residues (*see* **Note 4**).
2. CUGBP1ab binding solution: 20 mM Tris-HCl, 100 mM KCl, 5 mM MgCl$_2$, 5 mM DTT, 10% glycerol, 1% BSA, pH 7.6.
3. Other solutions are the same as those described in **Subheading 2.2.**

3. Methods

3.1. miRNA Detection Array Application and Analysis

The discovery that miRNA can act as transcriptional and translational regulators is an important event in recent biological science. miRNAs are small, noncoding RNA sequences recently found across various organisms and species *(36)*. miRNAs are generated from pre-miRNA by endogenous processing enzymes, such as DROSH, DICER, and RISC in humans. The length of pre-miRNA known thus far is mostly about 70 residues or longer and that of its product, miRNA, is 17–26 residues *(36)*. Although miRNA has only a few years of history in the literature *(43,44)*, the ubiquitous existence of this family of molecules has been established in many eukaryotic species and miRNA expression can be associated with cell development, tumorgenesis, and a growing list of important cellular activities. Therefore, it is of great interest to understand the expression of miRNA associated with different cell- or tissue-types and at different cell-states or disease-stages. As of June 2005, miRBase (a comprehensive database of miRNA) *(36)* contains 2909 entries of miRNA from living species and 321 entries of miRNA from humans.

Microarrays have been proven to be powerful tools for the rapid profiling and analysis of a large number of RNA transcripts. However, array applications for miRNA (pri-miRNA, pre-miRNA, and mature-miRNA) profiling and analysis are not straight forward extensions of those of mRNA expression profiling. A few unique properties are associated with small RNA detection, which are: (1) their low abundant presence (approx 0.01%) in total RNA; (2) mature miRNAs do not have a 3′-poly-A tail (polyadenylated); (3) some miRNA sequences are highly homologous (paralogs) with only slight disparity at 3′-end; (4) their short lengths (~20 residues); (5) no flexibility in probe sequence choices. As a result, these small RNA molecules present a wide range of hybridization strengths (reflected in the heterogeneous distributions of their T_m and CG composition). And their detection probes on the chip cannot be optimized as in the case of probe design for mRNA detection, where probes of different sequences and variant lengths can be selected from the target mRNA for optimal hybridization. Furthermore, these small RNA molecules cannot be labeled using oligothymine-based reverse transcription and dye-NTP (or modified NTP) incorporation as commonly used for cDNA preparation. The analysis of miRNA on a genome-scale is thus a new challenge in terms of sensitivity and specificity.

Recently, a number of oligonucleotide array-based miRNA profiling methods have been reported *(45–56)*. In general, the steps of miRNA profiling consist of: (1) fractionation of small RNA from total RNA, usually the cut-off is approx 200 residues or in some cases is approx 70 residues. It is possible to analyze small RNA from the total RNA sample, but this option requires vigorous evaluation with more data points, (2) direct labeling, (3) indirect labeling of small RNA molecules, and (4) detection of labeled molecules using hybridization. The array experimental results are validated by Northern blot, real time PCR, and occasionally by cloning experiments. Literature information on large scale profiling of miRNA is quickly mounting and the following discussion will focus on the methods of miRNA labeling and detection on arrays based on total RNA samples as reported in the literature *(45,46)*.

3.1.1. Fractionation of Small RNA Molecules

Several commonly used separation methods for nucleic acids can be used to obtain small RNA fractions from total RNA samples: (1) polyacrylamide electrophoretic gel separation using 15% acrylamide and 8 *M* Urea Tris acetate electrophoresis gel and extracted in water. Microgel devices are now available for small RNA isolation from various vendors, (2) polyethylene glycol precipitation of high-molecular weight RNA with 12.5% polyethylene glycol-8000 and 1.25 *M* NaCl, (3) size-filtration separation using, for instance, Microcon YM-100 Fractionation, and (4) commercial kits, such as *mirVana* miRNA isolation Kit. Separation and the use of such kits might require additional adjustments in the procedure used for optimal results.

3.1.2. Methods of Labeling

Although a simple method reported *(50,51)* is to use short, random oligonucleotides (six to eight residues) for reverse transcription of small RNA to cDNA and fluorescent dye incorporation in the mean process, a number of new methods specially designed for miRNA detection (miRNA has 3′-OH) might become favorable choices.

1. Adapter ligation and cDNA- or cRNA-based detection *(49,52)* (**Fig. 1**). Adapter oligonucleotides (e.g., 5′-adapter is a 5′-DNA-RNA hybrid sequence and 3′-adapter is a RNA–DNA or a DNA sequence) are ligated using T4 RNA ligase to the 5′-phosphate and 3′-OH of small RNA, the adapter-RNA sequences are reverse-transcribed, and cDNA are amplified by PCR according to the procedures described *(57,58)*. The 3′-OH of the 3′- adapter should be modified so that no ligation is possible at this position. It is also possible to first add just 3′-A tail to the small RNA sequences, followed by reverse transcription to generate first strand cDNA, and then adding the 3′-T tail to the cDNA using terminal deoxynucleotidyl transferase, SenseAmp Plus. cRNA hybridization sample is obtained through transcription initiated by the T7 RNA polymerase promoter sequence, as a portion of the 3′ adapter, (5′-GCG-TAATACGACTCACTATA), where 5′-GCG might be other residues and will be transcribed.

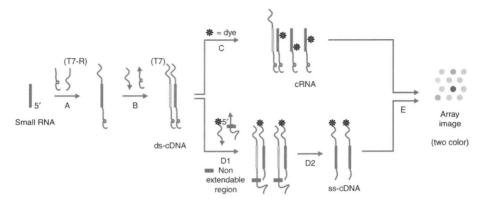

Fig. 1. Adapter ligation and cDNA or cRNA based detection of small RNA. The proto-cols involve several major steps. **(A)** Small RNA ligation with adapter sequences (T7-R sequence is needed when cRNA is used for detection). **(B)** Reverse transcription and PCR to generate ds-cDNA. **(C)** T7 transcription and incorporation of Cy3- or Cy5-labeled CTP or UTP to generate cRNA to give detection signals. Alternatively, **(D1)** PCR and **(D2)** gel purification to generate single-stranded cDNA (ss-cDNA) sequences containing detection signals. **(E)** The cRNA or ss-cDNA samples, each of which has two detection signal colors, cohybridize on miRNA array chip for detection of the sense strand of miRNA or other customized sequences.

Typical T7 RNA transcription reaction kits used for preparation of labeled cRNA for array hybridization are used for the cRNA preparation from the cDNA duplexes of small RNA, such as the Low-Input Linear Amplification kit. The reac-tion is carried out in the presence of Cy3- or Cy5-CTP using the protocol provided by the vendor to give labeled cRNA sample *(49)*. Other methods of labeling, such as incorporation of aminoallyl UTP or biotin UTP, are also suitable for cRNA labeling at this stage. The amino tag subsequently reacts with labeling fluores-cence dye molecules in the NHS-ester form through CyDyes, AlexaFluor 555, 594, 647, and others to generate labeled cRNA samples for hybridization after fil-tration purification to remove free dye molecules by using RNeasy MinElute Cleanup kit.

The duplex cDNA products might also be used for a second PCR using Cy3/Cy5-labeled primer for the sense strand (assuming the probes are for the sense strands) *(52)* (**Fig. 1**). The mixture contains 0.22 nM cDNA duplexes, a 5′-labeled sense strand primer and an antisense strand primer with an extension of a C18 spacer and dA$_{20}$. The PCR duplex products are separated from a denaturing polyacry-lamide gel (6% polyacrylamide, 8 M urea) to give Cy3/Cy5-labeled ss-cDNA sequences for hybridization.

2. DNA polymerase extension-based detection (2 µg or more of small RNA fraction used) *(46)*. The detection probes are built and placed on surface through a 5′-oligothymidine spacer to allow 3′-extension of miRNA of the known sequences

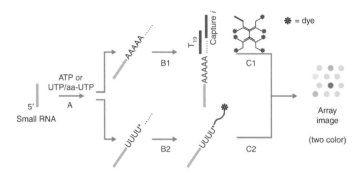

Fig. 2. Polyadenylate polymerase extension-based detection. (**A**) 3′-Extension of the poly A tail or oligo(dU/aa-dU) tail; (**B1**) Oligo-T hybridization and bridging a capture oligonucleotide (Genisphere) followed by the ligation, and (**C1**) Sample hybridization to chip followed by fluorescent dye dendrimer binding to the capture oligonucleotide (dye-unique). Alternatively, (**B2**) Labeling of the amino group of the aa-U with fluorescent dye (e.g., Cy3 or Cy5) (Ambion), and (**C2**) sample hybridization.

after hybridization. After hybridization, exonuclease I (in fresh buffer at pH 7.5, 4 U/μL) is used to digest single-stranded (unhybridized) probes for 3 h at 27°C. Hybridized miRNA and probe duplexes on the chip are treated with Exo(–) Klenow polymerase (in 1X DNA polymerase buffer, 0.15 U/μL) in the presence of dATP and biotin-7-dATP (4 μ*M*) for 1 h at 27°C. The detection is with streptavidin-conjugated Alexa-fluor-547 (15 ng/μL) at 25°C.

3. Poly A polymerase extension based detection (**Fig. 2**) (*48,59*). As described by *mirVana miRNA* Labeling kit or by 3DNA Array900 miRNA Direct, a small RNA sample is reacted with *E. coli* PAP (1 U for 100 μL reaction in PAP buffer) in the presence 2.5 m*M* MnCl$_2$ and ATP (1 m*M*) or UTP/aminoallyl dUTP (aaU) at 37°C for about 1 h for 3′-extension of poly A tail or oligo(U/aaU, less efficient incorporation by the enzyme) tail. After cleaning up of the reaction, the amino groups attached to the oligo (U/aaU) of the small RNA sequences are reacted with NHS-ester dye molecules to produce the labeled sample (*48*). Alternatively, signal amplification is achieved by hybridization of oligo(dT) to the 3′-A-tail to give overhang oligo(dT), followed by hybridization of a 5′-oligo(A)-capture oligonucleotide sequence and ligation of this sequence with the 3′-A-tail. The capture oligonucleotide is unique to a dendrimer decorated with fluorescence dye. After hybridization of the tagged small RNA sample, detection is achieved by staining the chip with the dendrimer-dye conjugate (*59*).

4. 3′-Modification-based detection with quantum dot/nanogold (**Fig. 3**) (*47*). The small RNA sample (from 90 μg total RNA) in 0.25 *M* NaAc solution, pH 5.6 (0.33 μg/μL) is mixed with one-fourth volume of 5 m*M* NaIO$_4$ at 25°C for 1.5 h in the dark. Two equivalences of Na$_2$S are added and allowed to react for 15 min, followed by addition of 1.5 equivalences of biotin-X-hydrazide and reaction at 37°C for 3 h. The biotinylated small RNA is recovered by ethanol precipitation and used

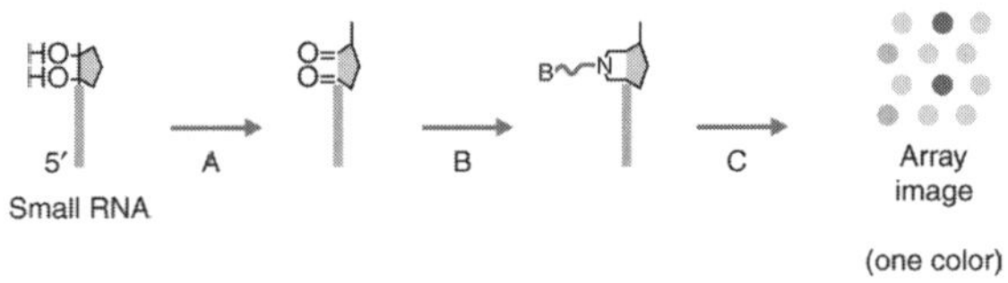

Fig. 3. 3′-Modification-based detection using quantum dot/nanogold. (**A**) The 2′- and 3′-diol of the small RNA 3′-terminal is oxidized to form a dialdehyde moiety. (**B**) Biotin labeling using biotin-X-hydrazide. (**C**) Hybridization and staining with streptavidin-quantum dot or streptavidin-gold sphere/silver enhancer (Sigma).

for hybridization. The hybridized miRNA array is stained with 10 μL (2 n*M*) Qdot 655 (excitation 633 nm, emission 655 nm) streptavidin conjugate for 1 h at room temperature. Alternatively, 10 μL gold-streptavidin is bound to the hybridized miRNA array followed by staining with the silver enhancer for 20 min. This colorimetric detection is performed using a commercial CCD camera mounted on a microscope.

5. Tyramide signal amplification based detection (**Fig. 4**) *(60)*. The small RNA (1 μg in 20 μL) is N7-labeled with fluorescein (FL) or biotin supplied in MICROMAX™ Accurate Sensitive and Precise (ASAP) miRNA Chemical Labeling kit and the two samples are combined after labeling. The FL and biotin tagged small RNA sample is hybridized to the chip. In the first dye labeling step, the anti-FL antibody conjugated with horse radish peroxidase (HRP) binds to FL-labeled, hybridized miRNA, and HRP catalyzes the labeling reaction of, for example, Cy3-Tyramide to produce Cy3 signals; the activated HRP is deactivated. In the following second dye labeling step, the streptavidin conjugated with HRP binds to biotin-labeled, hybridized miRNA and HRP catalyzes the labeling reaction of, for example, Cy5-Tyramide, to produce Cy5 signals.

3.1.3. Hybridization

Two miRNA detection chips, μParaflo™ MicroRNA, and two biological samples (e.g., mouse brain and thymus, 500 ng small RNA after fractionation) with each divided into two equal portions and labeled in either Cy3 or Cy5 as described in **Subheading 3.1.2.** In the first hybridization, the sample pair is Cy3 mouse brain and Cy5 mouse thymus; in the second hybridization, the sample pair is Cy3 mouse thymus and Cy5 mouse brain. In this color reversal experiment, each pair of the samples is present in the hybridization solution in a microtube, with two holes punctured in the cap for insertion of the liquid circulation tubes.

Caution: keep the chip wet all times and store at 4°C with the chip inlet and the outlet closed when not in use. All solutions are filtered through a 0.45-μm sterile syringe filter. All solutions loaded onto the chip are centrifuged at 13,000 rpm (16,000*g*), 1 min with an Eppendorf desktop centrifuge before use.

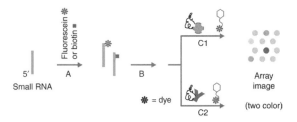

Fig. 4. Tyramide signal amplification detection (Perkin Elmer). **(A)** Direct tag-labeling of small RNA with MICROMAX Accurate Sensitive and Precise (*ASAP*) fluorescein or biotin labeling agent. **(B)** Cohybridization of tagged small RNA. **(C1)** Streptavidin HRP conjugate binding, followed by staining with Cy3- or Cy5-Tyramide. The activated HRP must be completely deactivated before the C2 step. **(C2)** Antifluorescein HRP conjugate binding, followed by staining with Cy5- or Cy3-Tyramide.

Temperature control for the chip is achieved through a contacting aluminum block placed on the heating/chilling plate.

3.1.3.1. Pre-Wash

1. Place a dummy chip in the adapter that is connected to the fluidic system (solution in a microtube, circulation tubing, heating/chilling plate, and the microperistaltic pump); the inlet tubing (withdrawing solution and delivering the solution to the chip) is immersed in the solution and the outlet tubing rests just above the liquid surface and the total volume for the circulation (excluding the chip) is approx 100 μL or larger.
2. Place a microtube containing 1 mL of the wash-1 solution (preheated to 95°C); the microtube is connected with the inlet and outlet tubing (*see* **Subheading 3.1.3.1., step 1**); alternating forward and reverse flow directions to completely remove the air bubbles in the chip (without further mentioning, this procedure is used in all cases where the solution circulation in a chip is being set up); circulate water at room temperature for 10 min.
3. Replace the microtube with 1 mL of water (preheated to 95°C) in a microtube; circulate at room temperature for 10 min.
4. Replace the microtube with 3 mL of water in a microtube and circulate at room temperature for 10 min.

3.1.3.2. Preparation for Hybridization

1. Connect the application chip to the fluidic system; the connected microtube contains 0.5 mL of water.
2. Circulate water at 500 μL/min, 35°C for 10 min.
3. Scan the chip image (no. 1 set of images).
4. Reconnect the chip to the fluidic system and circulate water as before to remove air bubbles.
5. Replace the microtube with 50 μL of the blocking solution in a microtube; circulate at 500 μL/min, 32°C for 10 min.

3.1.3.3. Hybridization

1. Preheat the labeled sample (10 µL) with 50 µL hybridization solution to 95°C for 5 min and microfuge briefly.
2. Replace the microtube with the sample solution in a microtube; circulate at 0.5 µL/min, 32°C for 6 h or more.
3. (Optional) Staining the chip by conjugation of dye molecules, signal amplification agents, or signal enhancers, as required by the labeling method used as discussed (*see* **Subheading 3.1.2.**).

3.1.3.4. OPTIMIZATION OF THE HYBRIDIZATION SPECIFICITY

1. Replace the microtube with 1 mL of the hybridization solution in a microtube; circulate at 100 µL/min, 32°C for 10 min.
2. Replace the microtube with 1 mL of wash-2 solution in a microtube; circulate at 100 µL/min, 32°C for 20 min.
3. Scan the chip image (no. 2 set of images) (tip: adjust focus to obtain uniform signals; obtaining images at different photomultiplier settings to cover both weak and strong signals in the linear intensity range).
4. If not satisfied with the results, continue the stringency wash as described above at an elevated temperature until a balance is reached for strong signals and low background (compared with the no. 2 set of images).
5. Replace the microtube with 1 mL of the 1X SSPE solution in a microtube; circulate at 100 µL/min and room temperature for 10 min.

3.1.4. Image Acquisition

This protocol is not limited to DNA oligonucleotide chips for hybridization but applies to all microfluidic chip experiments discussed in this chapter. The hybridized chip filled with 1X SSPE solution is mounted on a $1 \times 3''$ slide adapter and is placed on the support platform of the GenPix 4000B scanner with the chip facing down. The scanning is adjusted for optimal focal distance (using the Cy3 channel only) based on the intensity and sharpness and flatness of the image and on the uniform intensities of the replicate spots around the four corners of the chip. Cy3 and Cy5 signals are collected at several PMT settings, such as PMT 400, 500, 600, 700, and 750, to obtain image sets useful for weak as well as strong signals (*see* **Note 5**). During image acquisition, a quick reading of high-signal spots against background spots (low-intensity spots) should give a ratio better than 30-fold.

3.1.5. Data Extraction

This protocol is not limited to DNA oligonucleotide chips for hybridization but applies to all microfluidic chip experiments discussed in this chapter. **Figure 5** shows a portion of the images acquired at 5 µm resolution from a pair of samples: Cy3 mouse brain and Cy5 mouse thymus. A grid of rectangular boxes (128 rows and 31 columns) is generated using Array-Pro Analyzer 4.0

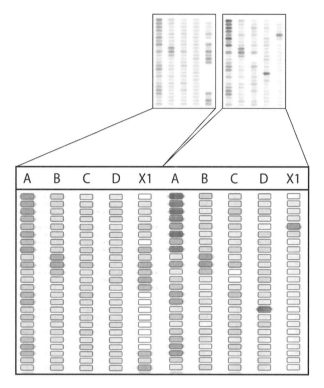

Fig. 5. Illustration of data extraction from the microfluidic chip hybridization images scanned at 5 μm resolution. The grid is built according to rows and columns of the spots on the chip and the total intensity and uniformity of each spot (average pixel intensities and standard deviation) are obtained (Array-Pro Analyzer, Media Cybernetics). The background signals are determined by the spots where no sequence was synthesized, such as the first spot under columns X1.

and each box contains more than 80 pixels. The regular shape and position of the microfluidic chip allow simple extraction of signal intensity and standard deviation into a data table. The background intensities are determined by the average signal of the spots where no sequence was synthesized.

3.2. Peptide Microarray Application and Analysis

Peptides are important ligand and messenger molecules of vast chemical diversity that are key components and regulators of biological processes through their interactions with proteins, nucleic acids, carbohydrates, and other biological molecules. However, our understanding of these interactions at the molecular level is rather scattered and limited. The studies and utilization of these interactions have increasingly become critical in both fundamental research and the development of applications based on these interactions.

In last few years, SPOT-based (a technique for peptide synthesis on membrane) peptide arrays have gained in popularity in high- throughput protein analysis *(28)*, a new high density, microfluidic-based peptide chip technology using *in situ* synthesis is coming to the stage of wide-spread applications *(27)*. These peptide arrays should complement random peptide libraries in the capability for further refining the consensus binding sequences at a much higher resolution by their quick readout of the results and discriminating at the level of a single amino acid. Moreover, synthetic peptide arrays greatly expand the sequence diversity owing to their ease for incorporating amino acid analogs or other chemical moieties compatible with the synthesis *(32,34,35)*.

We describe below the use of three types of microfluidic peptide chips: (1) epitope peptide chips for antibody binding **(Fig. 6A)**, (2) kinase substrate peptide chips for profiling kinases and their reactivities **(Fig. 6B)**, (3) phosphopeptide chips for mapping the binding motifs of phosphoprotein binding proteins. In some practical aspects, since the peptide chips and oligonucleotide chips are based on the same technology platform, they share many experimental requirements. These include: the fluidic system for chip experiments (a micropump, chip adapter, connecting tubing, and connectors), the flow control, the temperature control, and the image scanning. The general features used in the peptide chip experimental protocols are given only at the first time encountered.

3.2.1. Epitope Peptide Chips for Sequence Mapping and Antibody-Binding Studies

These experiments use a monoclonal Anti-FLAG® M2 antibody with the known epitope sequence DYKDDDDK, but similar procedures are applicable to other epitope-antibody-binding experiments (*see* **Note 6**) **(Fig. 6A)**. These applications are for epitope screening and antibody specificity assays and also for discovery of high affinity antibody-binding sequences for potential vaccine reagents or drug molecules. The high affinity-binding reagents are also suitable as candidates for sensing molecules in clinical or environmental diagnostic applications.

The peptide chip protocols have many similarities to those of oligonucleotides but with different target molecules and reaction solutions. For protein solutions, a major caution is that the circulation of a solution through the chip should generally be slow because proteins tend to precipitate under high-flow conditions and periodically reversing the flow direction (e.g., every 20 min) is also a good practice.

3.2.1.1. EXPERIMENTAL PREPARATION

1. Scan the chip image (no. 1 set of images, the first set of images).
2. The chip binding experiment uses the same fluidic system described in the DNA chip protocol (*see* **Subheading 3.1.3.1.**).

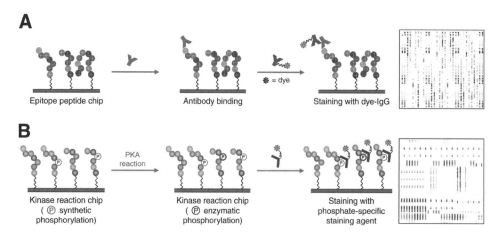

Fig. 6. Illustration of peptide chip applications and the fluorescence images of peptide chips. (A) Antibody-epitope peptide chip and the protocol for epitope-screening and specific antibody (ANTI-FLAG® M2 antibody AFM2) binding. The inverted fluorescence image illustrates the binding results. (B) Protein kinase A (PKA) substrate peptide chip and the protocol for the enzymatic reaction and detection. The inverted fluorescence image illustrates selective phosphorylation of PKA on the peptide chip containing sequences, such as Ac-RRXSLG and Ac-RKXSLG (X = one of the 20 amino acids), detected through phosphate-specific staining reagent.

3. Soak chip adapter parts and tubing in washing solution-1, ultrasonicate for 30 min, rinse the parts and tubing thoroughly with water.
4. Assemble the cleaned parts and the connection tubing and push 10 mL of water through the chip.
5. Connect the chip and a microtube containing 1 mL of water to a P625 peristaltic pump.

3.2.1.2. BLOCKING THE BINDING SURFACE

1. Circulate 1.5 mL of TBS to remove any bubble or small insoluble particles from chip by first discarding the 100 μL collected volume and circulate while reversing the flow direction at 50 μL/min for 20 min.
2. Replace the microtube with 2 mL of TBS-T in a microtube; circulate at 50 μL/min for 10 min.
3. Replace the microtube with 2 mL of the blocking solution in a microtube; circulate at 25 μL/min for 2 h at 4°C.
4. Scan the chip images (no. 2 set of images, the second set of images); no precipitation should be observed on the image.

3.2.1.3. STAINING BEFORE BINDING AS A NEGATIVE CONTROL FOR THE SECONDARY ANTIBODY BINDING

1. Connect the chip and a microtube containing 2 mL of the binding solution to the fluidic system; circulate the solution, and remove air bubbles.

2. Replace the microtube with 1 mL of the Cy5-anti-IgG (or Cy3-anti-IgG) staining solution (100 ng/mL) in a microtube; discard the first 100 µL collected at the flow outlet and circulate the staining solution in the dark at 25 µL/min for 1 h at room temperature.
3. Replace the microtube with 1 mL of the binding solution in a microtube and circulate at 50 µL/min for 20 min.
4. Scan the chip image (no. 3 set of images, dye staining control).

3.2.1.4. AFM2 ANTIBODY BINDING

1. Connect the chip and a microtube containing 1 mL of the binding solution to the fluidic system; circulate the solution at 50 µL/min for 20 min at room temperature.
2. Replace the microtube with 0.4 mL of AFM2 antibody in the binding solution (20 ng/mL) in a microtube (*see* **Note 7**). Discard the first 100 µL solution collected; circulate the antibody binding solution at 25 µL/min for 2 h at room temperature.

3.2.1.5. SECONDARY ANTIBODY-DYE CONJUGATE STAINING

1. Replace the microtube with 1 mL of the binding solution in a microtube and circulate at 50 µL/min for 20 min.
2. Replace the microtube with 0.4 mL of the Cy5-anti-IgG staining solution; discard the first 100 µL solution collected; circulate the staining solution at 25 µL/min for 1 h, 4°C.
3. Replace the microtube with 1 mL of the binding solution in a microtube and circulate the solution at 50 µL/min for 20 min.
4. Scan the chip images (no. 4 set of images). A typical image of the antibody-epitope chip binding is shown in **Fig. 6A**.

3.2.1.6. STRIPPING

1. Connect a chip (previously used for binding) and a microtube containing 1 mL of water with the fluidic system and circulate the water at 50 µL/min to establish uniform flow in the chip. Reverse the flow direction a few times when necessary.
2. Replace the microtube with 1 mL of the stripping solution in a microtube and circulate at 50 µL/min for 10 min at room temperature.
3. Place the chip in close contact with an aluminum block on a heating/chilling plate and circulate the stripping solution at 50 µL/min for 1 h at 40°C.
4. Scan the chip image (usually 80% of the fluorescence signals on the chip are washed away). If the results are not satisfactory, **step 3** is repeated.
5. Connect the chip and a microtube containing 5 mL of 20% (v/v) CH$_3$OH/water with the fluidic system and circulate the solution at 50 µL/min for 10 min.
6. Replace the microtube with 1 mL of TBS-T in a microtube; collect 0.5 mL solution at the outlet; close the inlet and the outlet using a coupler joining both ends; store the chip at 4°C.

3.2.2. Kinase Profiling Using Substrate Peptide Chip

A protein kinase reaction on a peptide chip is illustrated for PKA phosphorylation of Ser and Thr residues (**Fig. 6B**). The chip also contains the synthetic counterpart of the enzymatically phosphorylated peptides serving as positive controls and peptide sequences which contain Ala at the phosphorylation positions as negative controls. These experiments identify the specific peptide sequences as kinase substrates and provide relative reactivities for kinases against different substrate sequences; with real time data point collection, enzymatic reaction parameters (V_{max} and K_M) can be derived. The established high reactivity systems involving specific kinases and substrate sequences may be models for kinase inhibitor screening.

3.2.2.1. EXPERIMENTAL PREPARATION AND BLOCKING THE REACTION SURFACE

1. *See* the corresponding sub-sections in **Subheading 3.2.1.**

3.2.2.2. STAINING BEFORE THE KINASE REACTION AS NEGATIVE CONTROL OF THE SECONDARY ANTIBODY BINDING

1. Connect the chip and a microtube containing 2 mL of water; circulate at 50 μL/min for 10 min at room temperature.
2. Replace the water with 0.5 mL of 1:10 diluted Pro-Q phosphoprotein/peptide staining solution in a microtube. Discard the first 100 μL collected volume and circulate the staining solution at 50 μL/min for 20 min at room temperature.
3. Replace the microtube with 1.5 mL of water in a microtube; circulate the solution at 50 μL/min for 10 min at room temperature.
4. Scan the chip images (no. 3 set of images).

3.2.2.3. DESTAINING BEFORE THE KINASE REACTION

1. Connect the chip and a microtube containing 1.5 mL of destaining solution recommended by the vendor. Circulate the solution at 50 μL/min for 5 min at room temperature and then reverse the flow direction and circulate the solution for 15 min.

3.2.2.4. PKA REACTION

1. Replace the microtube with 1.5 mL of the fresh PKA kinase reaction solution in a microtube; circulate the solution at 50 μL/min for 5 min at 30°C.
2. Stop the flow and replace the microtube with 60 μL of the PKA kinase in the reaction solution in a microtube. Use the shorter connection tubing to directly load 25 μL of the solution to the chip (*see* **Note 8**); stop the flow and incubate the chip at 30°C for 15 min and load an additional 25 μL of the solution; incubate the chip at 30°C for 15 min.
3. Replace the microtube with 1.5 mL of the reaction solution in a microtube. Circulate the solution at 50 μL/min for 20 min at room temperature and the reverse the flow direction every 10 min.
4. Repeat the staining step using Pro-Q (*vide supra*).

5. Repeat the destaining step (*vide supra*).
6. Scan the chip image (no. 4 set of images). A typical image of the PKA reaction peptide chip is shown in **Fig. 6B**.

3.2.3. Protein Binding on a Phosphopeptide Chip

The binding of the phosphoprotein-binding motif, SH2 (N-terminal to SHP2 phosphatase enzyme) with an N-GST fusion tag, to phosphotyrosine-containing peptides on a chip is illustrated herein. The peptides on the chip include the positive binding sequences and their counter-parts of nonphosphorylated peptides as the negative controls. Similar experimental protocols are applicable to other phosphopeptide binding proteins. These applications allow the identification of high-affinity-binding sequences and specific residues which have significant roles in binding specificity. The results of the experiments reveal phosphoprotein–protein interactions, which are important fundamental elements of cellular regulation and targets for therapeutics.

3.2.3.1. EXPERIMENTAL PREPARATION AND BLOCKING THE REACTION SURFACE

1. *See* the corresponding subsections in **Subheading 3.2.1.**

3.2.3.2. STAINING BEFORE BINDING AS A NEGATIVE CONTROL FOR THE BINDING AND SECONDARY ANTIBODY BINDING

1. Connect the chip and a microtube containing 0.5 mL of the anti-GST mouse IgG antibody (primary antibody) solution to the fluidic system and circulate the solution at 25 μL/min for 1 h at room temperature.
2. Replace the microtube with 0.5 mL of the binding solution in a microtube and circulate at 50 μL/min for 15 min at room temperature.
3. Replace the microtube with 0.5 mL of the Cy5-anti-IgG antibody (secondary antibody) staining solution in a microtube and circulate at 25 μL/min for 1 h at room temperature.
4. Replace the microtube with 0.5 mL of the binding solution in a microtube and circulate at 50 μL/min for 15 min at room temperature.
5. Scan the chip image (no. 3 set of images).

3.2.3.3. SH2 MOTIF PROTEIN BINDING

1. Connect the chip and a microtube containing 1 mL of the binding solution to the fluidic system; circulate the solution at 25 μL/min for 10 min at room temperature.
2. Replace the microtube with 0.5 mL of the GST-NSH2 solution in a microtube; discard the first 100 μL collected volume; circulate the sample solution at 25 μL/min for 1 h at room temperature.

3.2.3.4. PRIMARY ANTIBODY BINDING AND SECONDARY ANTIBODY-DYE CONJUGATE STAINING

1. Repeat **steps 1–4** in **Subheading 3.2.3.2.**
2. Scan the chip images (no. 4 set of images).

3.3. Oligonucleotide Microarray for Protein Binding

Protein-DNA and protein-RNA binding are important events in gene activation, suppression, or expression. There have been intense interests in defining the binding sites (promoter sites, *trans*-splicing regulation sites, and so on) and the specific sequences on a genome scale, such as that reflected in the recent surge in using chip-to-chip (chromatin immunoprecipitation combined with the use of an array chip) to decode genomic transcription activation sites for the known and the speculated transcription factors in the scope of proteome. In these applications, oligonucleotide arrays are used to hybridize with the immunoprecipitated chromatin DNA sequences and thereby, verify the protein-binding sites. However, the potential of oligonucleotide arrays as powerful in vitro protein-binding platforms for mapping specific binding sites, defining binding sequences, and measurements of relative binding affinities has only slowly gained recognition *(61–67)*. The use of oligonucleotide arrays to screen artificial binding sequences for proteins or other targets is largely unexplored *(68,69)*. This delayed development of a clearly powerful high-throughput tool mostly to the unavailability of technologies, were attributed. But this will not necessarily continue to be a road block for interested researchers.

We describe herein the use of DNA oligonucleotide arrays for uncovering the binding properties of a previously known RNA-binding only protein motif (CUGBP1ab) containing 2X RBD (tandem RNA binding domain) from CUGBP1. The protein contains an N-terminal His6 tag. The protocol presented should be generally applicable to other recombination proteins in probing their DNA (DNA oligonucleotide chips) or RNA (RNA oligonucleotide chips)-binding properties.

3.3.1. Experimental Preparation and Blocking the Reaction Surface

1. *See* the corresponding sub-sections in **Subheading 3.2.1.**

3.3.2. Staining Before Binding as a Negative Control of Binding and Secondary Antibody Binding

1. Connect the chip and a microtube containing 0.5 mL of the anti-His mouse IgG antibody in the binding solution to the fluidic system. Discard the first 100 μL collected volume and circulate the solution at 25 μL/min for 1 h at room temperature.
2. Replace the microtube with 0.5 mL of the binding solution in a microtube and circulate at 50 μL/min for 15 min at room temperature.
3. Replace the microtube with 0.5 mL of the Cy5-anti-IgG antibody (primary antibody) staining solution in a microtube and circulate at 25 μL/min for 1 h at room temperature.
4. Replace the microtube with 0.5 mL of the binding solution in a microtube and circulate at 50 μL/min at room temperature for 15 min.
5. Scan the chip image (no. 3 set of images).

3.3.3. CUGBP1ab Protein Binding

1. Connect the chip and a microtube containing 1 mL of the binding solution to the fluidic system and circulate the solution at 25 µL/min for 10 min at room temperature.
2. Replace the binding solution with 0.5 mL of the His6-CUGBP1ab solution in a microtube. Discard the first 100 µL collected volume and circulate the sample solution at 25 µL/min at room temperature for 1 h.

3.3.4. Primary Antibody Binding and Secondary Antibody-Dye Conjugate Staining

1. Repeat **steps 1–4** in **Subheading 3.3.2.**
2. Scan the chip images (no. 4 set of images).

4. Notes

1. *General comments:* it is now possible to fully take advantage of arrays containing not only DNA but also RNA and peptide/peptide analogs to accelerate the studies of biomolecular interactions. The effective use of customizable arrays would require the strong support of bioinformatics, utilization of the existing databases, biostatistics for experimental design and analysis, and the experimental results that are carefully designed and carried out. These important aspects are not discussed herein but their importance cannot be overemphasized.

   The protocols discussed here are representative of typical in vitro binding procedures and these protocols outline major steps for achieving successful results. Proteins and antibodies are molecules of "strong personality," and thus optimization in the different systems studied is desirable. The application of serum, cell-lysate, or other biological samples will be more challenging because of their complex compositions, which result in high background noise and nonspecific interactions. Additionally, as these protocols are developed from our microfluidic chip, they would need modification if different types of arrays are used.

2. *miRNA detection microarrays:* miRNA and small RNA analyses are emerging fields of genomics that will continue to benefit from technology advancement. Currently, there exist uncertainties in several steps of these experiments. For instance, in small RNA isolation from total RNA, the total RNA sample might already have lost the small RNA fraction (<200 residues). Thus, it is important that the samples for small RNA analysis be analyzed before the fractionation and labeling experiments. Although, various options are available for the labeling step, direct labeling of small RNA is more reliable rather than using a sequence amplification protocol in detecting genuine miRNA sequences. Consideration should be given to the availability of the sample as a basis for the labeling method choices. If the sequence amplification method is used, the efficiency, the potential sequence bias of the enzymatic reactions and the effect of dye incorporation on hybridization stability should be evaluated. The adaptor ligation and cDNA- or cRNA-based detection—involving ligation and amplification, profiling, can only be

semiquantitative unless properly calibrated because of sequence-bias in these nucleic acid processing processes. It is desirable to use the two-color hybridization method to generate the ratio of coregulated RNA transcripts. This would require two different dyes, such as Cy3 and Cy5, postlabeled to the hybridization samples, or different tags are used so that a different tag can be labeled with a different dye and the labeling efficiency of different tags needs to be the same or else artifact interference would be generated. For achieving detection sensitivity and specificity, the final stringency wash and the stability of the dye molecules are important. The protocols described herein for the microfluidic chips prevent air-oxidation of dye molecules as it is a closed system and allow for additional cycles of stringency wash after inspection of the images. These are the practical aspects of a more controllable process for obtaining high quality and reproducible results.

3. *Peptide microarrays:* a major difference between peptide arrays and oligonucleotide arrays is the diverse application of the former against hybridization applications of the latter. An oligonucleotide array analysis usually focuses on coregulated sequences by clustering of signal intensities or ratios; but a peptide array analysis often examines the sequence patterns of the binding/reaction activities. Few software applications for peptide array analysis are available, but this situation is improving with the increasing use of peptide arrays and chips.

4. *Protein binding oligonucleotide microarrays:* in several aspects, the technical issues associated with these applications are similar to those of protein-binding peptide arrays. These include the binding and detection procedures and sequence-specific binding data obtained. Yet, the peptides and oligonucleotides are different chemical moieties and their binding behavior might be influenced by different factors and by different sets of oligonucleotides, which might be nature or modified residues, used against specific protein targets. One example is that, binding temperature may cause dramatic differences in protein-binding results as oligonucleotides could undergo hairpin to duplex transition or transitions involving other folding forms as a function of temperature. Because this is a fairly new area of array applications, much of the experimental details for reliable and effective use of the array-based in vitro protein binding experiments will continuously evolve.

5. Detailed instruction should be referred to the instrumental manual of Genepix™ 4000B microarray scanner and Genepix™ Pro 4.0 Array Acquisition and Analysis Software user's Guide.

6. Binding conditions (buffer type, pH, temperature, concentration, incubation time, and so on) depend on the individual antibody/protein.

7. Alternatively, Cy3/Cy5 dye-conjugated anti-AFM2 antibody is used. The chip image of the antibody binding can be directly obtained without using a dye-conjugated secondary antibody.

8. The total volume inside a peptide chip is 9.6 μL and the volume of a shorter end of tubing is 3 μL. Including the circulation tubing, the minimal amount of the solution in a microtube should be approx 100 μL.

Acknowledgments

This research is supported by grants by National Institutes of Health and the R. A. Welch Foundation (E-1270). The authors thank Mr. Christopher Collins for the careful reading of the manuscript.

References

1. Gao, X., Gulari, E., and Zhou, X. (2004) *In situ* synthesis of oligonucleotide microarrays. *Biopolymers* **73,** 579–596.
2. Zhou, X., Cai, S., Hong, A., et al. (2004) Microfluidic PicoArray synthesis of oligodeoxynucleotides and simultaneous assembling of multiple DNA sequences. *Nucleic Acids Res.* **32,** 5409–5417.
3. Gao, X., Pellois, J. P., Na, Y., Kim, Y., Gulari, E., and Zhou, X. (2004) High density peptide microarrays. *In situ* synthesis and applications. *Mol. Divers* **8,** 177–187.
4. Hinds, D. A., Stuve, L. L., Nilsen, G. B., et al. (2005) Whole-genome patterns of common DNA variation in three human populations. *Science* **307,** 1072–1079.
5. Ekins, R. P. (1989) Multi-analyte immunoassay. *J. Pharm. Biomed. Anal.* **7,** 155–168.
6. Mirzabekov, A. and Kolchinsky, A. (2002) Emerging array-based technologies in proteomics. *Curr. Opin. Chem. Biol.* **6,** 70–75.
7. Schena, M., Shalon, D. D., Davis, R. W., and Brown, P. O. (1995) Quantitative monitoring of gene expression patterns with a complementary DNA microarray. *Science* **270,** 460–467.
8. Lockhart, D. J., Dong, H., Byrne, M. C., et al. (1996) Expression monitoring by hybridization to high-density oligonucleotide arrays. *Nat. Biotech.* **14,** 1675–1680.
9. Ekins, R. P. (1998) Ligand assays: from electrophoresis to miniaturized microarrays. *Clin. Chem.* **44,** 2015–2030.
10. Stoll, D., Templin, M. F., Bachmann, J., and Joos, T. O. (2005) Protein microarrays: applications and future challenges. *Curr. Opin. Drug. Discov. Devel.* **8,** 239–252.
11. The chipping forecast. (1999) *Nat. Genet. Suppl* **21,** 3–60.
12. Fodor, S. P., Leighton, P. A. J., Pirrung, M. C., Stryer, L., and Solas, D. (1991) Light-directed spatially addressable parallel chemical synthesis. *Science* **251,** 767–773.
13. Maskos, U. and Southern, E. M. (1992) Parallel analysis of oligodeoxyribonucleotide (oligonucleotide) interactions. I. Analysis of factors influencing oligonucleotide duplex formation. *Nucleic Acids Res.* **20,** 1675–1678.
14. Blanchard, A. P., Kaiser, R. J., and Hood, L. E. (1996) High-density oligonucleotide arrays. *Biosens. Bioelectron* **11,** 687–690.
15. Blanchard, A. P. and Hood, L. E. (1996) Sequence to array: probing the genome's secrets. *Nat. BioTechnol.* **14,** 1649.
16. Gao, X., Yu, P. Y., LeProust, E., Sonigo, L., Pellois, J. P., and Zhang, H. (1998) Oligonucleotide synthesis using solution photogenerated acids. *J. Am. Chem. Soc.* **120,** 12,698–12,699.
17. Singh-Gasson, S., Green, R. D., Yue, Y., et al. (1999) Maskless fabrication of light-directed oligonucleotide microarrays using a digital micromirror array. *Nat. Biotech.* **17,** 974–978.

18. Gao, X., LeProust, E., Zhang, H., et al. (2001) Flexible DNA chip synthesis gated by deprotection using solution photogenerated acids. *Nucleic Acids Res.* **29,** 4744–4750.

19. Hughes, T. R., Mao, M., Jones, A. R., et al. (2001) Expression profiling using microarrays fabricated by an ink-jet oligonucleotide synthesizer. *Nat. Biotechnol.* **19,** 342–347.

20. Bulter, J. H., Cronin, M., Anderson, K. M., et al. (2001) *In situ* synthesis of oligonucleotide arrays by using surface tension. *J. Am. Chem. Soc.* **123,** 8887–8894.

21. McGall, G. H. and Fidanza, J. A. (2001) DNA Microarrays: photolithographic synthesis of high-density oligonucleotide arrays: in *Methods and Protocols in Molecular Biology*, vol. 170 (Rampal, J. B., ed.), Humana, Totowa, NJ, pp. 71–101.

22. Luebke1, K. J., Balog, R. P., and Garner, H. R. (2003) Prioritized selection of oligodeoxyribonucleotide probes for efficient hybridization to RNA transcripts. *Nucleic Acids Res.* **31,** 750–758.

23. Tesfu, E., Maurer, K., Ragsdale, S. R., and Moeller, K. D. (2004) Building addressable libraries: the use of electrochemistry for generating reactive Pd(II) reagents at preselected sites on a chip. *J. Am. Chem. Soc.* **126,** 6212–6213.

24. Srivannavit, O., Gulari, M., Gulari, E., et al. (2004) Design and fabrication of microwell array chips for a solution-based, photogenerated acid-catalyzed parallel oligonucleotide DNA synthesis. *Sensors Actuators A* **116,** 150–160.

25. Frank, R. (2002) The SPOT-synthesis technique. Synthetic peptide arrays on membrane supports—principles and applications. *J. Immunol. Methods* **267,** 13–26.

26. Reimer, U., Reineke, U., and Schneider-Mergener, J. (2002) Peptide arrays: from macro to micro. *Curr. Opin. Biotechnol.* **13,** 315–320.

27. Lam, K. S. and Renil, M. (2002) From combinatorial chemistry to chemical microarray. *Curr. Opin. Chem. Biol.* **6,** 353–358.

28. Panse, S., Dong, L., Burian, A., et al. (2004) Profiling of generic anti-phosphopeptide antibodies and kinases with peptide microarrays using radioactive and fluorescence-based assays. *Mol. Divers* **8,** 291–299.

29. Pease, A. C., Solas, D., Sullivan, E. J., Cronin, M. T., Holmes, C. P., and Fodor, S. P. (1994) Light-generated oligonucleotide arrays for rapid DNA sequence analysis. *Proc. Natl. Acad. Sci. USA* **91,** 5022–5026.

30. Holmes, C. P., Adams, C. L., Kochersperger, L. M., Mortensen, R. B., and Aldwin, L. A. (1995) The use of light-directed combinatorial peptide synthesis in epitope mapping. *Biopolymers* **37,** 199–211.

31. Pellois, J. P., Wang, W., and Gao, X. (2000) Peptide synthesis based on t-Boc chemistry and solution photogenerated acids. *J. Comb. Chem.* **2,** 355–360.

32. Pellois, J. P., Zhou, X., Srivannavit, O., Zhou, T., Gulari, E., and Gao, X. (2002) Individually addressable parallel peptide synthesis on microchips. *Nat. Biotechnol.* **20,** 922–926.

33. Komolpis, K., Srivannavit, O., and Gulari, E. (2002) Light-directed simultaneous synthesis of oligopeptides on microarray substrate using a photogenerated acid. *Biotechnol. Prog.* **18,** 641–646.

34. Li, S., Bowerman, D., Marthandan, N., et al. (2004) Photolithographic synthesis of peptoids. *J. Am. Chem. Soc.* **126,** 4088–4089.

35. Li, S., Marthandan, N., Bowerman, D., Garner, H. R., and Kodadek, T. (2005) Photolithographic synthesis of cyclic peptide arrays using a differential deprotection strategy. *Chem. Commun. (Camb)* **2005,** 581–583.

36. http://microrna.sanger.ac.uk/cgi-bin/sequences/browse.pl.

37. http://www.protocol-online.org/prot/Molecular_Biology/RNA/RNA_Extraction/Total_RNA_Isolation/.

38. Sun, X. J., Crimmins, D. L., Myers, M. G. Jr., Miralpeix, M., and White, M. F. (1993) Pleiotropic insulin signals are engaged by multisite phosphorylation of IRS-1. *Mol. Cell Biol.* **13,** 7418–7428.

39. Hers, I., Bell, C. J., Poole, A. W., et al. (2002) Reciprocal feedback regulation of insulin receptor and insulin receptor substrate tyrosine phosphorylation by phosphoinositide 3-kinase in primary adipocytes. *Biochem. J.* **368,** 875–884.

40. Lehr, S., Kotzka, J., Herkner, A., et al. (2000) Identification of major tyrosine phosphorylation sites in the human insulin receptor substrate Gab-1 by insulin receptor kinase in vitro. *Biochemistry* **39,** 10,898–10,907.

41. Lehr, S., Kotzka, J., Herkner, A., et al. (1999) Identification of tyrosine phosphorylation sites in human Gab-1 protein by EGF receptor kinase in vitro. *Biochemistry* **38,** 151–159.

42. Yokote, K., Mori, S., Hansen, K., et al. (1994) Direct interaction between Shc and the platelet-derived growth factor beta-receptor. *J. Biol. Chem.* **269,** 15,337–15,343.

43. Lagos-Quintana, M., Rauhut, R., Lendeckel, W., and Tuschl, T. (2001) Identification of novel genes coding for small expressed RNAs. *Science* **294,** 853–858.

44. Lee, R. C. and Ambros, V. (2001) An extensive class of small RNAs in Caenorhabditis elegans. *Science* **294,** 862–864.

45. Thomson, J. M., Parker, J., Perou, C. M., and Hammond, S. M. (2004) A custom microarray platform for analysis of microRNA gene expression. *Nat. Methods* **1,** 47–53.

46. Nelson, P. T., Baldwin, D. A., Scearce, L. M., Oberholtzer, J. C., Tobias, J. W., and Mourelatos, Z. (2004) Microarray-based, high-throughput gene expression profiling of microRNAs. *Nat. Methods* **1,** 155–161.

47. Liang, R. Q., Li, W., Li, Y., et al. (2005) An oligonucleotide microarray for microRNA expression analysis based on labeling RNA with quantum dot and nanogold probe. *Nucleic Acids Res.* **33,** E17.

48. Shingara, J., Keiger, K., Shelton, J., et al. (2005) An optimized isolation and labeling platform for accurate microRNA expression profiling. *RNA* **11,** 1461–1470.

49. Barad, O., Meiri, E., Avniel, A., et al. (2004) MicroRNA expression detected by oligonucleotide microarrays: system establishment and expression profiling in human tissues. *Genome Res.* **14,** 2486–2494.

50. Liu, C. G., Calin, G. A., Meloon, B., et al. (2004) An oligonucleotide microchip for genome-wide microRNA profiling in human and mouse tissues. *Proc. Natl. Acad. Sci. USA* **101,** 9740–9744.

51. Babak, T., Zhang, W., Morris, Q., Blencowe, B. J., and Hughes, T. R. (2004) Probing microRNAs with microarrays: tissue specificity and functional inference. *RNA* **10,** 1813–1819.

52. Baskerville, S. and Bartel, D. P. (2005) Microarray profiling of microRNAs reveals frequent coexpression with neighboring miRNAs and host genes. *RNA* **11,** 241–247.

53. Miska, E. A., Alvarez-Saavedra, E., Townsend, M., et al. (2004) Microarray analysis of microRNA expression in the developing mammalian brain. *Genome Biol.* **5,** R68.

54. Sioud, M. and Rosok, O. (2004) Profiling microRNA expression using sensitive cDNA probes and filter arrays. *Biotechniques* **37,** 574–576, 578–580.

55. Sun, Y., Koo, S., White, N., et al. (2004) Development of a micro-array to detect human and mouse microRNAs and characterization of expression in human organs. *Nucleic Acids Res.* **32,** E188.

56. Krichevsky, A. M., King, K. S., Donahue, C. P., Khrapko, K., and Kosik, K. S. (2003) A microRNA array reveals extensive regulation of microRNAs during brain development. *RNA* **9,** 1274–1281.

57. Lau, N. C., Lim, L. P., Weinstein, E. G., and Bartel, D. P. (2001) An abundant class of tiny RNAs with probable regulatory roles in *Caenorhabditis elegans*. *Science* **294,** 858–862.

58. See Protocol listed in http://www.protocol-online.org/prot/Molecular_Biology/RNA/microRNA/microRNA_Cloning/ and http://web.wi.mit.edu/bartel/pub/protocols/miRNAcloning.pdf.

59. http://www.genisphere.com/pdf/array900_mirna_direct_manual_12_15_04.pdf.

60. http://las.perkinelmer.com; http://las.perkinelmer.com/content/manuals/mps545.pdf.

61. Bulyk, M. L., Gentalen, E., Lockhart, D. J., and Church, G. M. (1999) Quantifying DNA-protein interactions by double-stranded DNA arrays. *Nat. Biotechnol.* **17,** 536–537.

62. Bulyk, M. L., Huang, X., Choo, Y., and Church, G. M. (2001) Exploring the DNA-binding specificities of zinc fingers with DNA microarrays. *Proc. Natl. Acad. Sci. USA* **98,** 7158–7163.

63. Krylov, A. S., Zasedateleva, O. A., Prokopenko, D. V., Rouviere-Yaniv, J., and Mirzabekov, A. D. (2001) Massive parallel analysis of the binding specificity of histone-like protein HU to single- and double-stranded DNA with generic oligodeoxyribonucleotide microchips. *Nucleic Acids Res.* **29,** 2654–2660.

64. Bulyk, M. L., Johnson, P. L., and Church, G. M. (2002) Nucleotides of transcription factor binding sites exert interdependent effects on the binding affinities of transcription factors. *Nucleic Acids Res.* **30,** 255–261.

65. Wang, J., Bai, Y., Li, T., and Lu, Z. (2003) DNA microarrays with unimolecular hairpin double-stranded DNA probes: fabrication and exploration of sequence-specific DNA/protein interactions. *J. Biochem. Biophys. Methods* **55,** 215–232.

66. Wang, J. K., Li, T. X., Bai, Y. F., and Lu, Z. H. (2003) Evaluating the binding affinities of NF-kappaB p50 homodimer to the wild-type and single-nucleotide mutant Ig-kappaB sites by the unimolecular dsDNA microarray. *Anal. Biochem.* **316,** 192–201.

67. Mukherjee, S., Berger, M. F., Jona, G., et al. (2004) Rapid analysis of the DNA-binding specificities of transcription factors with DNA microarrays. *Nat. Genetics* **36,** 1331–1339.

68. Yamamoto-Fujita, R. and Kumar, P. K. (2005) Aptamer-derived nucleic acid oligos: applications to develop nucleic acid chips to analyze proteins and small ligands. *Anal. Chem.* **77,** 5460–5466.

69. Collett, J. R., Cho, E. J., Lee, J. F., et al. (2005) Functional RNA microarrays for high-throughput screening of antiprotein aptamers. *Anal. Biochem.* **338,** 113–123.

20

Application of ProteinChip Array Profiling in Serum Biomarker Discovery for Patients Suffering From Severe Acute Respiratory Syndrome

Timothy T. C. Yip, William C. S. Cho, Wai Wai Cheng,
Johnny W. M. Chan, Victor W. S. Ma, Tai-Tung Yip,
Christine N. B. Lau Yip, Roger K. C. Ngan, and Stephen C. K. Law

Summary

A new strain of coronavirus has caused an outbreak of severe acute respiratory syndrome (SARS) from 2002 to 2003 resulting in 774 deaths worldwide. By protein chip array profiling technology, a number of serum biomarkers that might be useful in monitoring the clinical course of SARS patients were identified. This book chapter describes how the protein chip array profiling was carried out for this study. Briefly, SARS patients' serum samples were first fractionated in Q Ceramic HyperD ion exchange sorbent beads by buffers at different pH. Serum protein fractions thus obtained were then bound onto a copper (II) immobilized metal affinity capture (IMAC30 Cu [II]) ProteinChip® Array or a weak cation-exchange (CM10) ProteinChip Array. After washing and addition of sinapinic acid, the chips were read in a Protein Biological System (PBS) IIc mass spectrometer. Ions were generated by laser shots and flied in a time of flight mode to the ion detector according to their mass over charge (m/z) ratio. The serum profiling spectra in SARS patients were acquired, baseline subtracted and analyzed in parallel with those from the control subjects by Ciphergen ProteinChip Software 3.0.2 with their peak intensities compared by a nonparametric two sample Mann–Whitney-U test. More than twelve peaks were differentially expressed in SARS patients with one at m/z of 11,695 (later identified to be serum amyloid A protein), which had increase in peak intensity correlating with the extent of SARS-coronavirus induced pneumonia as defined by a serial chest X-ray opacity score. The remaining biomarkers could also be useful in the study of other clinical parameters in SARS patients.

Key Words: Pneumonia; ProteinChip array profiling; SARS-Coronavirus (CoV); serum amyloid A protein (SAA); severe acute respiratory syndrome (SARS); surface-enhanced laser desorption/ionization time of flight mass spectrometry (SELDI-TOF-MS).

From: *Methods in Molecular Biology, vol. 382: Microarrays: Second Edition: Volume 2*
Edited by: J. B. Rampal © Humana Press Inc., Totowa, NJ

1. Introduction

A new strain of coronavirus (CoV) caused a pandemic outbreak of severe acute respiratory syndrome (SARS) in China, Hong Kong, Singapore, Toronto, and Taiwan from 2002 to 2003, resulting in 8098 individuals being infected and 774 deaths *(1)*. As compared to other upper viral respiratory infections *(2)*, SARS-CoV can rapidly induce pneumonia. The accompanying adult respiratory distress syndrome can quickly progress resulting in rapid death of the patients *(3,4)*. The most commonly used laboratory diagnostic tests for SARS-CoV-infection are reverse transcription (RT)-PCR and anti-SARS-CoV antibody serological tests *(5–7)*, but their roles in monitoring the extent of pneumonia are rather limited. Although, serial chest radiography is helpful for monitoring the progress of disease *(8,9)*, it is also limited by its subjectivity and variability in the interpretation of the imaging results and occasionally suboptimal quality of portable films taken in isolation ward to avoid the spread of the disease. In a recent article, the discovery of 12 up- or downregulated serum biomarkers were reported in SARS patients by a novel protein chip array profiling approach *(10–14)* with one biomarker appears to be useful in monitoring the extent of pneumonia *(15)*. In this chapter, how this protein chip array profiling technique is carried out is described. The technique involves four steps with the first step being serum fractionation in ceramic ion exchange sorbent beads. This step separates serum proteins into several fractions aiming at reducing their complexity before protein chip profiling is performed. The second step involves the binding of serum protein biomarkers onto protein chip array surfaces. The third step is desorption of bound serum biomarkers by laser shots in the mass spectrometer generating a time of flight spectrum for each sample. The fourth step concerns with data acquisition, data processing and statistical analyses of the spectra in SARS patients in comparison with the control groups. These procedures will be described in full details in the following chapter and summarized in the flow chart diagram as in **Fig. 1**.

2. Materials

2.1. SARS Patients

1. 28 Patients with confirmed SARS-CoV infection managed in the Department of Medicine, Queen Elizabeth Hospital, Hong Kong.
2. 87% of the patients have abnormal chest radiographs on admission.
3. 13% of the patients present early thus requiring high-resolution thoracic computer tomography scanning to confirm the presence of pulmonary involvement.

2.2. Serum Samples

1. 44 Serum samples from 24 SARS patients for differential mapping of SARS associated biomarkers.

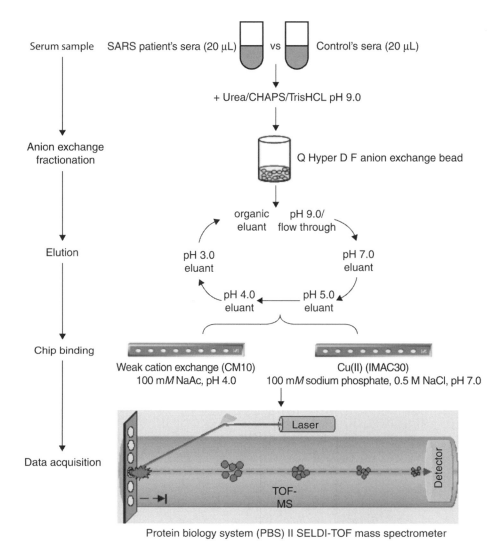

Fig. 1. Sample fractionation, chip binding, and reading in protein chip array profiling technique. Serum samples from SARS patients and various control groups were fractionated in Q Ceramic Hyper D F anion exchange sorbent beads. The eluted fractions were in turn spotted onto two types of protein chips, namely, copper IMAC30 Cu(II) ProteinChip array and weak cation-exchange (CM10) ProteinChip array. Each sample bound protein chip was in turn inserted into a Protein Biology System II SELDI-TOF-mass spectrometer and the protein biomarkers were ionized and desorbed by laser beam. The ions flied to the ion detector generating a biomarker peak intensity spectrum according to their mass over charge values (m/z).

2. 45 Serial serum samples from four additional SARS patients with comprehensive clinical follow-up for longitudinal correlation with clinical manifestation.
3. 10 Sera from 10 apparently healthy individuals (from a familial cancer screening clinic) as normal controls.
4. 72 Sera from 51 patients suffering from other viral or bacterial infections (influenza A virus [$n = 12$], influenza B virus [$n = 8$], respiratory adenovirus [$n = 12$], respiratory syncytial virus [$n = 10$], hepatitis B virus [$n = 10$], *Mycobacterium tuberculosis* [$n = 10$], other bacteria [$n = 10$]) as control patients for comparison with SARS patients.

2.3. Ion Exchange Fractionation of Serum

1. BioSepra Q Ceramic Hyper D F ion exchange sorbent beads (Ciphergen Biosystems Incorporation, Fremont, CA).
2. 96-Well sample plates (Nalge Nunc International, Rochester, NY).
3. 96-Well Silent Screen filter plates with Loprodyne membrane filter (0.45 μm pore size; Nalge Nunc International).
4. 96-Well collection plates (Nalge Nunc International).
5. 50 m*M* Tris-HCl buffer at pH 9.0 (W1 ion exchange bead washing buffer).
6. 50 m*M* Tris-HCl buffer containing 9 mol/L urea and 20 g/L CHAPS (3,3-cholmido-propyl-dimethylammonio-1-propanesulfonate) (D1 sample denaturing buffer) at pH 9.0.
7. 50 m*M* Tris-HCl buffer containing 1 mol/L urea and 2.2 g/L CHAPS at pH 9.0 (Q1 bead equilibration buffer).
8. 50 m*M* Tris-HCl buffer containing 1 g/L *N*-octyl-β-D-glucopyranoside (OGP) at pH 9.0 (E1 elution buffer).
9. 100 m*M* Sodium phosphate containing 1 g/L OGP at pH 7.0 (E2 elution buffer).
10. 100 m*M* Sodium acetate containing 1 g/L OGP at pH 5.0 (E3 elution buffer).
11. 100 m*M* Sodium acetate containing 1 g/L OGP at pH 4.0 (E4 elution buffer).
12. 50 m*M* Sodium citrate containing 1 g/L OGP at pH 3.0 (E5 elution buffer).
13. 33.3% Isopropanol, 16.7% acetonitrile, and 0.1% trifluoroacetic acid (E6 elution buffer).
14. Micromix 5-01 shaker (Euro/DPC Ltd., Gwynedd, UK).

2.4. Chip Pretreatment, Sample Binding, and Analysis

1. Copper (II) immobilized metal affinity capture [IMAC30 Cu(II)] ProteinChip® Array (Ciphergen Biosystems Inc., Fremont, CA).
2. Weak cation-exchange (CM10) ProteinChip Array (Ciphergen Biosystems, Inc.).
3. 96-Well ProteinChip bioprocessor (Ciphergen Biosystems, Inc.).
4. 100 m*M* Copper sulfate (CuSO4; Sigma-Adrich, St. Louis, MO).
5. 100 m*M* Sodium phosphate containing 0.5 mol/L NaCl (IMAC30 chip binding buffer).
6. 100 m*M* Sodium acetate buffer pH 4.0 (serves as both IMAC30 chip neutralizing buffer and CM10 chip binding buffer).

7. 50% Sinapinic acid (Ciphergen Biosystems Inc.) in 500 mL/L acetonitrile and 5 mL/L trifluoroacetic acid (Sigma-Adrich).
8. Protein Biological System (PBS) IIc mass spectrometer reader (Ciphergen Biosystems).
9. All-in-1 peptide molecular mass standard (Ciphergen Biosystems Inc.).
10. Ciphergen ProteinChip Software 3.0.2 (Ciphergen Biosystems Inc.).

3. Methods

3.1. Treatment of SARS Patients

1. Patients with fever and other symptoms of respiratory infection are initially managed with broad-spectrum antibiotics and supportive therapy.
2. After the diagnosis of clinical SARS is made and if there is no response to antibiotic therapy, combination therapy with ribavirin and systemic steroids is initiated.
3. Intravenous pulse methylprednisolone is initiated when the clinical condition, radiological presentation, or oxygen saturation status of the SARS patients further deteriorates.
4. The clinical characteristics, detailed management plan, and treatment regimen of this cohort of patients can be referred to a previous publication from our hospital *(3)*.

3.2. Serial Chest Radiographic Score

The extent of pneumonia in the SARS patients with longitudinal follow-up is assessed by a radiologist according to a serial chest radiographic score, modified from a score system initially proposed to assess computer tomography of the chest *(16)* and summarized as in **Fig. 2**.

1. After X-ray chest radiography is taken, divide the frontal chest X-ray radiograph into six lung zones, namely, left upper zone, left middle zone, left lower zone, right upper zone, right middle zone, and right lower zone.
2. The upper zone (left or right) represents area above carina (including the apex).
3. The middle zone (left or right) represents area from carina to the level of inferior pulmonary veins.
4. The lower zone (left or right) represents area from the lower margin of middle zone to the lung base.
5. Score the opacity in each lung zone by a "coarse semiquantitative method" with a five points' scale of grades 0–4.
6. Grade 0 represents no opacity involved area.
7. Grade 1 represents 5–24% opacity involved areas.
8. Grade 2 represents 25–49% opacity involved areas.
9. Grade 3 represents 50–74% opacity involved areas.
10. Grade 4 represents 75–100% opacity involved areas.
11. Add the grading from each of the six lung zones to provide a 0–24 point summation scale for assessing the extent of pneumonia.

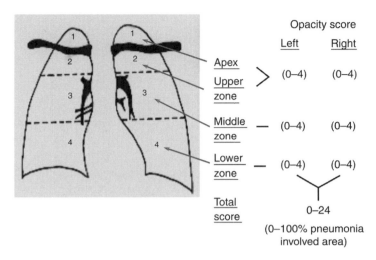

Fig. 2. Assessing the extent of pneumonia by chest X-ray opacity score. The opacity in each zone (apex+upper, middle, and lower zones) from the left and right regions of the lung was scored by a "coarse semiquantitative method" with a 5 points' scale of grades 0–4 representing involved areas of 0, 5–24, 25–49, 50–74, and 75–100%, respectively. A total score was then calculated by adding up all the grades in the six zones (Apex and upper zones were considered as one zone) to provide a 0–24 points' scale with the higher number representing more severe pneumonic disease.

3.3. Serum Preparation

1. 8 mL Blood is collected from each patient or control subject.
2. Incubate the blood for approx 2 h at 4°C to clot the blood cells.
3. Separate the serum portion from the clotted cells by centrifugation at 230g for 15 min.
4. Aliquot the sera for various routine laboratory diagnostic tests.
5. Freeze the remaining sera at −70°C for protein chip array profiling analysis.

3.4. Ion Exchange Fractionation of Serum Proteins

Owing to the protein complexity in the serum sample, an initial fractionation in Q Ceramic Hyper D F ion exchange sorbent beads is performed to generate serum fractions for chip binding experiments (**Fig. 1**) (*10,11*).

3.4.1. Washing of Ion Exchange Beads

1. Wash Q Ceramic Hyper D Fanion exchange sorbent beads three times each with five bed volumes of W1 bead washing buffer.
2. Drain the washing buffer each time in vacuum after washing.
3. Keep beads in 50% suspension in W1 buffer at RT.

3.4.2. Denaturation of Serum Proteins

1. Thaw serum from –70°C freezer immediately before serum fractionation.
2. Aliquot 20 µL of serum to each well in a 96-well sample plate.
3. Add 30 µL of D1 denaturing buffer to the well containing the serum to denature the serum proteins.
4. Shake the sample plate on a Micromix 5-01 shaker vigorously (in an amplitude of 7 and form of 20 Hg) for 20 min at 4°C to mix the serum with the D1 buffer well.

3.4.3. Equalibration of Ion Exchange Beads and Binding of Serum Samples

1. Add 180 µL of 50% suspension of ion exchange beads to each well of a 96-well Silent Screen filter plate and drain off the W1 bead-washing buffer by vacuum.
2. Wash the ion exchange beads in each well three more times by adding 200 µL of Q1 bead equilibration buffer to the beads and draining off the buffer by vacuum.
3. After the last vacuum drain, transfer 50 µL of the denatured serum sample from the well in the sample plate to the corresponding well in the filter plate containing the equilibrated ion exchange beads.
4. Add 50 µL of Q1 buffer to the well in the sample plate where serum was originally placed and rinse the residual serum by pipetting up and down five times.
5. Transfer the 50 µL of the serum rinse to the well in the filter plate containing the mixture of denatured serum sample and ion exchange beads.
6. Shake the filter plate vigorously on the Micromix shaker (in an amplitude of 7 and form of 20 Hg) for 30 min at 4°C to mix the serum and beads well.

3.4.4. Elution of Serum Fractions

1. After shaking, place a 96-well F1 collection plate under the filter plate and centrifuge at 1600*g* for 1 min to collect the flow-through fraction in the wells of the F1 collection plate.
2. Add 100 µL of E1 elution buffer (at pH 9.0) to each well in the filter plate containing the serum bound beads and shake vigorously again on orbital shaker (speed and time as before) for 10 min at RT.
3. Collect the eluate in the same collection plate F1 by centrifugation at 1600*g* for 1 min again.
4. This represents fraction 1, which contains both the flow-through fraction at pH 9.0 and the pH 9.0 eluate.
5. Add 100 µL of E2 elution buffer (at pH 7.0) to each well in the filter plate and again shake at the same speed and time.
6. Collect the eluate in another new F2 collection plate by centrifugation as before.
7. Add 100 µL of E2 elution buffer again to each filter well and elute a second time by shaking and centrifugation as before onto F2 collection plate again.
8. This represents fraction 2, which contains the pH 7.0 eluate.

9. Elute the serum protein bound beads similarly in turn with E3, E4, and E5 elution buffers as above resulting in fractions 3, 4, and 5 containing pH 5.0, pH 4.0, and pH 3.0 eluates, respectively, in three new F3, F4, and F5 collection plates.
10. Final elution is achieved by adding E6 organic solvent elution buffer and centrifugation at 2000g for 5 min giving rise to fraction 6 containing the organic solvent eluate in collection plate F6.
11. Freeze fractions 1–6 in collection plates E1-6 at –70°C until the chip binding protocol proceeds.

3.5. Chip Pretreatment

The ProteinChip Array Profiling technique is also called surface enhanced laser desorption and ionization time of flight mass spectrometry (SELDI-TOF-MS, Ciphergen Biosystems Inc., Fremont, CA). The chip array used is a 10 mm wide × 80 mm long metal chip having eight 2-mm spots with a specific chromatographic surface for binding of biomarkers of interest (*see* **Fig. 1**). For IMAC30 Cu(II) ProteinChip Array, a pretreatment procedure is required for loading copper ions onto the chip for binding protein biomarkers with affinity to copper ions. This metal loading step is, however, not required for CM10 ProteinChip Array, which is impregnated with carboxylate ions as weak cation exchanger for biomarker binding. CM10 chips only requires buffer washing instead. Both types of chips are tested to be ideal for binding serum proteins/peptides for the study.

3.5.1. IMAC30 Chip Pretreatment

1. Assemble the 96-well bioprocessor for chip pretreatment by placing 12 strips of protein chips in the base clamp assembly and then putting a 96-hole rubber gasket sheet and a 96-well bioprocessor reservoir on top of the chips.
2. Clamp this sandwich tight with the 96 incubation wells in the reservoir placed directly on top of the 96 chip spots.
3. Add 50 µL of 100 mM CuSO4 solution into each well in the reservoir making direct contact with each IMAC30 chip spot.
4. Get rid of air bubbles if present.
5. Shake the chips in the bioprocessor for 5 min at the same speed and time as before in the orbital shaker at RT to allow copper ions from CuSO4 solution to bind to the chip surface.
6. Pipet away the CuSO4 solution.
7. Rinse the chip spot in each well with 100 µL of Milli-Q grade of water and shake again under the same condition for 1 min at RT.
8. Rinse each well with 100 µL of sodium acetate buffer at pH 4.0 (IMAC30 chip neutralizing buffer) for 5 min with shaking at RT.

9. Remove the neutralizing buffer after shaking.
10. Rinse the well again with 100 µL of Milli-Q water by shaking for 1 min at room temperature.
11. Add 150 µL of 100 m*M* sodium phosphate buffer containing 0.5 mol/L NaCl (IMAC30 chip binding buffer) onto each well and shake for 5 min at RT to wash the chip spot.
12. Remove the buffer after shaking.
13. Wash with the same IMAC30 chip-binding buffer one more time.
14. Remove the buffer after washing.
15. Proceed immediately to **Subheading 3.6.1.** for sample binding without letting the chip surface to dry.

3.5.2. CM10 Chip Pretreatment

1. After assembling the CM10 chips onto the 96-well bioprocessor (*see* **steps 1–2** in **Subheading 3.5.1.**), add 150 µL of 100 m*M* sodium acetate buffer at pH 4. (CM10 chip binding buffer) onto each reservoir well and shake vigorously on micromix shaker again as before for 5 min at RT to wash the chip spot.
2. Remove the buffer from the well.
3. Wash with the same chip-binding buffer one more time.
4. Remove the buffer.
5. Proceed immediately to **Subheading 3.6.2.** for sample binding without letting the chip surface to dry.

3.6. Serum Fraction Binding on Chips

3.6.1. IMAC30 Chips

1. Add 80 µL of the IMAC30 chip-binding buffer into each well in the bioprocessor set up containing the IMAC30 chips.
2. Add 20 µL of each ion exchange bead eluate (serum fractions 1–6) to the well for sample binding.
3. Mix well by shaking for 30 min at RT.
4. Remove the serum fractions.
5. Wash the chip spot in the well by adding 150 µL of IMAC30 chip-binding buffer onto each well and shake for 5 min at RT
6. Remove buffer.
7. Repeat washing (**steps 5** and **6**) once more.
8. Rinse with water two times by adding 200 µL of Milli-Q water to each well and discard immediately.
9. Remove the chips from bioprocessor and air-dry the chips for 5 min.
10. Add 1 µL of 50% sinapinic acid (which is an energy absorbing molecule solution) to each chip spot.
11. Air-dry the chips for about 10 min.
12. Add 1 µL of 50% sinapinic acid and air-dry again.
13. The chip is ready to be read in the PBS IIc mass spectrometer reader.

14. Remember to process at least one reference control serum concurrently with the patients' samples on each chip for quality control of chip-to-chip variability.

3.6.2. CM10 Chips

1. Add 90 μL of 100 m*M* sodium acetate buffer at pH 4.0 (CM10 chip binding buffer) into each well in the bioprocessor set up containing the CM10 chips.
2. Add 10 μL of each ion exchange bead eluate (sample fractions 1–6) to the well for sample binding.
3. Mix well by shaking for 30 min at RT.
4. Remove the sample fractions.
5. Wash the chip spot in the well by adding 150 μL of CM10 chip binding buffer again into each well and shake for 5 min at RT.
6. Remove buffer.
7. Repeat washing (**steps 5** and **6**) once more.
8. Rinse with water two times by adding 200 μL of Milli-Q water to each well and discard immediately.
9. Remove the chips from bioprocessor and air-dry the chips for 5 min.
10. Add 1 μL of 50% sinapinic acid to each chip spot.
11. Air-dry the chip for around 10 min.
12. Add 1 μL of 50% sinapinic acid and air-dry again.
13. The chip is ready to be read in the PBS IIc mass spectrometer reader.
14. Process at least one reference control serum concurrently with the patients' samples on each chip for quality control of chip-to-chip variability as before.

3.7. Chip Reading and Data Acquisition

The PBS IIc SELDI-TOF mass spectrometer reader is a laser desorption ionization time of flight mass spectrometer equipped with a pulsed ultraviolet nitrogen laser source. When the laser activates the bound serum biomarkers on the chip surface, the biomarkers become desorbed and ionized. Ionized molecules fly along the mass spectrometer to the ion detector in a time of flight manner according to their mass over charge ratio (m/z). When the ion signal is detected in the ion detector, signal processing is accomplished by high-speed analog-to-digital converter linking to a computer. Detected protein biomarkers are displayed in spectral, map or gel view formats by the Ciphergen ProteinChip software 3.0.2. The following steps show how the chips are read in the mass spectrometer.

1. Click the "Sample Exchange Dialog" button from the manu bar of the Ciphergen ProteinChip software 3.0.2 in the PBS IIc SELDI-TOF mass spectrometer reader.
2. Click "Open Lid" button in the "Sample Exchange Dialog" box to open the lid of the chip chamber.
3. Place the sample treated chips into the slot of the chip chamber.
4. Click "Close Lid" button.

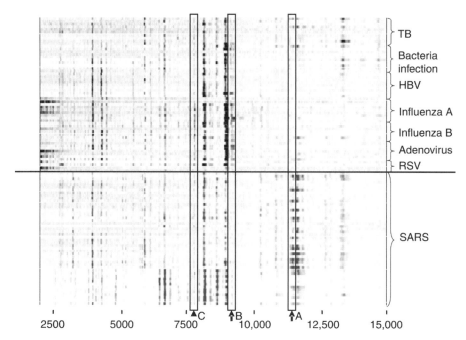

Fig. 3. Gel view of partial protein chip array profiling results in SARS patients' sera vs control infection groups' sera. Part of the profiling results of fraction 1 from the ion exchange serum eluate in SARS patients vs the control patients in m/z range from 2000 to 15,000 was illustrated. HBV, hepatitis B virus; TB, *M. tuberculosis*; RSV, respiratory syncytial virus. The three arrows showed three clusters of biomarkers, **A–C** at m/z of 11695, 9159, and 7784, respectively, with cluster **A** being significantly increased in SARS patients and the last two clusters (**B** and **C**) significantly decreased.

5. The chip is automatically inserted into the chip chamber.
6. In the "Sample Exchange Dialog" box, input the serial number, chip type, and chip format.
7. Click "Chip" button from the "Sample Exchange Dialog" box.
8. Input sample name and sample group.
9. Click "OK" button.
10. Click "Chip Protocol" button from the menu bar.
11. Input spectrum tag (which is the sample name) and the following spot protocols:
    a. Starting laser intensity to 165.
    b. Starting detector sensitivity to 9.
    c. Highest mass of detection to 200,000 Da.
    d. Optimal mass range of detection from m/z of 2000 to 20,000 (signals from m/z of 0–2000 are not analyzed as artifacts can be produced by energy-absorbing molecules or other chemical contaminants at this mass range).

Table 1
Serum Biomarkers Significantly Increased or Decreased in SARS Patients vs Seven Control Groups of Patients With Other Infections and Healthy Individuals

Biomarker number	Mass over charge (*m/z*)	Differential expression	*p*-value[a]
1	4922	Increased	1.1×10^{-4}
2	5104	Increased	7.1×10^{-8}
3	5215	Increased	6.4×10^{-4}
4	5833	Increased	3.1×10^{-9}
5	7784	Decreased	4.9×10^{-8}
6	8416	Decreased	1.6×10^{-7}
7	9159	Decreased	1×10^{-10}
8	10,867	Increased	3.6×10^{-7}
9	11,508	Increased	1.3×10^{-5}
10	11,695	Increased	5.4×10^{-6}
11	11,871	Increased	8.4×10^{-4}
12	14,715	Increased	3.9×10^{-9}

[a]*p*-values were obtained in Mann–Whitney *U*-test by comparing the normalized peak intensities of the biomarkers in 44 sera from 24 SARS patients vs those in 72 sera from 51 control patients with various viral or bacterial infections (*see* Materials for individual control group) plus 10 sera from 10 apparently healthy individuals.

     e. Focus lag time at 800 ns.
     f. Mass deflector to Auto.
     g. Data acquisition method to SELDI quantitation.
     h. 338 laser shots per sample.
12. Click "Start Running" button from the menu bar.
13. By the time 338 laser shots are fired, the collected data is automatically saved.
14. It is important to first carry out external calibration of the equipment using the All-In-1 Peptide molecular mass standard (according to the instruction sheet from the manufacturer) to ensure mass accuracy of the spectra before running the samples.
15. Carry out baseline subtraction and calibration for the spectra.
16. Generate labeled peak groups (clusters) across multiple spectra by the Biomarker Wizard mode.
17. Compare peak groups detected in the SARS patients with those of the controls by nonparametric two sample Mann–Whitney *U*-test in the Biomarker Wizard mode.
18. Biomarker Wizard operates in two passes, with the first pass uses low sensitivity settings to detect obvious and well-defined peaks and the second pass uses higher sensitivity settings to search for smaller peaks.
19. This will identify both strong and weak peaks that are significantly increased or decreased in the patients vs the controls.

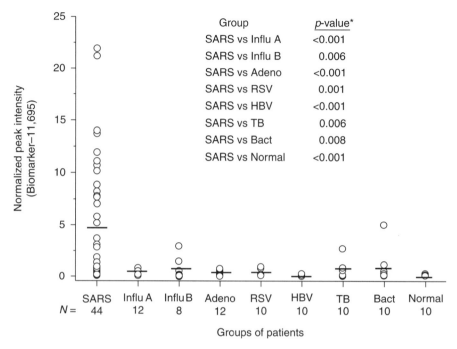

Fig. 4. Comparison of normalized peak intensities of the biomarker at m/z of 11,695 in sera from SARS patients and control groups. The normalized peak intensities of this marker in SARS patients' sera were compared with those of the other seven control infection groups and the normal individuals by Mann–Whitney *U*-test. A *p*-value of 0.05 or less demonstrated a statistically significant difference between the two groups. Adeno, adenovirus; Bact, bacterial culture positive; HBV, hepatitis B virus; Influ A, influenza A virus; Influ B, influenza B virus; Normal, apparently normal subjects; RSV, respiratory syncytial virus; TB, *M. tuberculosis*.

3.8. Data Analyses

1. As an example, **Fig. 3** illustrates a comparison of multiple biomarkers in serum fraction 1 (pH 9.0 eluate) of SARS patients vs all the other controls in a gel view format in the m/z range of 2000–15,000.
2. Highlighted in three boxes (A, B, and C) with arrows are biomarkers at m/z of 11,695, 9159, and 7784, respectively with biomarker at 11,695 being significantly increased in the SARS patients vs the controls and biomarkers 9159 and 7784 being significantly decreased.
3. Nine biomarker were found within m/z range of 4900–15,000 with highly significant increase in their normalized peak intensities in the SARS patients vs those of the controls and three biomarkers with highly significant decrease (**Table 1**).

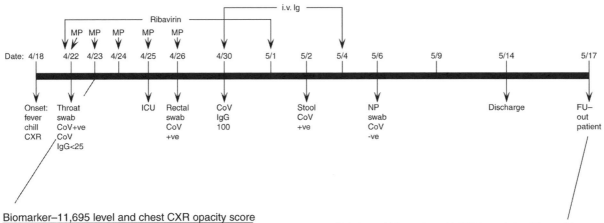

Clinical profile

i.v. Ig

Ribavirin

MP MP MP MP MP

Date: 4/18   4/22 4/23 4/24 4/25 4/26   4/30   5/1   5/2   5/4   5/6   5/9   5/14   5/17

Onset: Throat   ICU   Rectal   CoV   Stool   NP   Discharge   FU–
fever   swab   swab   IgG   CoV   swab   out
chill   CoV+ve   CoV   100   +ve   CoV   patient
CXR   CoV   +ve   -ve
         IgG<25

Biomarker–11,695 level and chest CXR opacity score

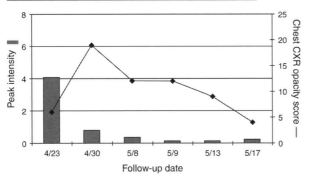

Peak intensity

Chest CXR opacity score —

8

6

4

2

0

25

20

15

10

5

0

4/23   4/30   5/8   5/9   5/13   5/17

Follow-up date

Gel view of biomarker–11,695 in protein chip profiling

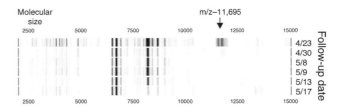

Molecular size

m/z–11,695

2500   5000   7500   10000   12500   15000

2500   5000   7500   10000   12500   15000

Follow-up date

4/23
4/30
5/8
5/9
5/13
5/17

4. Separating the controls into individual groups, a highly significant increase of bio-marker at 11,695 is also observed in SARS patients vs each control group (**Fig. 4**).
5. As pneumonia is a frequent life threatening clinical manifestation in the SARS patients, we further monitor the biomarker—11,695 level in a SARS patient longi-tudinally (**Fig. 5**) and correlate the biomarker level with the extent of pneumonia by chest X-ray opacity score (**Fig. 2**).
6. **Figure 5** illustrates, in this patient, an elevation of the biomarker—11,695 level preceding the development of pneumonia at the onset of disease (as indicated by an increasing opacity score) followed by a rapid drop of the biomarker level down to the background on recovery from the illness after stringent antiviral and steroid treatments (*see* **Note 1** for details of clinical presentation of the patients).
7. On the other hand, **Fig. 6** shows low biomarker—11,695 level during the whole monitoring period in another SARS patient who has low X-ray scores and whose clinical course is relatively mild (*see* **Note 2**).
8. By means of peptide mapping and tandem MS/MS mass spectrometry, this bio-marker is identified to be Serum Amyloid A protein (SAA), which is an acute phase reactant frequently and rapidly induced in abundance at pneumonia (*see* **ref. 15** for the protein ID results and **ref. 17–20** for the techniques adopted).
9. This serial study demonstrates the potential of the biomarkers discovered by protein chip array profiling in monitoring the disease manifestation in SARS patients.

4. Notes

1. *Clinical and protein chip monitoring of the first SARS patient* (see *Fig. 5*): one day after SARS-CoV infection was diagnosed, the radiographic score of this patient was increased from 6 to a peak value of >16 and then dropped to 12, 9, and finally

---

Fig. 5. *(Opposite page)* Correlation of the level of serum biomarker—11,695 with the extent of pneumonia in a SARS patient under clinical follow-up. Longitudinal fol-low-up of the clinical profile of this patient was illustrated in the top panel. Monitoring of the biomarker—11,695 level and chest X-ray opacity score (an indicator of the extent o pneumonia) was shown in the lower left panel. The lower right panel illustrates the biomarker—11,695 in a gel view measured at different time points. CoV IgG <25 and 100, Serum SARS-Coronavirus IgG antibody titers of <1/25 and 1/100; CXR, Lung consolidation as shown by chest X-ray imaging; FU outpatient, During follow-up as outpatient; ICU, Admitted to Intensive care unit; i.v. Ig, Intravenous injection of immune globulin; MP, Daily treatment by 500 mg × 2 of methylprednisolone; NP Swab CoV –ve, RT-PCR negative for SARS Coronavirus in nasopharyngeal swab; Rectal Swab CoV +ve, RT-PCR positive for SARS Coronavirus in rectal swab; Ribavirin, Daily treatment by 400 mg Ribavirin; Stool CoV +ve, RT-PCR positive for SARS Coronavirus in the stool; Throat Swab CoV +ve, RT PCR positive for SARS Coronavirus in throat swab.

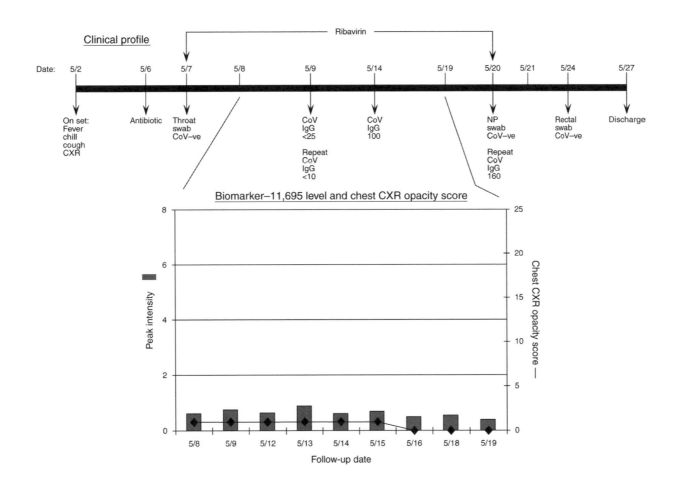

328

to 4 demonstrating a progressive recovery. The biomarker—11,695 level by protein chip array profiling study showed an elevation that peaked earlier than the radiographic score but it then gradually subsided along with the score to a nadir when the patient was discharged.

2. *Clinical and protein chip monitoring of the second SARS patient* (see *Fig. 6*): the second patient had all the typical clinical symptoms of SARS-CoV infection with left upper lobe consolidation in chest radiographs. Despite negative SARS-CoV RT-PCR, titer elevation in paired anti-SARS CoV serum antibody tests performed in two laboratories confirmed the viral infection. He was managed as SARS on the next day. The clinical course was uneventful and uncomplicated. His pneumonia had never been extensive and hence his radiographic score showed just a low level of lung involvement from May 8th to 15th and then totally subsided from May 16th onwards. Biomarker—11,695 level was essentially low throughout the monitoring period.

Acknowledgments

The authors thank our Kowloon Central Cluster/Queen Elizabeth Hospital Chief Executive, Dr. C. T. Hung and the SARS Management Committee for their support in this project. We also express our appreciation to Dr. Dominic N. C. Tsang, Dr. John K. C. Chan, Dr. King-Chung Lee, Dr. Angel Chan, Dr. Cesar S. C. Wong, Mr. K. H. Leung, Mr. C. C. Tiu, Mr. Gene G. T. Lau, and Miss Elena S. F. Lo in the Pathology Department of Queen Elizabeth Hospital for the provision of SARS patients' sera and their help in this project. The authors thank Dr. Wilina W. L. Lim for the provision of control sera from patients with upper respiratory infection and Dr. Ting-Lok Kwan for the provision of the serial chest X-ray score. We express our gratitude to Mr. Cadmon K. P. Lim in his technical assistance in this project. This work was supported in part by grants from Queen Elizabeth Hospital and Hong Kong Jockey Club. Part of the contents in this book chapter has been reproduced with permission from Yip et al., 2005 *see* **ref. 15** (Copyright 2004 AACC).

Fig. 6. *(Opposite page)*   Correlation of the level of serum biomarker—11,695 with the extent of pneumonia in the second SARS patient under clinical follow-up. Longitudinal follow-up of the clinical profile of this patient was illustrated in the top panel. Monitoring of the biomarker—11,695 level and chest X-ray opacity score was shown in the lower panel. Annotations were similar to those as tabulated in **Fig. 5**. Antibiotic, Treatment with conventional antibiotics; CoV IgG <10 and 160, Serum SARS-Coronavirus IgG antibody titers of <1/10 and 1/160; NP Swab CoV –ve, RT-PCR negative for SARS Coronavirus in nasopharyngeal swab; Rectal Swab CoV –ve, RT-PCR negative for SARS Coronavirus in rectal swab; Throat Swab CoV –ve, RT PCR negative for SARS Coronavirus in throat swab.

References

1. World Health Organization. (2003) WHO issued consensus document on the epidemiology of SARS. *WHO Wkly Epidemiol. Rec.* **78,** 373–375.
2. Engel, J. P. (1995) Viral upper respiratory infections. *Semin. Respir. Infect.* **10,** 3–13.
3. Chan, J. W. M., Ng, C. K., Chan, Y. H., et al. (2003) Short term outcome and risk factors for adverse clinical outcomes in adults with severe acute respiratory syndrome (SARS). *Thorax* **58,** 686–689.
4. Donnelly, C. A., Ghani, A. C., Leung, G. M., et al. (2003) Epidemiological determinants of spread of casual agent of severe acute respiratory syndrome in Hong Kong. *Lancet* **361,** 1761–1766.
5. Yam, W. C., Chan, K. H., Poon, L. L., et al. (2003) Evaluation of reverse transcriptase-PCR assays for rapid diagnosis of severe acute respiratory syndrome associated with a novel coronavirus. *J. Clin. Microbiol.* **41,** 4521–4524.
6. Xu, G., Lu, H., Li, J., et al. (2003) Primary investigation on the changing mode of plasma specific IgG antibody in SARS patients and their physicians and nurses. *Beijing Da Xue Xue Bao* **35,** 23–25.
7. Poon, L. L. M., Wong, O. K., Luk, W., Yuen, K. Y., Peiris, J. S. M., and Guan, Y. (2003) Rapid diagnosis of a coronavirus associated with severe acute respiratory syndrome (SARS). *Clin. Chem.* **49,** 953–955.
8. Grinblat, L., Shulman, H., Glickman, A., Matukas, L., and Paul, N. (2003) Severe acute respiratory syndrome: radiographic review of 40 probable cases in Toronto, Canada. *Radiology* **228,** 802–809.
9. Muller, N. L., Ooi, G. C., Khong, P. L., and Nicolaou, S. (2003) Severe acute respiratory syndrome: radiographic and CT findings. *AJR Am. J. Roentgenol.* **181,** 3–8.
10. Fung, E. T., Thulasiraman, V., Weinberger, S. R., and Dalmasso, E. A. (2001) Protein biochips for differential profiling. *Curr. Opin. Biotechnol.* **12,** 65–69.
11. Wright, G. L. Jr. (2002) SELDI proteinchip MS: a platform for biomarker discovery and cancer diagnosis. *Expert Rev. Mol. Diagn.* **2,** 549–563.
12. Weinberger, S. R., Dalmasso, E. A., and Fung, E. T. (2002) Current achievements using ProteinChip Array technology. *Curr. Opin. Chem. Biol.* **6,** 86–91.
13. Issaq, H. J., Veenstra, T. D., Conrads, T. P., and Felschow, D. (2002) The SELDI-TOF MS approach to proteomics: protein profiling and biomarker identification. *Biochem. Biophys. Res. Commun.* **292,** 587–592.
14. Vorderwulbecke, S., Cleverley, S., Weinberger, S. R., and Wiesner, A. (2005) Protein quantification by the SELDI-TOF-MS-based ProteinChip® system. *Nat. Method* **2,** 393–395.
15. Yip, T. T. C., Chan, J. W. M., Cho, W. C. S., et al. (2005) protein chip array profiling analysis in patients with severe acute respiratory syndrome identified serum amyloid a protein as a biomarker potentially useful in monitoring the extent of pneumonia. *Clin. Chem.* **51,** 47–55.
16. Ng, C. S., Desai, S. R., Rubens, M. B., Padley, S. P. G., Wells, A. U., and Hansell, D. M. (1999) Visual quantitation and observer variation of signs of small airways disease at inspiratory and expiratory CT. *J. Thorac. Imag.* **14,** 279–285.

17. Cleveland, D. W., Fischer, S. G., Kirschner, M. W., and Laemmli, U. K. (1977) Peptide mapping by limited proteolysis in sodium dodecyl sulfate and analysis by gel electrophoresis. *J. Biol. Chem.* **252,** 1102–1106.

18. Baudys, M., Foundling, S., Pavlik, M., Blundell, T., and Kostka, V. (1988) Protein chemical characterization of Mucor pusillus aspartic proteinase amino acid sequence homology with the other aspartic proteinases disulfide bond arrangement and site of carbohydrate attachment. *FEBS Lett.* **235,** 271–274.

19. Fournier, I., Chaurand, P., Bolbach, G., Lutzenkirchen, F., Spengler, B., and Tabet, J. C. (2000) Sequencing of a branched peptide using matrix-assisted laser desorption/ionization time-of-flight mass spectrometry. *J. Mass Spectrom.* **35,** 1425–1433.

20. Perkins, D. N., Pappin, D. J., Creasy, D. M., and Cottrell, J. S. (1999) Probability-based protein identification by searching sequence databases using mass spectrometry data. *Electrophoresis* **20,** 3551–3567.

21

Volumetric Mass Spectrometry Protein Arrays

Dobrin Nedelkov, Urban A. Kiernan, Eric E. Niederkofler, Kemmons A. Tubbs, and Randall W. Nelson

Summary

Affinity mass spectrometry is a proteomics approach for selectively isolating target proteins from complex biological fluids for mass spectrometric analysis. When executed in high through-put mode through affinity pipets, the resulting volumetric mass spectrometry arrays enable rapid protein assaying from hundreds of samples. Furthermore, in combination with postcapture prote-olytic degradation, this top-down proteomics approach can reveal structural features (i.e., modi-fications) in the protein sequences that are result of posttranslational modifications and/or point mutations. Described here in greater detail are the individual steps of the high throughput com-bination of affinity protein capture in antibody-derivatized affinity pipets, protein elution, and protein processing through enzyme-derivatized mass spectrometry targets.

Key Words: Affinity-capture; high throughput; mass spectrometry; proteins; plasma; enzy-matic digestion.

1. Introduction

In the last several years, proteomics and its subdiscipline clinical proteomics have been engaged in the discovery of the next generation protein biomarkers *(1–3)*. As the effort and the intensive debate it has sparked continue, it is becom-ing apparent that a paradigm shift is needed in proteomics in order to start grasping the complexity of the human proteome and assess its subtle variations among individuals. Current mass spectrometry (MS)-based proteomics tech-nologies and approaches are not simple or robust enough to allow for repro-ducible and repetitive protein analysis from hundreds-to-thousands of samples. Furthermore, MS bottom-up approaches seem to significantly complicate pro-teomic analyses by increasing the complexity of the starting sample, through generation of tens of peptides from a single protein. Furthermore, by relying on only few peptides for protein identification, a large portion of the protein

From: *Methods in Molecular Biology, vol. 382: Microarrays: Second Edition: Volume 2*
Edited by: J. B. Rampal © Humana Press Inc., Totowa, NJ

sequence remains "unassayed." And throughput remains a significant bottleneck with most of the existing proteomic instrumentation. Although high-density functional protein arrays have become viable, miniaturized, and high- throughput alternatives to classical immunoassays, they do not offer insight into the protein structure the way MS does.

Recently, a hybrid array-format methodology was developed that can readily be used to study protein modifications across and within populations. This top-down proteomics approach combines protein affinity-extraction with rigorous characterization using MALDI-TOF mass spectrometry. Protein affinity extraction (from a biological sample) is achieved with the help of affinity pipets, which are made of pipet tips fitted with porous silicon-based microcolumns derivatized with polyclonal antibodies toward a specific protein. The affinity capture is followed by elution of the protein onto arrayed proteolytic enzymes derivatized MALDI targets for subsequent digestion. The MS analyses of the resulting peptide fragments reveal structural features (i.e., modifications) in the protein sequences that are result of posttranslational modifications and/or point mutations. High-throughput protein analyses are achieved through the use of robotics that enables parallel manipulation of 96 affinity pipets. The core technologies of the methodology were conceptualized in the mid-1990s *(4–7)*, but the enabling components were perfected only recently *(8–10)*. In the most recent applications of the methodology, we have developed several new protein assays *(11–14)*, performed comparative protein phenotyping *(15)* and expression profiling *(16)* across a small set of sample cohorts, and demonstrated highly reproducible and high-throughput proteomics analyses of proteins from hundreds of samples per day *(17)*. Here, the individual steps of the high throughput combination of arrayed affinity protein capture, elution, and on-target enzymatic processing are described in greater detail **(Fig. 1)**.

2. Materials

1. Normal (nonreactive) human heparin–plasma sample (ProMedDX, Norton, MA).
2. 1,1′-Carbonyldiimidazole (CDI)-activated affinity pipets (Intrinsic Bioprobes Inc., Tempe, AZ) (*see* **Note 1**).
3. Rabbit antihuman polyclonal antibody to transthyretin (DakoCytomation, Carpinteria, CA; cat. no. A0002, 3.9 g/L solution).
4. Antibody coupling buffer: 10 mM acetate, pH 5.0, prepared fresh.
5. Blocking solution: 1 M ethanolamine (ETA) (Sigma-Aldrich, St. Louis, MO), pH 8.5. Solution stored at 4°C.
6. 60 mM HCl, stored at room temperature.
7. HBS physiological buffer (HEPES-buffered saline): 10 mM HEPES, pH 7.4, 150 mM NaCl. Stored at 4°C.
8. Sterile water (American Bioanalytical, Natick, MA; cat. no. AB02120-01000).
9. Bradykinin and ACTH fragment 18–39 (Sigma).

10. Organic rinse solution: 2 *M* ammonium acetate–acetonitrile (3:1 v/v) mixture, prepared fresh.
11. Normalization rinse solution: 10 m*M* N-octylglucoside (NOG) (Roche Diagnostics, Mannheim, Germany; cat. no. 10281722). Stored at 4°C.
12. MALDI matrix: α-cyano-4-hydroxycinnamic acid (ACCA) (Aldrich, Milwaukee, WI; cat. no. 47,687-0), which is further processed by powder-flash recrystallization from a low-heat saturated acetone solution of the original stock. The working matrix solution is prepared fresh daily by dissolving ACCA (at 6 g/L) in aq solution containing 33% (v/v) acetonitrile and 0.4% (v/v) trifluoroacetic acid (TFA).
13. Digests-buffering solution: 25 m*M* Tris-HCl, pH 9.1. Stored at 4°C.
14. Digests rehydrating solution: 0.8% TFA (American Bioanalytical, cat. no. AB02010). Stored at 4°C.
15. 96-Well format trypsin-derivatized MALDI target (Bioreactive Probe, BRP, Intrinsic Bioprobes Inc.), made with sequencing-grade trypsin (Promega Co., Madison, WI; cat. no. V511A).
16. Beckman Multimek Automated 96-Channel Pipetor (Beckman Coulter, Fullerton, CA).
17. Bruker Biflex MALDI-TOF mass spectrometer (Bruker Daltonics, Billerica, MA).
18. Proteome Analyzer Software (Intrinsic Bioprobes Inc.).
19. 96-Well polypropylene microplates, V-shape bottom (Greiner, Frickenhausen, Germany; cat. no. 651201).
20. Windows-based PC using Proteome Analyzer (Intrinsic Bioprobes, co-developed with Beavis Informatics, Winnipeg, Canada).
21. PAWS-sequence display and manipulation software (Proteometrics LLC, New York, NY).

3. Methods

There are six main steps in the analysis:

1. Derivatization of the affinity pipets with antibody.
2. Protein affinity extraction from the biological samples.
3. Protein elution with matrix solution.
4. Enzymatic on-plate protein digestion and preparation for MS.
5. Mass spectrometry analysis.
6. Data analysis.

3.1. Derivatization of Affinity Pipetes With Antibody

1. All solutions used in the derivatization of the affinity pipets are aliquoted into 96-well microplates. The transthyretin (TTR) antibody is diluted 100-fold into the antibody coupling acetate buffer, and 200-µL aliquots are dispensed into each well of a 96-well microplate. The ETA, HCl, and HBS buffer solutions are similarly dispensed into separate microplates at 200 µL per well (two microplates of each HBS and HCl are prepared).
2. Ninty-six underivatized affinity pipets are loaded onto the Beckman Multimek 96-channel pipetor head.

3. A program macro is started on the workstation controlling the Multimek pipetor to automatically run the derivatization process.
4. In the first step, the 96 affinity pipetes are immersed into the antibody solution microplate, and 50 µL volumes are passed (through repetitive aspiration and dispensing) through the pipets 400 times, allowing for the covalent attachment of the antibody to the activated microcolumn within each affinity pipet.
5. After the coupling, the pipets are moved to the microplate containing the ethanolamine solution, and ETA is repetitively aspirated and dispensed 50 times, in 100 µL volumes, to block the unreacted active sites on the microcolumns.
6. The pipets are then moved to the two HCl-containing microplates, and 100 µL volumes are aspirated and dispensed 30 times from each of the two microplates, consecutively.
7. Finally, the pipets are moved to the HBS-containing microplates, and 100 µL volumes are aspirated and dispensed 30 times from each of the two microplates, consecutively.
8. Once reconditioned with the final HBS buffer rinse, the antibody-derivatized affinity pipets are ready for use (*see* **Note 2**).

3.2. Affinity Extraction

1. All solutions used in the protein affinity capture are aliquoted into 96-well microplates. The human plasma sample is diluted 10-fold into the HBS buffer, and 200-µL aliquots are dispensed into each well of the 96-well microplate. Two microplates containing HBS, three with H_2O, and one with the organic rinse solution (each at 200 µL/well) are prepared (**Fig. 1**).
2. The 96 anti-TTR derivatized affinity pipets are loaded onto the Beckman Multimek 96-channel pipetor head (*see* **Note 3**).
3. A program macro is started on the workstation controlling the Multimek pipetor to automatically run the affinity extraction process.
4. In the first step, the 96 affinity pipets are immersed into the microplate containing the HBS buffer, and 100 µL volumes are passed (through repetitive aspiration and dispensing) through the pipets 10 times, to (re)condition the affinity pipets.
5. The pipets are then moved to the microplate containing the plasma samples, and 100 µL volumes are aspirated and dispensed 70 times to allow for the affinity capture of transthyretin by the immobilized antibody.
6. The pipets are then moved to the second HBS-containing microplate, and 100 µL volumes are aspirated and dispensed 10 times to remove any loosely retained components.
7. The pipets are next rinsed with water, by aspirating and dispensing (five times) 100 µL-volumes from the water-containing microplate.
8. The pipets are then subjected to an organic rinse, by aspirating and dispensing (10 times) 100 µL volumes from the organic rinse containing microplate (*see* **Note 4**).
9. The pipets are then moved to the two H_2O-containing microplates, and 100 µL volumes are aspirated and dispensed 10 times from each of the microplates,

Affinity pipetes   Bioreactive MS targets   HT processing

+   +

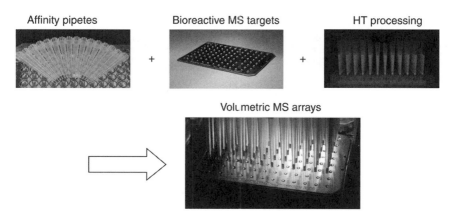

Volumetric MS arrays

Fig. 1. Illustration of the major assay components.

consecutively. The protein loaded-affinity pipets are now ready for elution, which should be carried out immediately after the affinity capture step.

3.3. Protein Elution Onto Bioreactive Probe

1. All the reagents required for the elution of the affinity-retained proteins are prepared fresh. This includes making 50 mL of the 10 m*M* NOG solution, and 50 mL of MALDI matrix solution, and pouring them into separate 200-mL volume basins that fit within the Multimek pipetting station (*see* **Note 5**). Also, the trypsin-derivatized MALDI target is positioned within the Multimek pipetor in preparation for eluates deposition.
2. Another program macro is started on the workstation controlling the Multimek pipetor to automatically run the elution process.
3. The protein-loaded 96 affinity pipets are immersed into the NOG-containing basin for a single 100-µL aspiration and dispense rinse, which uniformly wets the solid support inside the affinity pipets (*see* **Note 6**).
4. The program is then paused, and the affinity pipets are blotted with a Kimwipe to remove any residual NOG solution.
5. Continuing the program, the pipets are next moved to the basin containing the matrix solution, and 6 µL volumes are aspirated within each pipet, enough to cover the microcolumns at the entrance of the pipets that contain the captured protein. The matrix serves as an elution solution (*see* **Note 7**).
6. The 96 affinity pipets containing the matrix solution are then moved and positioned vertically to within 0.5–1 mm above the trypsin-derivatized MALDI target. After a 10 s delay, the 6 µL volumes inside the pipets (the protein-containing eluates) are directly dispensed onto the 96-well formatted target surface. The trypsin-derivatized target is thus imprinted with protein samples in the same 8 × 12 array format as the one in which the proteins were captured from the samples.
7. After deposition of the eluates, the affinity pipets can be discarded.
8. The MALDI target is allowed to dry, at which point it can undergo MS analysis (*see* **Note 8**) or be subjected to the subsequent digestion step.

3.4. On-Target Digestion and Preparation for MS

1. The digestions of the protein on the trypsin-derivatized MALDI target is initiated by placing 10-µL aliquots of 25 m*M* Tris-HCl, pH 9.1 buffer onto each of the 96 spots (*see* **Note 9**).
2. The drops-containing trypsin-derivatized target is then placed into a humidified enclave at 50°C for optimal digestion conditions.
3. To keep the samples solvated, one 10-µL aliquot of water is added to each spot at approx 15 min into the digestion.
4. Digestions are terminated after 25 min by air-drying the bioreactive probe.
5. In preparation for MS, the sample spots are rehydrated with 5-µL aliquots of 0.8% TFA, and allowed to air-dry again (*see* **Note 10**).

3.5. MALDI-TOF MS Analysis

1. Following matrix recrystallization, the MALDI target is inserted into the mass spectrometer for analysis.
2. MS analysis is performed on a Bruker *Biflex* III MALDI-TOF mass spectrometer in a reflectron mode using an instrument setting of full accelerating potential of 19.35 kV, an ion-mirror voltage of 20 kV, draw-out pulse voltage of 2.7 kV and a 600 ns delay. The instrument is calibrated using external mass calibrants such as Bradykinin and ACTH fragment 18–39.

3.6. Data Analysis

The data from all 96 parallel analyses was retrieved from the Bruker workstation and analyzed on a Windows-based PC using Proteome Analyzer. The Proteome Analyzer is mass spectra visualization program that, when used in conjunction with PAWS-sequence display and manipulation software, allows for rapid identification of protein sequence modifications. High-resolution spectra (m/Δm > 5000) in the range of 1000 to 4000 Da were acquired from each of the 96 sample spots. **Figure 2** shows eight mass spectra obtained from column 2 of the target plate (A2 through H2). Database search using Profound (with monoisotopic mass tolerance of 150 ppm, maximum of four missed cleavages and all other parameters including species left wide open), returned human TTR as the best match for all 96 data sets. Characteristic of all 96 spectra is the presence of 14 tryptic fragments that allow TTR sequence mapping with 100% coverage (**Fig. 3**).

4. Notes

1. Affinity pipets are available with a variety of active supports. The CDI-activated pipets were found to work best for assaying plasma proteins and were utilized in this example. The coupling protocol and reagents described here are specific for the CDI-activated pipets.

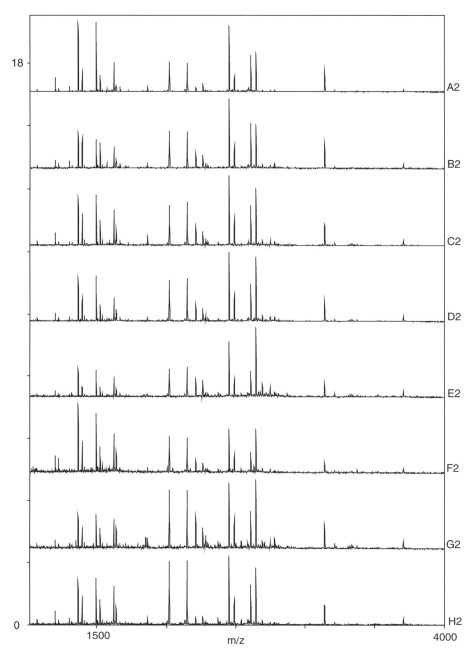

Fig. 2. An example of eight digests mass spectra from column 2.

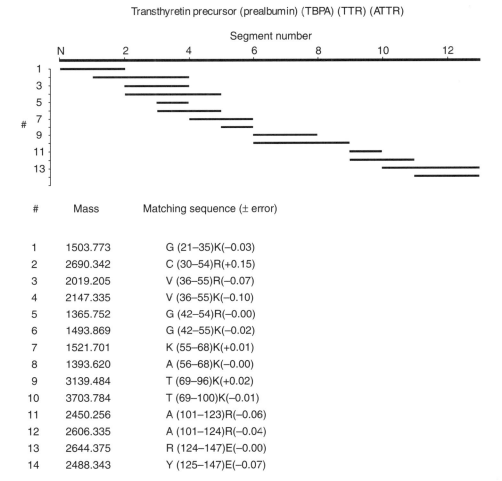

Fig. 3. TTR sequence coverage obtained from the 14 commonly observed digest peptides.

2. Immediate use of the antibody-derivatized pipets is recommended for optimal results. However, if needed, the derivatized pipets can be stored (wet, in HBS buffer) for up to 6 mo at 4°C.

3. Before loading the affinity pipets onto the Multimek head, any residual buffer that may be contained within them should be removed because the Beckman pipetor is unable to purge any initial volume of fluid. Such residual fluid might aspirate into the instrument head mechanism and contaminate the robotic system.

4. Because the two components of the organic rinse mixture (acetonitrile and ammonium acetate) are not highly miscible, the stock rinse solution should be agitated and aliquoted just before use in the analysis. At that point, the robotic method applied will pause to allow for the insertion of the organic mix microplate into the rinse series.

5. Before running the elution program macro, the matrix solution should be left at room temperature for 5 min to allow any matrix sediment to settle at the bottom of the basin. The MALDI matrix is aspirated from a large container so that the matrix molecules are uniformly dispersed. Aspirating the matrix from a 96-well microplate can result in poor uniformity and sample/matrix crystal formation owing to matrix precipitation within the individual wells of the microplate.

6. Different amounts of residual water are present in the pipets after the final water rinse of the affinity extraction step, which, owing to the slight differences in capillary action, can result in inconsistencies in the amount of matrix eluant solution drawn into each pipet. Hence, a final equilibrating rinse that either completely dries or wets the affinity pipets is needed to achieve uniform matrix aspiration and elution. For most protein assays, NOG (which uniformly wets the affinity pipets surfaces), or acetone (which rapidly evaporates and leaves the pipets uniformly dry) can be used as an equilibrating rinse. Both are nondetrimental to the overall analysis (NOG more so than acetone), and do not disrupt the affinity interaction between the immobilized antibodies and the target proteins, so that there is no loss of protein during this rinse step.

7. A major concern regarding the elution solution is the possible introduction of elution compounds and/or conditions onto the bioreactive probe that might destroy the enzymatic activity, or ultimately interfere with the MALDI process. Because most antibody–antigen interactions can be broken at extreme pH (i.e., pH ~2.0 or ~12.0), the natural choice for elution solution would be volatile acids and bases. However, it was found that the low-pH MALDI matrix is the optimal elution solution, because of its low pH and composition. The matrix has the apparent effect of denaturing the eluted proteins, which in turn yields highly reproducible digests. Moreover, the presence of matrix compounds does not have a negative influence on the trypsin enzymatic activity, and the matrix could be readily buffered up to the pH optimal for trypsin digestion. The elution with the matrix is beneficial in one more way: the matrix molecules can serve as a good indicator of the pH of digestion; at neutral pH (7.0–9.0, the pH of trypsin digestion) the ACCA matrix molecules have yellow appearance.

8. By doing MS of the native protein before digestion, structural changes that exist as a result of point mutations and/or posttranslational modifications can be rapidly uncovered through observation of additional signals in the mass spectra (usually in the vicinity of the native protein peaks). The MS acquisition has no detrimental effect on the ensuing digestion and peptide MS analysis. Actually, when screening large number of samples, it is beneficial to first uncover which samples contain modified proteins and then to perform enzymatic digestion only on those samples.

9. The matrix/protein eluates have a pH of about 2.0 (the pH of the matrix solution) and need to be buffered up to pH of about 8.0 for optimal trypsin digestion. To bring the pH up, a small aliquot of Tris-HCl buffer is added to the dried–out matrix/protein spots. The amount of Tris-HCl needed for efficient buffering is low because most of the TFA in the eluates evaporates during the drying process. This reduced amount of Tris-HCl is beneficial to the MS acquisition process and results

in improved signal quality in the mass spectra. For the digestion of disulfide bonds containing proteins, the Tris-HCl buffer can be fortified with 1 mM dithiothreitol (DTT). This concentration of DTT is effective in reducing all of the disulfides and is readily tolerated during the MALDI-TOF MS analysis.

10. The digestion-termination and preparation for MALDI-TOF MS are critical part of the experiment. Highly consistent and homogeneous samples sites can be obtained only when the Tris/matrix/protein digest mixture is first air-dried and then rehydrated with 5-μL aliquots of 0.8% TFA. The addition of surplus, volatile TFA converts the already present matrix molecules on the sample spots from the salt form (at neutral pH) to the MALDI-desired acid form. This conversion can be monitored by the disappearance of the neutral pH yellow color of the ACCA matrix. On the other hand, rehydration of the Tris/matrix/protein digest mixture with more matrix solution is not desirable as it results in overall matrix excess on the sample spots and yields lower quality mass spectra.

Acknowledgments

This publication was supported in part by grant number 5 R44 CA099117-03 and contract number N44ES-35511, from the National Institutes of Health. Its contents are solely the responsibility of the authors and do not necessarily represent the official views of the National Institute of Health.

References

1. Petricoin, E. F., Zoon, K. C., Kohn, E. C., Barrett, J. C., and Liotta, L. A. (2002) Clinical proteomics: translating benchside promise into bedside reality. *Nat. Rev. Drug Discov.* **1**, 683–695.
2. Diamandis, E. P. (2004) Mass spectrometry as a diagnostic and a cancer biomarker discovery tool: opportunities and potential limitations. *Mol. Cell. Proteomics* **3**, 367–378.
3. Coombes, K. R., Morris, J. S., Hu, J., Edmonson, S. R., and Baggerly, K. A. (2005) Serum proteomics profiling-a young technology begins to mature. *Nat. Biotechnol.* **23**, 291–292.
4. Nelson, R. W., Krone, J. R., Bieber, A. L., and Williams, P. (1995) Mass-spectrometric immunoassay. *Anal. Chem.* **67**, 1153–1158.
5. Dogruel, D., Williams, P., and Nelson, R. W. (1995) Rapid tryptic mapping using enzymatically active mass spectrometer probe tips. *Anal. Chem.* **67**, 4343–4348.
6. Krone, J. R., Nelson, R. W., and Williams, P. (1996) Mass spectrometric immunoassay, *SPIE.* **2680,** 415–421.
7. Nelson, R. W. (1997) The use of bioreactive probes in protein characterization. *Mass Spectrom. Rev.* **16**, 353–376.
8. Niederkofler, E. E., Tubbs, K. A., Gruber, K., et al. (2001) Determination of beta-2 microglobulin levels in plasma using a high-throughput mass spectrometric immunoassay system. *Anal. Chem.* **73**, 3294–3299.

9. Kiernan, U. A., Black, J. A., Williams, P., and Nelson, R. W. (2002) High-throughput analysis of hemoglobin from neonates using matrix- assisted laser desorption/ionization time-of-flight mass spectrometry. *Clin. Chem.* **48,** 947–949.

10. Kiernan, U. A., Tubbs, K. A., Gruber, K., et al. (2002) High-throughput protein characterization using mass spectrometric immunoassay. *Anal. Biochem.* **301,** 49–56.

11. Kiernan, U. A., Tubbs, K. A., Nedelkov, D., Niederkofler, E. E., and Nelson, R. W. (2003) Detection of novel truncated forms of human serum amyloid A protein in human plasma. *FEBS Lett.* **537,** 166–170.

12. Kiernan, U. A., Tubbs, K. A., Nedelkov, D., Niederkofler, E. E., and Nelson, R. W. (2002) Comparative phenotypic analyses of human plasma and urinary retinol binding protein using mass spectrometric immunoassay. *Biochem. Biophys. Res. Commun.* **297,** 401–405.

13. Niederkofler, E. E., Tubbs, K. A., Kiernan, U. A., Nedelkov, D., and Nelson, R. W. (2003) Novel mass spectrometric immunoassays for the rapid structural characterization of plasma apolipoproteins. *J. Lipid Res.* **44,** 630–639.

14. Kiernan, U. A., Nedelkov, D., Tubbs, K. A., Niederkofler, E. E., and Nelson, R. W. (2004) Proteomic characterization of novel serum amyloid P component variants from human plasma and urine. *Proteomics* **4,** 1825–1829.

15. Kiernan, U. A., Tubbs, K. A., Nedelkov, D., Niederkofler, E. E., McConnell, E., and Nelson, R. W. (2003) Comparative urine protein phenotyping using mass spectrometric immunoassay. *J. Proteome Res.* **2,** 191–197.

16. Kiernan, U. A., Nedelkov, D., Tubbs, K. A., Niederkofler, E. E., and Nelson, R. W. (2004) Selected expression profiling of full-length proteins and their variants in human plasma. *Clin. Proteomics. J.* **1,** 7–16.

17. Nedelkov, D., Tubbs, K. A., Niederkofler, E. E., Kiernan, U. A., and Nelson, R. W. (2004) High-throughput comprehensive analysis of human plasma proteins: a step toward population proteomics, *Anal. Chem.* **76,** 1733–1737.

22

Microarray Data Classified by Artificial Neural Networks

Roland Linder, Tereza Richards, and Mathias Wagner

Summary

Systems biology has enjoyed explosive growth in both the number of people participating in this area of research and the number of publications on the topic. The field of systems biology encompasses the *in silico* analysis of high-throughput data as provided by DNA or protein microarrays. Along with the increasing availability of microarray data, attention is focused on methods of analyzing the expression rates. One important type of analysis is the classification task, for example, distinguishing different types of cell functions or tumors. Recently, interest has been awakened toward artificial neural networks (ANN), which have many appealing characteristics such as an exceptional degree of accuracy. Nonlinear relationships or independence from certain assumptions regarding the data distribution are also considered. The current work reviews advantages as well as disadvantages of neural networks in the context of microarray analysis. Comparisons are drawn to alternative methods. Selected solutions are discussed, and finally algorithms for the effective combination of multiple ANNs are presented. The development of approaches to use ANN-processed microarray data applicable to run cell and tissue simulations may be slated for future investigation.

Key Words: Artificial neural network; comparison of methods; data analysis; microarray; multicategory classification; bibliometry.

1. Introduction

Systems biology is an emerging discipline focused on tackling the enormous technical and intellectual challenges associated with devising generally applicable methods of interpreting data in a way that will shed light on the complex relationships between multiple genes and their products in order to generate comprehensive understanding of how organisms are built and run. The term *systems biology* includes for example computer assisted mathematical modeling and data analysis. The latter encompasses dealing with high-throughput analyses

From: *Methods in Molecular Biology, vol. 382: Microarrays: Second Edition: Volume 2*
Edited by: J. B. Rampal © Humana Press Inc., Totowa, NJ

such as DNA microarrays. A DNA microarray is a two-dimensional array, typically mounted on a glass, filter, or silicon wafer, upon which genes or gene fragments otherwise referred to as complementary DNA or cDNA, are deposited or synthesized in a predetermined spatial order, which quantitatively measures corresponding mRNA sequences. These signals (expression rates) may be analyzed and are characterized by few observations and potentially thousands of predictor variables per microarray *(1)*. The likely outcome of the analysis can be subdivided in three ways *(2)*. This may be done by:

α. The detection of differences in the expression rates among diverse groups/populations.
β. Cluster analysis of genes or samples in order to detect groups or structures (*unsupervised learning*) *(3)*.
γ. The classification of diseased entities (*supervised learning*) *(4)*.

Concering α: it may be preferable to make the distinction between tumorous as opposed to sound tissue or to make a comparison of the expression rates of patients as against healthy controls. The rationale is the detection of features associated with the pathogenesis or manifestations of complex diseases as the point of departure for causal therapy concepts *(5)*. This comparison of treated cell cultures against untreated may potentially help in improving therapy. Additionally, it is possible to improve therapy by comparing treated cell cultures against untreated.

Concering β: the cluster analysis can be useful in identifying groups or subgroups of genes that produce similar expression patterns *(6,7)*. For example new tumor subtypes can be detected *(8–10)*.

Concering γ: the classification of diseased entities may be one of the most promising possibilities of the microarray analysis. The classification of varied tumor subtypes can improve the diagnostic quality and so inform the selection of the most acceptable therapy. Researchers are continuously exploring gene expression profiles that are designed to give a pretherapeutic indicator concerning the success of certain chemotherapy *(11)*. Another way in which this may be applied is in the categorization of breast tumor patients into high risk and low risk groups again with the objective of optimizing therapy *(12)*. With respect to microbiology, classification may help to more accurately classify strains of bacteria *(13,14)* or to identify new strains *(15)*. An advantage for the patient is that his or her infection can be identified more quickly and therefore treated more specifically *(16,17)*.

The objectives previously outlined can be analyzed by artificial neural networks (ANN). ANNs (also termed "neural nets" or "connectionist models") are a series of nonlinear, interconnected mathematical equations, which tangentially resemble biological neuronal systems and which are used to calculate an output variable on the basis of independent input variables. Neural network analysis is derived from artificial intelligence, however, it differs from expert systems in that, instead of being rule-based with preprogrammed constraints, rules, or conditions, it consists of neural networks, which "learn" and progressively develop meaningful reliable relationships between input and output variables. Unlike classical statistical models and correlative methods, neural networks comprise multiple indirect interconnections between input and output variables and employ nonlinear mathematical equations and statistical techniques to successively minimize the variance between actual and predicted outputs. This consequently produces a model, which can be subsequently applied to an independent data set, and in turn produces predicted outputs that reliably correspond to the actual observed values.

This work focuses on multilayered feed-forward networks using ANN as a synonym for this type of network. Even if this type of ANN is feasible to classify unsupervised networks (so-called *bottleneck networks) (18)* its primary effectiveness is the supervised classification as outlined above.

2. Methods

The history of ANNs dates back to as early as 1890, when a model was developed by an American psychologist to explain the capabilities of the brain in making associations *(19)*. In 1958, a two-layered ANN known as "perceptron" was described *(20)*. In 1986, a powerful learning algorithm was introduced: this algorithm could also handle hidden neurons *(21)*, representing the starting point for the triumphant progress of ANN technology. Of no less importance to medicine, ANNs are used for supporting decisions in diagnosis, classification, early detection, prognosis, and quality control.

2.1. Bibliometric Analysis

As revealed by a MEDLINE search for the MeSH term "Neural Networks Computer," there is a growing interest in ANNs, with the search having produced more than 7000 articles dealing with this topic, most of which focus on feed-forward networks (**Fig. 1**). Searching MEDLINE for "Neural Networks" (Computer) [MeSH] and "Microarray Analysis" [MeSH] retrieved 60 hits (as of 4th May, 2005), indicating that ANNs have been established for the analysis of microarray data (**Fig. 2**). The use of MeSH headings may be controversially discussed *(22)*.

In light of the aforementioned results a simplified analysis of abstracts (without searching MeSH terms) was carried out to analyze the representation of the present topic in abstracts of the biomedical body of literature. The current approach was chosen because abstracts appear to be more frequently used to obtain a quick overview on a given topic *(23)*. The bibliometric overview of

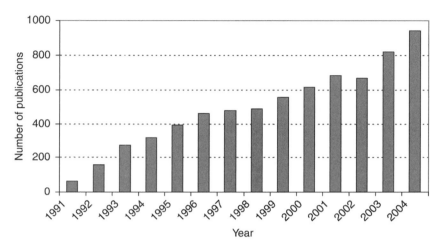

Fig. 1. MEDLINE search for the MeSH term "Neural Networks Computer" (as of April 7th, 2005).

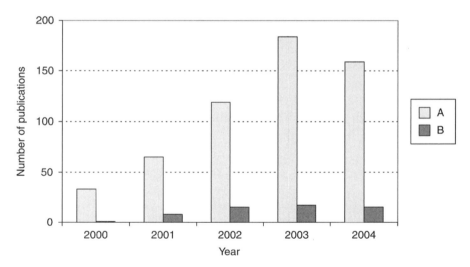

Fig. 2. MEDLINE search (as of May 5th, 2005) on methods involved in microarray analysis. Reviews have been omitted from the results of this search. Conventional statistical methods **(A)** vs ANN **(B)** are displayed; search algorithm for **(A)**: "Microarray Analysis[MeSH] AND "Medical Informatics" [MeSH] NOT "Neural Networks (Computer)" (MeSH); search algorithm for **(B)**: "Neural Networks (Computer)"[MeSH] AND "Microarray Analysis" [MeSH].

ANN and microarray was initiated by interrogating PubMed using the search strategies "artificial neural network" and microarray; "neural network" and microarray; and ANN and microarray (as of 3rd May, 2005). A total of 28 abstracts were found of which 23 (82.14%) were considered relevant on perusal of the abstracts. Analysis of this data set, albeit a small one, gives a synoptic overview of how the abstracts serve to indicate the status of research activity and therefore give a microcosmic perspective of the state of research in that field as it is represented without including MeSH terms. The data set shows that of the 23 abstracts, one was published in 1999, whereas 22 (95.65%) came out between 2000 and 2005, and all corresponding articles had been written in English.

Lotka *(24)* states that the number of authors making n contributions to the literature is about $1/n^2$ of those making one contribution. Using this premise as a point of departure, the data set shows that there are 127 distinct authors who contributed to the articles and of this number, 112 authors (88.18%) contributed to one article; 13 authors (10.23%) contributed to two articles; and 2 authors (1.57%) contributed to three articles.

The articles from which the abstracts have been analyzed were published across 18 journals and the level of productivity of these can best be assessed by considering Bradford *(25)*, which contends that, if a set of articles is divided into three approximately equal "zones" of productivity such that the ratio $1:n:n^2$ will hold, where, 1 is the number of journals in the first zone and n is a proportional multiplier, then there is always a small "core" of journals, which contains a large number of the articles, usually about one-third of the total, a second larger group which accounts for another third and the last very large group of journals containing the final third. In applying Bradford's data set there are 16 journals (88.88%) having only one article each, one journal (5.55%) with three articles, and one journal (5.55%) with four articles. In this case the very small core consists of two journals *Bioinformatics* (four articles) and *BMC Bioinformatics* (three articles). Although mathematically the precise formulation might not "hold" in this case (which is a contention bibliometricians often cite in critiquing Bradford's Law), in principle it quite accurately represents the basic premise, that in most fields of research it is only a small number of "core" journals that are the most productive with decreasing productivity over the secondary and tertiary level journals.

There are eight countries which contributed to the research output (as represented by the abstracts excluding the MeSH terms) (**Fig. 3**).

Consistent with the literature of the sciences, journals are the primary vehicle of publication, with the small data set indicating the 23 articles were published in 18 different journals and with the small core of two journals being responsible for 7 of the 23 articles which is 30.43%. This approx ⅓ or 33% that Bradford's Law postulates that the first zone of "core" journals will contain. Increasing research interest is demonstrated, with over 95% of the articles having been published just for the period 2000–2005. A notable pattern of collaboration is shown through coauthorship of articles (**Fig. 4**).

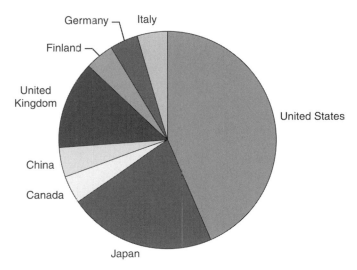

Fig. 3. Geographical location of research groups contributing relevant abstracts to MEDLINE in terms of ANN usage in microarray data analysis (as of May 3rd, 2005).

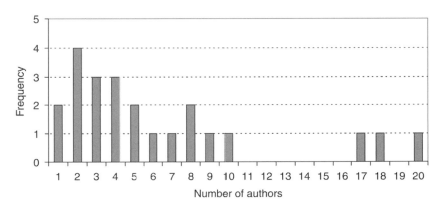

Fig. 4. Quantitative analysis of authors contributing to an ANN based analysis of microarray data (abstracts found in MEDLINE; as of May 3rd, 2005).

The mean number of authors collaborating per article is eight. There is a notable lack of articles written in languages other than English. This may be attributable to the assertion that it is usually more difficult to get cited when published in non-English language journals, resulting in English being the first language of choice, in which to get articles printed, as researchers will undoubtedly aspire to get their research results accepted for publication and so increase the likelihood of their articles being "acknowledged" by the academy and consequently cited in the literature.

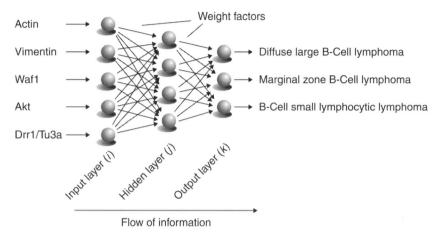

Fig. 5. An example for the structure of a feed-forward artificial neural network: Different types of lymphoma are distinguished by the expression rates of selected genes.

2.2. How Does an ANN Work?

Generally, feed-forward networks consist of three or more layers of artificial neurons each with data entered in the input layer and further processed in the hidden and output layers. Through a learning process, which consists of a "training phase" and a "recall phase" ANNs use nonlinear mathematical equations to successively develop meaningful relationships between input and output variables. The relationships between the different input variables and the output variable(s) are established through adaptations of the weight factors assigned to the interconnections between the layers of the artificial neurons, and this takes place in the training phase. This adaptation is based on rules that are set in the learning algorithm. At the end of the training phase, the weight factors are fixed. Data from patterns not previously interpreted by the network are entered in the recall phase, and an output is calculated based on the previously mentioned and now fixed weight factors (**Fig. 5**). Output neurons usually produce what may be loosely referred to as "activity" between 0 and 1 when this occurs a pattern is assigned to the class with the highest neuronal activity which equates to the "winner takes all" rule.

Preprocessing (*see* **Note 1**) and gene selection (*see* **Note 2**) precede the training phase which will be initiated with random weights at the connections between the neurons. A training data set with known outcome is then entered at the input neurons. The ANN compares its own output values with the known outcome (e.g., 1 for "diffuse large B-cell lymphoma," 0 otherwise), calculates an error value, which will change as the weights at the connections change, with the ANN attempting to minimize the error by adjusting the weights according

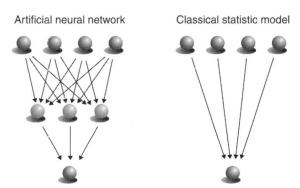

Fig. 6. This schematic comparison as described in **ref. 87** reveals differences in architecture between a three layered artificial neural network and a classical statistic model.

to the learning algorithm. This process is repeated for a predefined number of times (or "epochs") and later, in the recall phase, the ANN can be tested on data with known outcome values (*see* **Note 3**).

2.3. What Are the Advantages of ANNs Compared to Alternative Methods?

As shown earlier (**Fig. 1**), ANNs are increasingly being used as an "intelligent" alternative to the conventional statistical multivariate analysis methodology. The impetus behind the increasing interest in ANNs are primarily the consideration given to possible nonlinear connections (1) between the predictor variables (genes) and the output variables as well as (2) nonlinear interdependencies between the predictor variables. These correlations are biological realities and ought not to be ignored. ANNs consider such nonlinear relationships based on nonlinear mathematics, namely by sigmoid activation functions *(26)* and the multilayer concept; are independent from the partially rigid guidelines as they are known from conventional approaches like the *linear discriminant analysis*; and do not need to fulfill the assumptions of any well-defined distributions (e.g., the *Gaussian distribution*). There are different concepts of ANNs and linear statistical models (**Fig. 6**).

Based on these advantages ANNs generally produce better overall results in classification *(27)* and to some degree better results when compared to experts in the corresponding fields *(28–33)*. ANNs may also outperform the *logistic regression* (LR), which is considered to be the *gold standard* in biomedical research (**Fig. 7**).

2.4. Statistical Approaches

Standard statistical approaches, such as the *linear discriminant analysis* and LR, are easily implementable but assume independence among all inputs

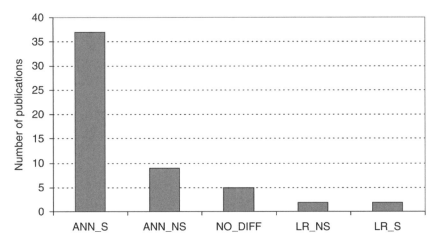

Fig. 7. MEDLINE search for "logistic regression neural network" comprising the years 2002 to 2004. In 11 publications the difference in accuracy was not significant or was not investigated regarding significance (ANN: $n = 9$, LR: $n = 2$). In 15 publications there was no comparison given (meta-analysis, reviews, and so on) and six publications were not available to the authors. The ANN classifies more accurately with significance (ANN_S) or without significance (ANN_NS). No difference between ANN and LR regarding accuracy (NO_DIFF). The LR classifies more accurately without significance (LR_NS) or with significance (LR_S).

and the existence of a linear relationship between input and output variables. Although *linear discriminant analysis* is incapable of handling interactions between predictor variables, using the LR approach, nonlinear input-output-relationships can be explicitly modeled and interactions between variables can be clearly defined. Based on the assumption of complexity in the relationships, the LR modeler might fail to consider all of the nonlinearities, although progress has been achieved through the modeling of nonlinear relationships through *fractional polynomials*, therefore these complexities as well as interactions are factors that can automatically be considered *(34,35)*.

2.4.1. Classification Trees

Classification trees implicitly perform a step-by-step variable selection and—assuming binary predictor variables—are easy to interpret. Trees may be bushy when genes are used as input and *multisplit functions* are employed. Moreover, classification trees have a tendency to be unstable and lacking in accuracy. Their accuracy can be greatly improved by aggregation (*bagging* or *boosting*; *see* **Subheading 2.6.**). However, some simplicity is lost by aggregating trees.

2.4.2. K-Nearest Neighbor Classifier

Nearest neighbor classifiers are simple, intuitive and have remarkably low error rates when compared to more sophisticated classifiers and as such they are able to handle interactions between genes, they do so using a "black-box" approach and give very little insight into the structure of the data. As a disadvantage predictor variables are not weighted based on their discriminatory power, especially when using the most widespread *Euclidian distance*. The introduction and use of alternative distance measures, for example, the *Malahanobis distance*, classification could possibly fail if the expression levels of the (few) observations that are made are not distributed normally or if any one class comprises more than one cluster.

2.4.3. Support Vector Machines

SVMs are receiving increased attention and have been applied successfully to microarray gene expression cancer diagnosis *(36–39)*, however they require more training than the methods previously discussed (e.g., choice of kernel function K and scale factor λ). Additionally, SVMs are difficult to interpret if there are diverse support vectors (as is usually the case) and to date could not consistently be proved to be more accurate than alternative classification methods or even to solve problems where other methods failed *(40)*.

In the case of DNA microarray analysis, a number of classification methods have been compared *(41)*, and it was concluded that—beside the *k-nearest neighbor—backpropagation neural networks* are the best classifiers for that purpose. ANNs have been demonstrated to be nearly perfect in distinguishing different sets of lymphoma patients or predicting the long-term survival of individuals suffering from various lymphoproliferative conditions *(42)*. When considered from a mathematical perspective the so-called "general function approximation theorem" would suggest that a three-layered ANN with appropriate weights could approximate any arbitrary nonlinear function *(43)*. For this reason ANNs can serve as universal approximators.

2.5. What Are the Disadvantages and How to Overcome These Problems?

The criticism most frequently proffered is that because ANNs converge slowly, the learning phase may be stuck in local minima and so achieve poor accuracy; numerous learning parameters as well as the topology of the network have to be optimized by an expert; and the trained ANN resembles a "black box," thereby, hampering any interpretation of the classifier's response *(44)*.

Since the late 1980s when ANNs began attracting increased attention, these problems have become well known and so various efforts have been undertaken to overcome these drawbacks. Acceleration of this convergence may be facilitated by, numerous new learning algorithms being described in **refs. *45–47***; second order methods being introduced *(26,48,49)*; the *standard logistic activation* function being replaced by the *tangent hyperbolic (50)*; and approaches to eliminate flat spots in the derivation of the *activation function (51)* or modifications of the error function, for example, *cross-entropy (52–55)*. To improve the generalization accuracy, local adaptive learning rates were established *(56–58)*, weight initialization was optimized *(59–61)*, and modular approaches were tested *(62)*. To facilitate more intense learning data were *jittered (63–65)* or ensembles of ANNs were trained for majority or weighted voting. The body of literature comprises reviews of all these improvements *(66–70)*. In the meanwhile most data analysis tools offer a neural network analysis, which considers some of the aforementioned learning strategies. A selection of these tools allows the user to implement further strategies *(71)* by programming additional routines. Even if one is not a computer-scientist there is a facet one can easily adjust in order to improve the accuracy, namely the size of the network structure, i.e., to increase considerably the number of hidden neurons. Although this may be considered time-consuming, 100 hidden neurons is determined a good choice, overcompensating for the network structure, however the current theory speaks to the efficacy of choosing the smallest admissible size that will provide a solution because many researchers believe that the simplest architecture is that which is most suitable for generalization. Notwithstanding this belief, several neural net empiricists have shown that surprisingly good generalization can be achieved with oversized feed-forward networks *(72–76)*. From the experience of the authors, overcompensating the network architecture leads to a marked increase in the generalization performance although the network structure does not require additional incremental adjustments. Moreover, the capacity is sufficient to face almost every classification problem. There is however, one condition that must be adhered to: overtraining the ANN must be avoided.

Overtraining or *overfitting* can occur during the learning phase, when the error in the training set decreases more or less steadily, while the error in unseen patterns begins to deteriorate. This usually occurs in the later stages of learning. Before reaching this point, the network learns the general characteristics of the classes; afterwards, it takes advantage of some idiosyncrasies in the training data that aggravate generalization performance (**Fig. 8**). Several theoretical studies have evaluated the optimal stopping time *(77–78)*. One common approach toward avoiding overfitting is *early stopping*, which consists of estimating the generalization performance during training (with an extra validation set removing some patterns from the training data) and stopping when it begins to decrease. This technique has been reported to be superior to other regularization methods in many cases *(79)*. However, the real situation is somewhat more complex, in that

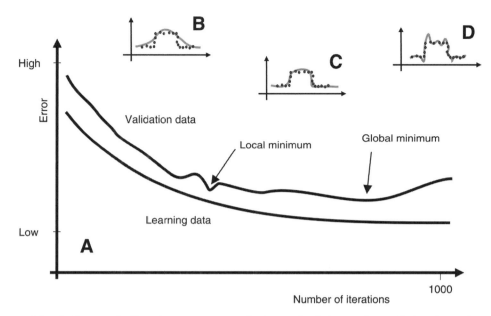

Fig. 8. Rectangle function as an example for *underfitting* and *overfitting*. Learning the rectangle function by an ANN (**A**). Approximation accuracy is still inadequate (**B**). The characteristic of the function has been learned (**C**). All sampling points have been precisely fitted but the generalization performance has worsened (**D**). The labeling of the axes applies to all graphs (**A–D**).

generalization curves almost always have more than one local minimum. In light of this, 14 different automatic-stopping criteria have been developed *(80)*.

In order not to waste valuable data for the extra validation set, an ensemble technique can be used, with each ANN consisting of an ensemble of five modules. In this technique, all training data are divided into five equally sized sets, designated as "A," "B," "C," "D," and "E." The first module is trained by sets A, B, C, and D, and set E serves as a validation set; the second module is trained by sets A, B, C, and E, and set D is the validation set; this is continued until all the possible combinations are complete (**Fig. 9**). To obtain a single common result for each class, the outputs of the five modules can be (weighted) averaged. Early stopping can therefore be performed using an extra validation set and all training patterns can effectively be used for training.

Of central importance to ANNs is that the weights at the connections are "learned" during training of the network. "Experience" in the trained network is stored in these interconnection weights *(32)*. To open the black box and therefore to give insight into the response behavior, one can either try to extract rules

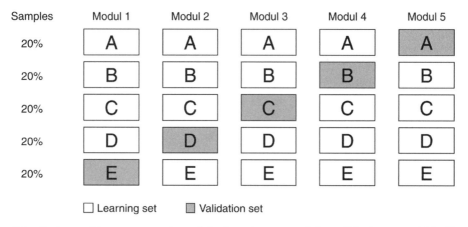

Fig. 9. A graphic representation of the learning set and the validation set within five modules.

from the weights *(81–83)* or perform so-called sensitivity analyses, i.e., systematically varying the input while observing the output. For instance, the ANN software tool *BrainMaker (84)* (California Scientific®, Nevada City, USA) offers a sensitivity analysis. Another approach, which is easily implementable and understood is the *Causal Index* introduced by Baba *(85)* or its modified version *MCI* *(86)* also considering the sigmoid curve progression of the activation function. The **Eqs. 1** and **2** indicate how the impact of an input neuron *i* regarding an output neuron *k* can be quantified as having a *Causal Index* or *MCI* value. Where, *j* denotes the hidden neurons, w_{ij} is the weight from neuron *i* to neuron *j* (**Fig. 5**).

$$CI_{ik} = \sum_{j=1}^{n} w_{ij} \times w_{jk} \tag{1}$$

$$MCI_{ik} = \sum_{j=1}^{n} \tanh(w_{ij}) \times \tanh(w_{jk}) \tag{2}$$

Irrespective of the common preconception regarding the black box character of ANNs it has been stated that "(…) ANNs are a superior tool for digesting microarray data both with regard to making distinctions based on the data and with regard to providing very specific reference as to which genes were most important in making the correct distinction in each case," *(42)* (*see* **Note 4**).

Furthermore, detailed discussions about the advantages and disadvantages of the ANNs have been published previously *(87–89)*.

2.6. How to Combine the Power of Multiple ANNs?

In order to make learning more robust as well as to achieve a more accurate classification *committee* or *ensemble techniques* may be employed *(90–91)*.

To attain this objective, several classifiers are specified for the same classification task and in this way produce one common result based on their collective voting. The ANNs involved can be differentiated based on their network architecture *(92)*, different initialization procedures used *(93)* or learning data permutated as outlined above. Applying ensembles is not only restricted to ANNs, other learning algorithms can be grouped as ensembles, for example, SVMs *(94)*. Different voting schemes can be used such as the *simple averaging method (95–97)*, the *weighted averaging (96,98,99)*, or the *majority voting scheme* or *plurality voting (100,101)*. Further voting rules are the *Maximum Vote*, the *Borda Count* or the *Nash Vote (102)*.

Moreover, generating perturbed versions of the learning set can be done by *bagging* and *boosting (103,104)*. In the *bootstrap aggregating* or *bagging* procedure *(105)*, perturbed learning sets of the same size as the original learning set are formed by forming bootstrap replicates of the learning set. In *boosting (106)* the data are resampled adaptively so that the weights in the resampling are increased for those cases most often misclassified. Hereby the aggregation of predictors is done by weighted voting.

Although an ensemble is a collection of different classifiers, for example, ANNs, specified for the same task, ANNs may also be trained for different subtasks. Those approaches are categorized into *mixture of experts*. For example, a two-level ANN has displayed superior performance over a single-level ANN *(107)*. The task was to differentiate chest radiographs with lung nodules from those without lung nodules. To concentrate ANN-learning on patterns difficult to assign, first an ANN differentiated between patterns being "possible a nodule" from those being "probably no nodule." In the second step—when the "probably no nodule" patterns have been sorted out—the second ANN focused on the remaining patterns which were difficult to classify (**Fig. 10**).

A two-level ensemble architecture distinguishes normal cells from four types of malignant cells (adenocarcinoma, squamous cell carcinoma, small cell carcinoma, large cell carcinoma) using concepts of varying complexity *(108)*. The first-level ANN ensemble classifies higher-level concepts (benign or malign), whereas the second-level ensemble is used to classify lower-level concepts (concrete malign types). Then a cell is only classified by the second ensemble if at least one individual ANN of the first ensemble concludes it is malign (**Fig. 11**).

Dealing with a multicategory classification problem comprising three classes, the same strategy can be employed. To predict the three possible states of the protein secondary structure (α-helix, β-strand, or coil) from protein sequences, an ensemble of different neural networks for state prediction can be used in a first step *(109)*. Positions where there was a full agreement between all ensemble members were taken as the final prediction. In case of no final outcome a second ANN obtained the final prediction (**Fig. 12**).

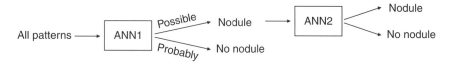

Fig. 10. Two-level architecture, two-class classification problem.

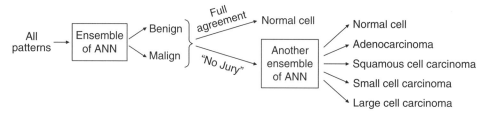

Fig. 11. Two-level architecture, classifying higher-level and lower-level concepts.

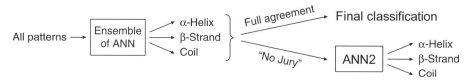

Fig. 12. Two-level architecture, multiclass classification problem.

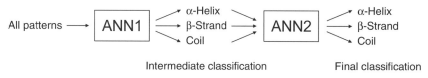

Fig. 13. Usage of output as input.

Another way to combine two ANNs is to simply use the output of the first ANN (protein secondary structure) as an input for a second ANN that finally makes the prediction (**Fig. 13**). Employing a consecutive structure–structure network is reported to result in a 2% improvement in prediction *(110)*.

The *one-vs-all* approach builds k (the number of classes) binary classifiers which distinguish one class from all other classes added together. A sample is assigned to the corresponding class label of the binary classifier achieving the greatest output activity (**Fig. 14**).

Similarly, the all-pairs approach builds $k(k - 1)/2$ binary classifiers. For each class there are $k-1$ relevant binary classifiers, which distinguish it from the other classes. The output activities of those $k-1$ binary classifiers are summed up, and

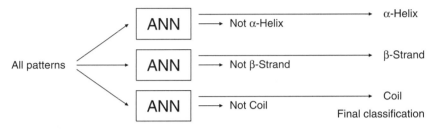

Fig. 14. *One-vs-all* approach.

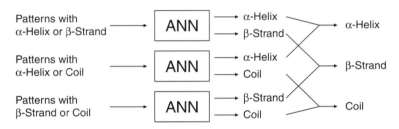

Fig. 15. *All-pairs* approach.

the class with the greatest overall activity is the winning class (**Fig. 15**). Moreover, there are more sophisticated approaches that somehow combine the one-vs-all approach with the all-pairs approach *(111)*.

The *Subsequent ANN (SANN)* approach *(112)* is oriented to the human decision making process. In this situation, a process of exclusion occurs which is the first step where preferred choices are selected and included in the "narrowed-down choice," after which the final decision is made in a succeeding step. This means that the classification made by the first ANN is interpreted as a preselection to be followed by a final categorization by a successive, second application of ANN. Classification is therefore concentrated on the two primary classes, i.e., the two most preferred classes with the highest activities of the corresponding output neurons (**Fig. 16**).

When comparing these amalgamations of experts (as the authors have published elsewhere *[112]*) the all-pairs method and the SANN approach clearly outperform the other methods, probably because they concentrate on classification problems comprising only of patterns of two adjacent classes. However, the all-pairs method leads to intense computing on both, classes which are difficult to separate, and classes which are readily separable. This shortcoming does not apply to the SANN approach calculating no. 2-class classification problems of classes easy to separate. Finally, it should be emphasized that mixtures of experts are—like the ensembles techniques—by no means restricted to ANNs. An example combining four different types of predictors has been reported previously *(113)*.

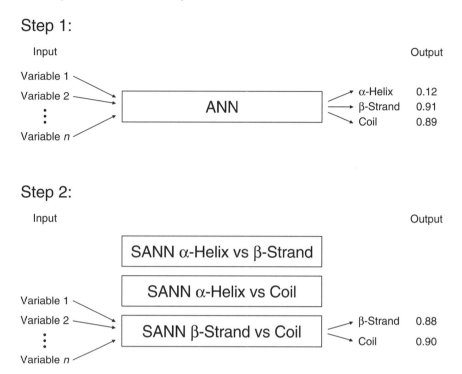

Fig. 16. Paradigmatic three-class problem. As an example (fictive values), the pattern, with parameter 1–*n,* will be primarily categorized as β-strand. A subsequent ANN (SANN) trained to discriminate between β-strand and coil conclusively classifies the pattern in the sense that the secondary structure will be a coil.

3. Closing Remarks

Whether the theoretical superiority of highly sophisticated methods as compared with more simple methods is really evident in practical applications is debatable. Comparisons of methods often do not consider important parameters such as population drift, sample selectivity bias, errors in class labels, arbitrariness in the class definition, or misleading optimization criteria and performance assessment *(114)*. This is observed particularly in the analysis of gene expression data, where possible errors may occur with the marking of the mRNA by fluorescent nucleotides, when producing the chip, during the hybridization procedure, scanning the arrays or in the course of image processing *(115)*. With this in mind, the importance in searching for only slightly more accurate algorithms makes this all relative. Deciding between two algorithms is mainly guided by the availability of a corresponding computer program and the experience to work with it, therefore, researchers who are experienced in applying a certain classification method will not really be affected by the views presented.

For researchers who are interested in a potent classification procedure, neural network technology might be a powerful alternative to other existing methods.

This also applies to the authors of the present report and other researchers in the field of systems biology who increasingly consider expression rates of genes as they are provided by the microarray analysis to modify their mathematical models *(116,117)*. Simulating cell behavior, various multicategory classification problems have to be faced. The present work will hopefully be supportive with regards to this approach.

4. Notes

4.1. Preprocessing

Preprocessing is as important a consideration as the subsequent classification procedure and includes a number of aspects. Although facets such as data cleaning, coding of nominal variables or feature extraction are less applicable to the analysis of microarray data, some essential steps are the imputation of missing data, the normalization, the logarithmic transformation, and the standardization of data.

4.1.1. Imputation of Missing Data

Some of the discriminatory methods are able to deal with missing data (e.g., *classification trees*); however, others require complete data and to this end there exists a comprehensive body of literature on how to input missing data *(118)*. Interestingly, even ANNs are suited to the substitution of missing values *(119,120)*. A very pragmatic approach is to use a simple k nearest neighbor algorithm, in which the neighbors are the genes and the distance between neighbors is based on their correlation *(121)*. For each gene with missing data: (1) compute its correlation with all other $n - 1$ genes, and (2) for each missing entry, identify the k nearest genes having complete data for this entry and impute the missing entry by the average of the corresponding entries for the k neighbors. Setting $k = 5$ has been proved to work successfully *(121)*.

4.1.2. Normalization

In order for the comparison of gene expression levels of different microarray to be made, the signals must be normalized. To achieve this, *housekeeping genes* can be employed *(122,123)* or the average intensity of all genes (or a subset of the genes) can be used for scaling *(124,125)*. Additionally, artificial transcripts can be added with a defined behavior suited for normalization *(2)*.

4.1.3. Logarithmic Transformation

It is customary that the expression signals are log-transformed and in that regard the base 10 logarithmic transformation *(17)* is as well used as the base 2 logarithm *(8,126)*.

4.1.4. Standardization

Conventional practice allows the use of the correlation between the gene expression profiles of two mRNA samples to measure their similarity *(8,126,127)*. Consequently, observations are standardized to have mean 0 and variance 1 across variables (genes). With the data standardized in this manner, the distance between two mRNA samples may be measured by their Euclidean distance, for example, important for the *k nearest neighbor algorithm*.

4.2. Gene Selection

When administering an ANN, it makes for greater efficiency if gene selection is also done using an ANN, for example, using the *Neural Net Clamping Technique (128)*. This technique exemplifies a kind of backward search, the process involves starting with *n* input neurons; training the ANN only once, including all features, yet testing the ANN *n* times; and setting (or clamping) the considered feature to its mean value over all test patterns. Should the feature with the smallest contribution to classification be omitted, the search continues with *n*-1 input neurons. This procedure is repeated until the best feature set is found and when compared to a standard backward search, the ANN only has to be trained once to leave out one feature.

Based on computational challenges experienced when starting the selection procedure with thousands of genes, a preselection has to be performed, for example, by a common *univariate method* (based on the *BSS/WSS criterion [121]*, *S2N ratio [17]*, *Wilcoxon test [129]*, *t-Statistic [121]*, or the *misclassification rate [130]*). Increasingly, *multivariate approaches* are being reported *(91,131,132)* that—in contrast to the *univariate methods*—also consider possible interdependencies between the genes. Reviews on this issue have been previously published *(133,134)*.

4.3. Statistical Evaluation

The disproportionately high number of possible predictor variables (genes) associated with only few observations may result in many false-positive findings. To avoid this, one should be careful to ensure that either multiple statistical testing can be performed *(135,136)* or validation is done by a *k*-fold cross-validation, which entails splitting the total amount of records into *k* equally sized and representative subsets. This method requires deriving ANN calculations using *(k-1)/k* records and evaluating the remaining *1/k* records, repeating this procedure *k* times to obtain a final result based on all of the records. The most commonly used values for *k* are 5 or 10. Setting *k = n*, cross-validation is also called the *"Leave One Out Method,"* which is suited for very small sample sizes. For *n* < 50, some researchers have used *bootstrapping* or its variants *(137)*. The most current variant is 632B+ *(138)*.

4.4. Integration of Microarray Data Into Cell or Tissue Simulations (Mathematical Models)

Pathway information based on microarray analyses may become increasingly vital for successful modeling of biological systems. Such integrated systems will greatly facilitate the constructive cycle of computational model building and experimental verification that lies at the heart of systems biology. One of the major issues is the modeling of transition, for example, normal cell to tumor cell. Mathematical models may be designed using the results of ANN-based microarray data analyses to describe transition. A simple equation can hence be formulated where x_t represents a given condition at starting point t, and x_{t+1} map to the condition at time point $t + 1$ while φ represents factors of influence (**Eq. 3**).

$$x_{t+1} = f(x_t) + \varphi \tag{3}$$

The ANN-based classification may help model the transition function f in which the corresponding microarray data are assigned to both conditions (x_t and x_{t+1}).

References

1. Schena, M. (ed.) (1999) *DNA Microarrays: A Practical Approach.* Oxford University Press, Oxford.
2. Victor, A., Klug, S., and Blettner, M. (2005) cDNA-microarrays—strategien zur bewältigung der datenflut. *Deutsches Ärzteblatt* **102,** 355–360.
3. Quackenbush, J. (2001) Computational analysis of microarray data. *Nat. Rev. Genet.* **2,** 418–427.
4. Ringner, M. and Peterson, C. (2003) Microarray-based cancer diagnosis with artificial neural networks. *Biotechniques Suppl.* 30–35.
5. Gu, C., Rao, D., Stormo, G., Hicks, C., and Province, M. (2002) Role of gene expression microarray analysis in finding complex disease genes. *Genet. Epidemiol.* **23,** 37–56.
6. Eisen, M., Spellman, P., Brown, P., and Botstein, D. (1998) Cluster analysis and display of genome-wide expression patterns. *Proc. Natl. Acad. Sci. USA* **95,** 14,863–14,868.
7. Tamayo, P., Slonim, D., Mesirov, J., et al. (1999) Interpreting patterns of gene expression with self-organizing maps: methods and application to hematopoetic differentiation. *Proc. Natl. Acad. Sci. USA* **96,** 2907–2912.

8. Alizadeh, A., Eisen, M., Davis, R., et al. (2000) Distinct types of diffuse large B-cell lymphoma identified by gene expression profiling. *Nature* **403,** 503–511.

9. Perou, C., Sørlie, T., Eisen, M., et al. (2000) Molecular portraits of human breast tumours. *Nature* **406,** 747–752.

10. Sørlie, T., Perou, C., Tibshirani, R., et al. (2001) Gene-expression patterns of breast carcinomas distinguish tumor subclasses with clinical implications. *Proc. Natl. Acad. Sci. USA* **98,** 10,869–10,874.

11. Chang, J., Wooten, E., Tsimelzon, A., et al. (2003) Gene expression profiling for the prediction of therapeutic response to docetaxel in patients with breast cancer. *Lancet* **362,** 362–369.

12. van de Vijver, M., He, Y., van't Veer, L., et al. (2002) A gene-expression signature as a predictor of survival in breast cancer. *New Engl. J. Med.* **347,** 1999–2009.

13. Broekhuijsen, M., Larsson, P., Johansson, A., et al. (2003) Genome-wide DNA microarray analysis of *Francisella tularensis* strains demonstrates extensive genetic conservation within the species but identifies regions that are unique to the highly virulent F. tularensis subsp. tularensis. *J. Clin. Microbiol.* **41,** 2924–2931.

14. Li, J., Chen, S., and Evans, D. (2001) Typing and subtyping influenza virus using DNA microarrays and multiplex reverse transcriptase PCR. *J. Clin. Microbiol.* **39,** 696–704.

15. Bekal, S., Brousseau, R., Masson, L., et al. (2003) Rapid identification of *Escherichia coli* pathotypes by virulence gene detection with DNA-microarrays. *J. Clin. Microbiol.* **41,** 2113–2125.

16. Fukushima, M., Kakinuma, K., Hayashi, H., Nagai, H., Ito, K., and Kawaguchi, R. (2003) Detection and identification of *mycobacterium* species isolates by DNA microarray. *J. Clin. Microbiol.* **41,** 2605–2615.

17. Golub, T., Slonim, D., Tamayo, P., et al. (1999) Molecular classification of cancer: class discovery and class prediction by gene expression monitoring. *Science* **286,** 531–537.

18. Tafeit, E., Möller, R., Sudi, K., and Reibnegger, G. (1999) The determination of three subcutaneous adipose tissue compartments in non-insulin-dependent diabetes mellitus women with artificial neural networks and factor analysis. *Artif. Intell. Med.* **17,** 181–193.

19. James, W. (1890) The principles of psychology, in *Neurocomputing: Foundations of Research,* (Anderson, J. and Rosenfeld, E., eds.), Henry Holt and Co. New York, NY, USA.

20. Rosenblatt, F. (1958) The percepton: a probabilistic model for information storage and organization in the brain. *Psycholog. Rev.* **65,** 386–408.

21. Rumelhart, D., Hinton, G., and Williams, R. (1986) Learning representations by back-propagating errors. *Nature* **323,** 533–536.

22. Jenuwine, E. and Floyd, J. (2004) Comparison of medical subject headings and text-word searches in MEDLINE to retrieve studies on sleep in healthy individuals. *J. Med. Libr. Assoc.* **92,** 349–353.

23. Kuller, A., Wessel, C., Ginn, D., and Martin, T. (1993) Quality filtering of the clinical literature by librarians and physicians. *Bull. Med. Libr. Assoc.* **81,** 38–43. Erratum in *Bull. Med. Libr. Assoc.* **81,** 233.

24. Lotka, A. (1926) Frequency distribution of scientific productivity. *Journal of the Washington Academy of Sciences* **16,** 317–325.
25. Bradford, S. (ed.) (1953) *Documentation.* 2nd ed., Crosby Lockwood, London.
26. Bishop, C. (ed.) (1995) *Neural Networks for Pattern Recognition.* Clarendon Press, Oxford.
27. Penny, W. and Frost, D. (1996) Neural networks in clinical medicine. *Med. Decis. Making* **16,** 386–398.
28. Baxt, W. and Skora, J. (1996) Prospective validation of artificial neural network trained to identify acute myocardial infarction. *Lancet* **347,** 12–15.
29. El-Solh, A., Hsiao, C. -B., Goodnough, S., Serghani, J., and Grant, B. (1999) Predicting active pulmonary tuberculosis using an artificial neural network. *Chest.* **116,** 968–973.
30. Bottaci, L., Drew, P., Hartley, J., et al. (1997) Artificial neural networks applied to outcome prediction for colorectal cancer patients in separate institutions. *Lancet* **350,** 469–472.
31. Burke, H., Goodman, P., Rosen, D., et al. (1997) Artificial neural networks improve the accuracy of cancer survival prediction. *Cancer* **79,** 857–862.
32. Geddes, C., Fox, J., Allison, M., Boulton-Jones, J., and Simpson, K. (1998) An artificial neural network can select patients at high risk of developing progressive IgA nephropathy more accurately than experienced nephrologists. *Nephrol. Dial. Transplant* **13,** 67–71.
33. Jiang, Y., Nishikawa, R., Wolverton, D., et al. (1996) Malignant and benign clustered microcalcifications: automated feature analysis and classification. *Radiology* **198,** 671–678.
34. Royston, P. and Sauerbrei, W. (2003) Stability of multivariable fractional polynomial models with selection of variables and transformations: a bootstrap investigation. *Stat. Med.* **22,** 639–659.
35. Royston, P. and Sauerbrei, W. (2004) A new approach to modeling interactions between treatment and continuous covariates in clinical trials by using fractional polynomials. *Stat. Med.* **23,** 2509–2525.
36. Statnikov, A., Aliferis, C., Tsamardinos, I., Hardin, D., and Levy, S. (2005) A comprehensive evaluation of multicategory classification methods for microarray gene expression cancer diagnosis. *Bioinformatics* **21,** 631–643.
37. Lee, Y. and Lee, C. (2003) Classification of multiple cancer types by multicategory support vector machines using gene expression data. *Bioinformatics* **19,** 1132–1139.
38. Ramaswamy, S., Tamayo, P., Rifkin, R., et al. (2001) Multiclass cancer diagnosis using tumor gene expression signatures. *Proc. Natl. Acad. Sci. USA* **98,** 15,149–15,154.
39. Furey, T., Cristianini, N., Duffy, N., Bednarski, D., Schummer, M., and Haussler, D. (2000) Support vector machine classification and validation of cancer tissue samples using microarray expression data. *Bioinformatics* **16,** 906–914.
40. Hearst, M. (1998) Support vector machines. *IEEE Intell. Syst.* **13,** 18–28.

41. Cho, S. -B. and Won, H. (2003) *Machine Learning in DNA Microarray Analysis for Cancer Classification* in Chen, Y. -P. (ed.). First Asia-Pacific Bioinformatics Conference (APBC 2003). Adelaide, Australia: CRPIT 19 Australian Computer Society 2003, pp. 189–198.

42. O'Neill, M. and Song, L. (2003) Neural network analysis of lymphoma microarray data: prognosis and diagnosis near-perfect. *BMC Bioinformatics* **4,** 13.

43. Hornik, K., Stinchcombe, M., and White, H. (1989) Multilayer feedforward networks are universal approximators. *Neural Networks* **2,** 359–366.

44. Benítez, J., Castro, J., and Requena, I. (1997) Are artificial neural networks black boxes? *IEEE Transactions on Neural Networks* **8,** 1156–1164.

45. Riedmiller, M. and Braun, H. (1993) A direct adaptive method for faster back-propagation learning, in *The RPROP Algorithm* (Ruspini, E., ed.), IEEE International Conference on Neural Networks. San Francisco, CA, pp. 586–591.

46. Zimmermann, H. and Neuneier, R. (1998) The observer-observation dilemma in neuro-forecasting, in *Advances in Neural Information Processing Systems* (Jordan, M. I., Kearns, M. J., and Solla, S. A., eds.), MIT Press, pp. 992–998.

47. Fahlman, S. and Lebiere, C. (1990) The cascade-correlation learning architecture, in *Advances in Neural Information Processing Systems,* (Touretzky, D., ed.), Morgan Kaufmann, pp. 524–532.

48. Battiti, R. (1992) First- and second-order methods for learning: between steepest descent and Newton's method. *Neural Computation* **4,** 141–166.

49. Shepherd, A. (ed.) (1997) *Second-Order Methods for Neural Networks.* Springer, New York.

50. LeCun, Y., Bottou, L., Orr, G., and Müller, K. -R. (1998) Efficient BackProp, in *Neural Networks: Tricks of the Trade* (Orr, G., & Müller, K. R., eds.), Springer, Berlin, pp. 9–50.

51. Fahlman, S. (1988) *An Empirical Study of Learning Speed in Backpropagation.* Carnegie Mellon University.

52. Humpert, B. (1994) Improving back propagation with a new error function. *Neural Networks* **7,** 1191–1192.

53. Oh, S. (1997) Improving the error backpropagation algorithm with a modified error function. *IEEE Trans. Neural Networks* **8,** 799–803.

54. Solla, S., Levin, E., and Fleisher, M. (1988) Accelerated learning in layered neural networks. *Complex Syst.* **2,** 625–639.

55. van Ooyen, A. and Nienhuis, B. (1992) Improving the convergence of the back-propagation algorithm. *Neural Networks* **5,** 465–471.

56. Tollenaere, T. (1990) SuperSAB: fast adaptive back propagation with good scaling properties. *Neural Networks* **3,** 561–573.

57. Jacobs, R. (1988) Increased rates of convergence through learning rate adaptation. *Neural Networks* **1,** 295–307.

58. Linder, R., Wirtz, S., and Pöppl, S. (2000) Speeding up backpropagation learning by the APROP algorithm in *Proceedings of the Second International ICSC Symposium on Neural Computation*, (Bothe, H. and Rojas, R., eds.), Berlin, Germany: ICSC Academic Press, Millet, pp. 122–128.

59. Weymaere, N. and Martens, J. (1994) On the initialization and optimization of mutilayer perceptrons. *IEEE Trans. Neural Networks* **5**, 738–750.

60. Yam, Y., Chow, T., and Leung, C. (1997) A new method in determining initial weights of feedforward neural networks for training enhancement. *Neurocomputing* **16**, 23–32.

61. Lehtokangas, M., Saarinen, J., Kaski, K., and Huuhtanen, P. (1995) Initializing weights of a multilayer perception by using the orthogonal least squares algorithm. *Neural Computation* **7**, 982–999.

62. Anand, R., Mehrotra, K., Mohan, C., and Ranka, S. (1995) Efficient classification for multiclass problems using modular neural networks. *IEEE Trans. Neural Networks* **6**, 117–124.

63. Rögnvaldsson, T. (1994) On Langevin updating in multilayer perceptrons. *Neural Computation* **6**, 916–926.

64. Murray, A. and Edwards, P. (1993) Synaptic weight noise during multilayer perceptron training: Fault tolerance and training improvements. *IEEE Trans. Neural Networks* **4**, 722–725.

65. Grandvalet, Y., Canu, S., and Boucheron, S. (1997) Noise injection: theoretical prospects. *Neural Computation* **9**, 1093–1108.

66. Barnard, E. and Holm, J. (1994) A comparative study of optimization techniques for backpropagation. *Neurocomputing* **6**, 19–30.

67. Alpsan, D., Towsey, M., Ozdamar, O., Tsoi, A., and Ghista, D. (1995) Efficacy of modified backpropagation and optimisation methods on a real-world problem. *Neural Networks* **8**, 945–962.

68. Orr, G. M. K. -R. (ed.) (1998) *Neural Networks: Tricks of the Trade*. Springer, New York.

69. Looney, C. (1996) Stabilization and speedup of convergence in training feedforward neural networks. *Neurocomputing* **10**, 7–31.

70. Linder, R. and Pöppl, S. (2001) ACMD: a practical tool for automatic neural net based learning. *Lect. Notes Comp. Sci.* **2199**, 168–173.

71. Stuttgarter *Stuttgarter Neuronale Netze Simulator*. http://www-ra.informatik.uni-tuebingen.de/SNNS (as of May 5th, 2005).

72. Amirikian, B. and Nishimura, H. (1994) What size network is good for generalization of a specific task of interest? *Neural Networks* **7**, 321–329.

73. Murata, N. (1996) An integral representation of functions using three-layered networks and their approximation bounds. *Neural Networks* **9**, 947–956.

74. Kröse, B. and van der Smagt, P. (ed.) (1993) *An Introduction to Neural Networks*. 5, University of Amsterdam,

75. Bartlett, P. (1993) Vapnik-Chervonenkis dimension bounds for two- and three-layer networks. *Neural Computation* **5**, 371–373.

76. Lewicki, M. and Sejenowski, T. (2000) Learning overcomplete representations. *Neural Computation* **12**, 337–365.

77. Amari, S., Murata, N., Müller, K. -R., Finke, M., and Yang, H. (1997) Asymptotic statistical theory of overtraining and cross-validation. *IEEE Trans. Neural Networks* **8**, 985–996.

78. Wang, C., Venkatesh, S., and Judd, J. (1995) Optimal stopping and effective machine complexity in learning. *Adv. Neural Inf. Processing Syst.* **6,** 303–310.
79. Finoff, W., Hergert, F., and Zimmermann, G. (1993) Improving model selection by nonconvergent methods. *Neural Networks* **6,** 771–783.
80. Prechelt, L. (1998) Automatic early stopping using cross validation: quantifying the criteria. *Neural Networks* **11,** 761–767.
81. Bologna, G. (1996) Rule extraction from the IMLP neural network: a comparative study. Proc. of the NIPS workshop of rule extraction from trained artifical neural networks. Snowmass, CO.
82. Setiono, R. and Liu, H. (1997) NeuroLinear: from neural networks to oblique decision rules. *Neurocomputing* **17,** 1–24.
83. Towell, G. and Shavlik, J. (1993) Extracting refined rules from knowledge based neural networks. *Machine Learning* **13,** 71–101.
84. Lawrence, J. and Frederickson, J. (eds.) (1993) *BrainMaker Professional User's Guide and Reference Manual*, 4*th*, California Scientific Software Press, Nevada City, CA.
85. Baba, K., Enbutu, I., and Yoda, M. (1990). Explicit representation of knowledge acquired from plant historical data using neural network in *Int. Joint Conf. on Neural Networks* (Caudill, M., ed.), San Diego, CA, pp. 155–160.
86. Linder, R., Theegarten, D., Mayer, S., et al. (2003) Der Einsatz eines Modifizierten Causal Index erleichtert die interpretation des Antwortverhaltens eines mit Daten einer Whole-Body Plethysmographie an einem Knock Out Mausmodell trainierten Artifiziellen Neuronalen Netzwerks (ANN). *Atemw. Lungenkrkh.* **29,** 340–343.
87. Chalfin, D. B. (1996) *Neural Networks: A New Tool for Predictive Models*, (Vincent, J. L, ed.) Springer, Berlin, Germany, pp. 816–829.
88. Tu, J. (1996) Advantages and disadvantages of using artificial neural networks versus logistic regression for predicting medical outcomes. *J. Clin. Epidemiol.* **49,** 1225–1231.
89. Dreiseitl, S. and Ohno-Machado, L. (2002) Logistic regression and artificial neural network classification models: a methodology review. *J. Biomed. Informa.* **35,** 352–359.
90. Dimopoulos, I., Tsiros, I., Serelis, K., and Chronopoulou, A. (2004) Combining neural network models to predict spatial patterns of airborne pollutant accumulation in soils around an industrial point emission source. *J. Air. Waste Manag. Assoc.* **54,** 1506–1515.
91. Liu, B., Cui, Q., Jiang, T., and Ma, S. (2004) A combinational feature selection and ensemble neural network method for classification of gene expression data. *BMC Bioinforma.* **5,** 136.
92. Rogova, G. (1994) Combining the results of several neural network classifiers. *Neural Networks* **7,** 777–781.
93. Doyle, H., Parmanto, B., Munro, P., et al. (1995) Building clinical classifiers using incomplete observations—a neural network ensemble for hepatoma detection in patients with cirrhosis. *Methods of Inf. Med.* **34,** 253–258.

94. Valentini, G., Muselli, M., and Ruffino, F. (2004) Cancer recognition with bagged ensembles of support vector machines. *Neurocomputing* **56,** 461–466.

95. Hansen, L. and Salamon, P. (1990) Neural networks ensembles. *IEEE Trans. Neural Networks* **12,** 993–1001.

96. Tumer, K. and Ghosh, J. (1995) Order statistics combiners for neural classifiers in *Worlds Congress on Neural Networks,* INNS Press, Washington, DC, pp. 31–34.

97. Munro, P. and Parmanto, B. (1997) Competition among networks improves committee performance, in *Advances in Neural Information Processing Systems,* (Mozer, M., Jordon, M., and Petsche, T., eds.), MIT Press, Cambridge, pp. 592–598.

98. Wolpert, D. (1992) Stacked generalization. *Neural Networks* **5,** 241–259.

99. Hashem, S. (1997) Optimal linear combinations of neural networks. *Neural Networks* **10,** 599–614.

100. Battiti, R. and Colla, A. (1994) Democracy in neural nets: voting schemes for classification. *Neural Networks* **7,** 691–707.

101. Lam, L. and Suen, C. (1995) Optimal combination of pattern classifiers. *Pattern Recognition Lett.* **16,** 945–954.

102. Wanas, N. and Kamel, M. (2001). Feature based decision fusion, in *ICAPR* (Singh, S., Murshed, N., and Kropatsch, W., eds.), Springer-Verlag, Berlin, Heidelberg, pp. 176–185.

103. Carney, J. and Cunningham, P. (1999) *The NeuralBAG* Algorithm: Optimizing Generalization Performance in Bagged Neural Networks in *Proceedings of the 7th European Symposium on Artificial Neural Networks* (Verleysen, M. ed.). pp. 3540.

104. Drucker, H., Schapire, R., and Simard, P. (1993). Improving Performance in Neural Networks Using a Boosting Algorithm, in *Advances in Neural Information Processing Systems* (Hanson, S., Cowen, J., and Giles, C. eds.), Morgan Kaufman, pp. 42–49.

105. Breiman, L. (1996) Bagging predictors. *Machine Learning* **24,** 123–140.

106. Schapire, R. (1990) The strength of weak learnability. *Machine Learning* **5,** 197–227.

107. Lin, J. -S., Lo, S. -C., Hasegawa, A., Freedman, M., and Mun, S. (1996) Reduction of false positives in lung nodule detection using a two-level neural classification. *IEEE Trans. Med. Imag.* **15,** 206–217.

108. Zhou, Z., Jiang, Y., Yang, Y. -B., and Chen, S. -F. (2002) Lung cancer cell identification based on artificial neural network ensembles. *Artif. Intell. Med.* **24,** 25–36.

109. Cuff, J. and Barton, G. (2000) Application of multiple sequence alignment profiles to improve protein secondary structure prediction. *Proteins* **40,** 502–511.

110. Qian, N. and Sejenowski, T. (1988) Predicting the secondary structure of globular proteins using neural network models. *J. Molec. Biol.* **202,** 865–884.

111. Yeang, C. -H., Ramaswamy, S., Tamayo, P., et al. (2001) Molecular classification of multiple tumor types. *Bioinformatics* **17,** 316–322.

112. Linder, R., Dew, D., Sudhoff, H., Theegarten, D., Pöppl, S., and Wagner, M. (2004) The "subsequent artificial neural network" (SANN) approach might bring more classificatory power to ANN-based DNA microarray analyses. *Bioinformatics* **20,** 3544–3552.

113. Kittler, J., Hatef, M., Duin, R., and Matas, J. (1988) On combining classifiers. *IEEE Trans. Pattern Anal. Machine Intell.* **20(3),** 226–239.

114. Hand, D. (2004) Academic obsessions and classification realities: ignoring practicalities in supervised classification, in *Classification, Clustering, and Data Mining Applications,* (Banks, D., House, L., McMorris, F., Arabie, P., and Gaul, W., eds.), Springer, Berlin, Germany pp. 209–232.

115. Nguyen, D., Arpat, A., Wang, N., and Carroll, R. (2002) DNA microarray experiments: biological and technological aspects. *Biometrics* **58,** 701–717.

116. Dutilh, B. and Hogeweg, P. (1999) *Gene networks from microarray data: analysis of data from microarray experiments, the State of the art in gene network reconstruction. Bioinformatics,* Utrecht University.

117. Holter, N., Maritan, A., Cieplak, M., Fedoroff, N., and Banavar, J. (2001) Dynamic modeling of gene expression data. *Proc. Natl. Acad. Sci. USA* **98,** 1693–1698.

118. Little, R. and Rubin, D. (eds.) (2002) *Statistical Analysis with Missing Data.* 2, Wiley-Interscience, New York.

119. Yoon, S. -Y. and Lee, S. -Y. (1999) Training algorithm with incomplete data for feed-forward networks. *Neural Processing Lett.* **10,** 171–179.

120. Personen, E., Eskelinen, M., and Juhola, M. (1998) Treatment of missing data values in a neural network based decision support system for acute abdominal pain. *AI in Med.* **13,** 139–146.

121. Dudoit, S., Fridlyand, J., and Speed, T. (2002) Comparison of discrimination methods for classification of tumors using gene expression data. *J. Am. Stat. Assoc.* **97,** 77–87.

122. Beissbarth, T., Fellenberg, K., Brors, B., et al. (2000) Processing and quality control of DNA array hybridization data. *Bioinformatics* **16,** 1014–1022.

123. Schuchhardt, J., Beule, D., Malik, A., et al. (2000) Normalization strategies for cDNA microarrays. *Nucleic Acids Res.* **28,** E47.

124. Schadt, E., Li, C., Ellis, B., and Wing, H. (2001) Feature extraction and normalization algorithms for high-density oligonucleotide gene expression array data. *J. Cell Biochem. Suppl.* **37,** 120–125.

125. Yang, Y. H., Dudoit, S., Luu, P., Lin, D., Peng, V., Ngai, J., and Speed, T. (2002), Normalization for cDNA microarray data: a robust composite method addressing single and multiple slide systematic variation. *Nucleic Acids Res.* **30,** E15.

126. Ross, D., Scherf, U., Eisen, M., et al. (2000) Systematic variation in gene expression patterns in human cancer cell lines. *Nat. Genet.* **24,** 227–234.

127. Perou, C., Jeffrey, S., van de Rijn, M., et al. (1999) Distinctive gene expression patterns in human mammary epithelial cells and breast cancers. *Proc. Natl. Acad. Sci. USA* **96,** 9212–9217.

128. Wang, W., Jones, P., and Partridge, D. (1998) Ranking pattern recognition features for neural networks, in *Advances in Pattern Recognition,* (Singh, S., ed.), Springer, Berlin, Germany pp. 232–241.

129. Park, P., Pagano, M., and Bonetti, M. (2001) A nonparametric scoring algorithm for identifying informative genes from microarray data. *Pac. Symp. Biocomput.* **6,** 52–63.

130. Ben-Dor, A., Bruhn, L., Friedman, N., Nachman, I., Schummer, M., and Yakhini, Z. (2000) Tissue classification with gene expression profiles. *J. Comput. Biol.* **7,** 559–583.

131. Tsai, C., Chen, C., Lee, T., Ho, I., Yang, U., and Chen, J. (2004) Gene selection for sample classifications in microarray experiments. *DNA Cell Biol.* **23,** 607–614.

132. Bo, T. and Jonassen, I. (2002) New feature subset selection procedures for classification of expression profiles. *Genome Biol.* **3,** Research0017.

133. Guyon, I. and Elisseeff, A. (2003) An introduction to variable and feature selection. *J. Machine Learning Res.* **3,** 1157–1182.

134. Cho, S. -B. and Won, H. -H. (2003) Data mining for gene expression profiles from DNA microarray. *Int. J. Software Eng. & Knowledge Eng.* **13,** 593–608.

135. Dudoit, S., Shaffer, J., and Boldrick, J. (2003) Multiple hypothesis testing in microarray experiments. *Stat. Sci.* **18,** 71–103.

136. Dudoit, S., Yang, Y., Callow, M., and Speed, T. (2002) Statistical methods for identifying differentially expressed genes in replicated cDNA microarray experiments. *Stat. Sinica* **12,** 111–139.

137. Efron, B. and Tibshirani, R. (eds.) (1993) *An Introduction to the Bootstrap.* Chapman and Hill, London, UK.

138. Efron, B. and Tibshirani, R. (1997) Improvements on cross-validation the 632+ Bootstrap Method. *J. Am. Stat. Assoc.* **92,** 548–560.

23

Methods for Microarray Data Analysis

Veronique De Bruyne, Fahd Al-Mulla, and Bruno Pot

Summary

This chapter outlines a typical workflow for micraorray data analysis. It aims at explaining the background of the methods as this is necessary for deciding upon a specific numerical method to use and for understanding and interpreting the outcomes of the analyses. We focus on error handling, various steps during preprocessing (clipping, imputing missing values, normalization, and transformation of data), statistic tests for variable selection and the use of multiple hypothesis testing procedures, various metrics and clustering algorithms for hierarchical clustering, principles, and results from principal components analysis and discriminant analysis, partitioning, self-organizing map, *K*-nearest neighbor classifier, and the use of a neural network and a support vector machine for classification.

Key Words: Bioinformatics; data mining; microarray analysis; numerical methods; statistic tests; algorithm; classification.

1. Introduction

The analysis of data sets produced by high-density gene arrays (microarrays) and gene chips is no longer a task that can be managed by a standard spreadsheet, but requires special tools capable of processing and analyzing many thousands of numerical values. There is no standard cookbook for microarray data analysis because different scientific cases generally require different combinations of data processing and analysis techniques. In many applications, the successful exploration of gene expression data relies on the ability to detect groups of genes (e.g., for the identification of genes that are coexpressed or that have similar expression patterns) or to associate certain clusters of genes with groups of arrays (e.g., for the discovery of genes with highly differentiating patterns for a set of tumor types). But there is more to microarray data analysis than finding coexpressed or differentially expressed genes, because a through

From: *Methods in Molecular Biology, vol. 382: Microarrays: Second Edition: Volume 2*
Edited by: J. B. Rampal © Humana Press Inc., Totowa, NJ

preprocessing of the data and a strategic data selection are prerequisites for obtaining scientifically meaningful results.

2. Data Management and Error Handling

In case of microarray experiments, data are more than just one set of expression values. Ideally the data set should originate from a number of repeated experiments. It is also relevant information contained in text fields like, for example, patient information or annotations from a public database.

The idea of using error estimates through the whole process of data reduction and analysis may not be so widespread. However, error estimates may be the companions that confirm a result or that urge one to be rather careful when drawing conclusions.

Smart data management requires that repeated experiments be stored in such a way that the arithmetic average of the values can be used in analyses. Original data should be kept, so that problems or questions can be traced back to an early stage in the reduction process. It is also improtant to keep the expression values extracted from one chip as a whole and to work with a selection of spots. This allows one to maintain an overview of what happened on the chip and at the same time to focus on the values of interest. Another important issue in data management is to be able to divide arrays and genes in groups based on various criteria. These groups are necessary for certain statistical tests like, for example, ANOVA, analyses, or classifiers and are also helpful for the visualization of results.

3. Data Preprocessing

As a general rule, hybridization quantifications and even expression values can only be used in analyses after some preprocessing. There are various operations that are considered as preprocessing actions like, for example, arithmetic operations, clipping, imputing missing values, normalization, integrating data from multiple arrays, combining color-flip data, combining repeated measurements, performing a log transform, the specific necessary actions depending on the transformations executed during the image processing stage.

3.1. Clipping

Clipping can be necessary when the data set contains data points with extreme values or with extreme error values. The need for clipping becomes immediately clear on regarding a histogram plot of the data and/or its errors. Clipping is the act of removing these data points by setting thresholds on the expression values or the error values of the expression values. Instead of removing the data points and hence introducing a missing value, one can immediately replace the outliers, for example, by the minimum allowed value if they had too low a value or by the maximum allowed value if they had too high a value, or by a value estimated from the neighboring data points (e.g., taking the average or the median).

3.2. Imputing Missing Values

If for the experiment there is no measurement for the expression level for a certain combination of an array and a gene; there will be a missing value in the extracted matrix. This poses a technical problem, which requires analysis techniques to complete filled data matrix. A replacement of the missing value can be estimated with several techniques. The simplest solution is to substitute missing values with a constant or a value calculated from the row or column in the expression matrix it belongs to. There are also solutions that rely on the fact that there are correlations within the data set and that try to come to a better substitution for the missing value. One such technique is based on *K*-nearest neighbors. An estimate for a missing value for a given gene on certain array is obtained from genes with a very similar profile to the one of the given gene. From the profiles with the smallest distance measure (often quantified using the Euclidean distance), the *K*-nearest neighbors of the given gene profile are used to obtain an estimate for the missing value. The expression values for the given gene on the *K*-nearest profiles are used to obtain a weighted mean that is used as substitution value for the missing one. The weighting depends on the distance between the individual profiles and the one of the given gene. The value of *K* is a free parameter of the algorithm. Apparently, the method works well if *K* ranges between 10 and 20 *(1)*.

3.3. Normalization and Transformation

Some variability in microarray data is systematic and does not carry any biological information as it has a technical origin. To reduce this undesired variability the data should be normalized, but unthoughtfull normalization can seriously distort the obtained results and introduce false signal. When the need for normalization first became clear, people started using total intensity normalization techniques. These techniques work on (log) expression ratios and are simple to implement. Nowadays it is widely accepted that intensity dependent normalization techniques, working on the two hybridization quantities from the two channels, are better suited for microarray data.

3.3.1. Total Intensity Normalization Techniques

Total intensity normalization techniques generally assume that the average expression ratio of (1) all genes in the array or (2) a subset of genes should be equal to one. For log expression values this translates to assuming that the average or median log expression ratio should be equal to zero. In case of (1), the complete data set is used to calculate the parameters for the normalization. In case of (2), only that particular subset is to be used, for example, housekeeping genes or control elements. Hence, total normalization techniques can apply a centering and a scaling. Centering is extracting an offset that is a good approximation of the population mean. Scaling aims at bringing the expression ratios into the same scale by dividing profiles by a certain value, often the standard deviation of its (log) expression ratios. Scaling is a necessary step before results based on a metric and from different experiments can be compared. It is known that centering of genes may distort the results, that scaling has the tendency to weaken strong signals and to amplify weaker signals, and that uncarefull normalization can create false-positives or -negatives.

3.3.2. Intensity-Dependent Normalization

In case of experiments where for each gene on an array pairs of fluorescence intensity (R,G) are available in a query sample and a reference sample, a plot of the ratio $\log(R/G)$ vs the intensity $\log(\sqrt{R \times G})$ is very helpful to detect intensity dependent effects. Any base for the log transform can be used, but log base 2 makes an elegant interpretation of the plot possible as it visualizes over- and underexpression in a symmetrical way. If the data are normal the plot will show data points equally scattered below and above a global average value of the ratio.

If this is not the case, a transformation that depends on the intensity of the data points can be used to obtain a normalized data set. The transformation can be noted as, $\log(R/G)' = \log(R/G) - c\log\sqrt{R \times G}$, where the function $c\log\sqrt{R \times G}$ stands for the intensity dependent normalization. A popular method to obtain such an intensity dependent normalization function is to perform a Lowess fit to the plot of the ratio vs the intensity.

The principle of a Lowess fit to a data set can be intuitively understood. For each point in the data set a local neighborhood is considered and the behavior of the data in that neighborhood is by a low degree (2 at most will do) polynomial, fit to the data using a weighted least squares method. More weight is attributed to the points near the original data point and less weight to the points further away. These local polynomial fits are then combined into one normalization function that applies a custom transformation to each array in the expression matrix. The extent of the neighborhood of a data point is determined by a so-called smoothing parameter and this is a crucial parameter for the fit. Larger local neighborhoods result in smoother polynomials, while smaller fitting regions result in functions that follow the data set more closely but that show more features on a smaller scale. Useful values for the smoothing parameter lie in the range between 0.25 and 0.5 for most applications; it is the purpose to capture the local variation in the plot without producing a rapidly varying function.

As an extension of this method, different Lowess normalization functions can be applied to each groups of elements obtained with a single print tip. It is not advisory to perform Lowess normalization if there are less than 1000 genes in the sample. In general, intensity dependent normalization schemes deliver better results than global normalization schemes *(2–5)*.

4. Variable Selection

There are several classical and well-known statistic tools that can be applied on profiles to select and filter-out relevant information from the thousands of entries that are probably irrelevant to the intended question. The most currently used statistics: (1) on one profile: Kolmogorov–Smirnov test for normality, (2) for two profiles: *t*-test for equal means, Wilcoxon signed-rank test for means of paired samples, Pearson correlation test, Spearman rank-order correlation test, (3) for one profile on which groups are defined: *t*-test for independent samples, Man-Witney test, (4) for one profile on which more than two groups are defined: ANOVA test (or parametric *f*-test), and Kruskal–Wallis test.

Some of these are parametric tests, they basically assume that the data follow a normal distribution; they generally make use of the values for the mean and the standard deviation of the distribution. On the other hand, nonparametric tests are commonly based on ranking of the data. These ranks are distributed uniformly; hence these tests are independent of any underlying distribution. On the other hand, an estimate of the significance is more complicated and often relies on approximations.

If the assumption of normality of the data is not fulfilled one cannot completely rely on the results of parametric tests. So before applying, for example, a *t*-test or ANOVA test (*see* **Subheading 4.1.**) one should check, which profiles are very unlikely to follow a normal distribution and exclude those from further analysis. For the other profiles, it is still not proven whether they do follow a normal distribution; there might be some slight chance that they will deviate from a normal distribution. The Kolmogorov–Smirnov test for normality can be applied to test how different the cumulative distribution of the sample is from the cumulative distribution of a model normal distribution, for which the mean value and standard deviation have been estimated from the sample.

4.1. Statistic Tests for a Profile With Groups Defined on It

If a gene profile contains data from arrays that are known to fall into groups, a plot of the gene profile presents a categorical and quantitative variable.

A categorical variable is a type of variable that divides a sample into separate categories or classes, for example, sex. A quantitative variable is a type of variable that can take continuous numerical values or binary values. The categorical variable splits the sample in a number of groups, while the quantitative variable describes a distribution within each group. This kind of data is called unpaired.

A typical question is whether the gene expression is significantly different among different groups. In case there are only two groups for the categorical variable, the parametric *t*-test or the nonparametric Mann–Whitney test can be applied. These tests can be used for testing whether a gene is differentially expressed within two groups of arrays. In most of the cases, the parametric *t*-test will be used. However, if the data is not normally distributed one should employ the Mann–Whitney test. If there are three or more groups for the categorical variable, the ANOVA test or the nonparametric Kruskal–Wallis test can be applied. These tests can be used for testing whether a gene is differentially expressed within three or more groups of arrays. The ANOVA test or parametric *f*-test is the most widely used one, whereas the Kruskal–Wallis test should be used if the data is not normally distributed. A more elaborate description of these tests is available (*6*).

5. Multiple Hypotheses Testing Procedures

The tests mentioned earlier are traditionally meant for testing one profile at the time. However, in microarray analysis these tests are mostly used within the context of identifying differentially expressed genes. This means that typically, many genes or hypothesis are tested simultaneously. In such situations, where many genes are involved, two types of errors can be distinguished: a type I error or false-positive, meaning that a gene is declared differentially expressed while it is not or a type II error or false-negative, meaning that a gene which is differentially expressed is not declared to be so.

In practice, a hypothesis is rejected if its *p*-value exceeds a threshold that is considered to be the maximum allowed probability of committing a type I error. When many hypotheses are tested simultaneously the chance of committing a type I error increases with the number of genes being tested at once. Clearly, increasing the number of genes in the data set increases the chance of finding genes with a very small *p*-value purely by chance. So when performing multiple hypotheses tests an appropriate type I error rate should be defined and a test procedure should be used to control this error rate. Accordingly, this control comes down to using an adjusted *p*-value for each hypothesis that is tested. Various testing procedures to achieve this are described in literature. A complete description of multiple testing procedures is beyond the scope of this chapter. In the context of multiple testing, the family-wise error rate is defined as the probability of committing at least one type I error. The false-discovery rate is defined as the expected proportion of type I error among the rejected hypotheses. An overview of procedures for multiple hypothesis testing can be found in *(7)* and references therein. Popular methods for family-wise error rate control are the Bonferroni procedure or the step-up Hochberg procedure. The Benjamini and Hochberg procedure is a multiple hypothesis testing procedure for controlling the false-discovery rate.

When using a multiple hypothesis testing procedure on the data set with an appropriate value for the level of error rate control, only a small selection of gene profiles will remain from most data sets. This is not a mild way to treat your data, but strictly speaking it is not scientifically defendable not to use multiple testing procedures.

6. Data Mining and Clustering Tools

There are numerous data analysis tools available that can be used to find gene or array profiles that have a similar behavior or that are significantly differently expressed *(8)*. For some researchers the problem with clustering tools is that there are many different data analysis tools that can be applied to a given data set and that the results of these analyses will rarely be exactly the same. Unfortunately, there is no way to ensure which technique gives the best result. In fact, the differences in outcome are only natural if one realizes that these methods differ in underlying mathematical techniques or in implicit assumptions that are made.

Grouping techniques that decide on the number of groups themselves (e.g., hierarchical clustering, principal components analysis [PCA]) can be very practical if there is no idea about the expected number of groups in the data set. On the other hand, for example, in the case of the hierarchical clustering, the method will propose certain clustering the data, while there is no guarantee that the result has any biological relevance. For clustering tools that take the number of expected groups as input (e.g., self-organizing map [SOM] and partitioning) it may be challenging to find a good estimate of this number of groups. In most cases, comparing the results from different grouping techniques, examining the average group profiles and performing some statistics, will lead to a satisfying and relevant result. The tools mentioned in this section are also referred to as unsupervised clustering techniques.

6.1. Hierarchical Clustering

This is a commonly used technique because it is easy to implement and the visualization as a series of nested clusters allows an easy interpretation, for example, use of different hierarchical clustering techniques were described in **refs. *9–11***. The clustering technique uses a metric (or similarity coefficient) to

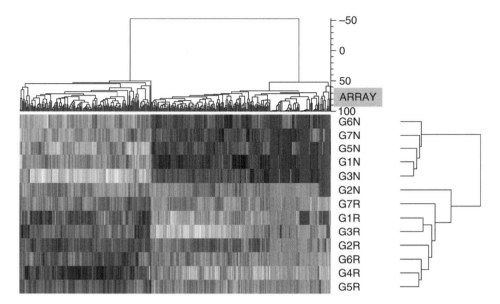

Fig. 1. A typical result of an unsupervised clustering of both genes and arrays is shown in this figure. The result of such a clustering may vary with the metric and clustering algorithm that is used.

measure the distance between all individual expression profiles. The relational distance table (a matrix of similarity values) obtained by the metric is used to come to a tree-like dichotomic order with closely resembling profiles in tighter groups and less related profiles in more loose groups (**Fig. 1**).

6.1.1. Metrics

There is a close interplay between the metrics are described earlier and the normalization of the data. If the data set is centered round the mean and scaled with the root mean square, the Euclidean distance, the Pearson correlation distance and the Cosine correlation distance will give the same result. If the data set is centered round the mean but not scaled, the Euclidean distance and Cosine correlation distance will give the same result. The Pearson correlation coefficient can be positive or negative as it gives the slope of the linear relation between two profiles. A Pearson correlation value of −1 means that one profile is horizontally mirrored compared to the other profile. In the context of gene expression profiles, this may as well be an indication for a strong relationship while the Pearson correlation distance will be high, hence the squared Pearson correlation can be of particular interest since it considers mirrored profiles as being strongly correlated.

6.1.1.1. EUCLIDEAN DISTANCE

The square of the Euclidean distance D between two profiles $x = (x_1, x_2, ..., x_n)$ and $y = (y_1, y_2, ..., y_n)$ is defined as $D^2 = \sum_{k=1}^{n} (x_k - y_k)^2$. This is the distance measure one would use to measure how close two physical locations are, in this example the vectors would have two elements: longitude and latitude.

This metric gives good results if the data has compact or isolated clusters. The metric is sensitive to the centering and scaling of the data. When using data that is not centered to a common mean and scaled to a common variance, the largest-scale features will dominate the distance measure. Also, linear relations between the profiles will distort the Euclidean distance.

6.1.1.2. PEARSON CORRELATION DISTANCE

The often used Pearson correlation coefficient r between two profiles $x = (x_1, x_2, ..., x_n)$ and $y = (y_1, y_2, ..., y_n)$ is defined as $r = \text{Cov}(x,y)/\sqrt{s_x s_y}$. The Pearson correlation distance $P = 1 - r$. These measures how similar the shapes of the profiles are, disregarding a constant offset or a difference in scaling between the two profiles. By its definition, centering and scaling are dealt with by the coefficient itself.

6.1.1.3. SQUARED PEARSON CORRELATION DISTANCE

This distance metric is defined as, $P_2 = 1 - r^2$, with r the Pearson correlation coefficient as explained earlier.

6.1.1.4. COSINE CORRELATION DISTANCE

The two points in n-dimensional space defined as $x = (x_1, x_2, ..., x_n)$ and $y = (y_1, y_2, ..., y_n)$ can be regarded as the end points of two vectors. The cosine of the angle between the two vectors can be used as metric. The Cosine correlation distance is $1 - C$. Clearly, the angle between the vectors x and y is equal to the angle between, for example, $2x$ and y and so this metric is independent of the scaling of the profiles. The Cosine correlation between x and y is also equal to the cosine correlation distance between $-x$ and y or x and $-y$, because the angles between these vectors have equal cosines. The metric is sensitive to the centering of the data.

6.1.2. Clustering Algorithms

The grouping of the profiles uses a specific clustering algorithm to produce rooted trees (e.g., UPGMA, single linkage) or unrooted trees (neighbor joining) *(9)*. For clustering of large data sets, as gene expression data, modified versions of UPGMA can be used. These methods bypass the calculation of a complete similarity matrix and are much faster on large data sets than traditional clustering algorithms. As such, these algorithms are very interesting for microarray data analysis.

The first step in a hierarchical clustering is to calculate the matrix of pair wise similarities using one of the metrics described earlier. Then the two profiles with smallest distance are merged into a cluster. Once this cluster is formed, there is need for a criterion to establish the distance between clusters in the evolving tree.

6.1.2.1. UPGMA

UPGMA stands for unweighted pair-group method using arithmetic averages and is a widely used method. The distance between two clusters is the average of the distances between all pairs of profiles that can be considered with one profile from the first cluster and the other profile form the second cluster. This method can be used when the operator has no idea about the distribution of the data points in advance. If the data are expected to contain naturally distinct groups, a complete linkage is better suited. If the data is expected to have a more chained structure, a single linkage is preferable.

6.1.2.2. SINGLE LINKAGE OR NEAREST NEIGHBOR CLUSTERING

The minimum of the distances between all pairs of profiles that can be considered with one profile from the first cluster and the other profile form the second cluster is taken as the distance between two clusters. This criterion works well if there are multiple equal minimum distances between clusters and generally produces a chained clustering.

6.1.2.3. COMPLETE LINKAGE OR FURTHEST NEIGHBOR CLUSTERING

The maximum of the distances between all pairs of profiles that can be considered with one profile from the first cluster and the other profile form the second cluster is taken as the distance between two clusters. This criterion works well if the data contains naturally distinct groups and generally produces compact clusters.

6.1.2.4. NEIGHBOR JOINING

This method subsequently joins clusters to minimize the sum of branch lengths of the whole tree and continues until two clusters are left. The result is an unrooted tree.

6.2. Principal Components Analysis and Discriminant Analysis

Unlike hierarchical clustering, a PCA does not produce a hierarchical structure. Instead, the result is a two-dimensional (2D) or three-dimensional plot, in which the profiles are spread according to their relatedness.

6.2.1. Principle of PCA

A large number of variables are involved in a data set consisting of a large number of arrays and a large number of tested genes. This forms a complex situation for analyses and moreover, some of these variables may contain redundant information. It seems then a logical progression to reduce the dimension of the problem and to try to construct a smaller set of new variables that account for most of the variation in the original set of variables that are useful for a practical approximation of these original variables.

This idea can be translated into a mathematical minimization problem. There is a very interesting and elegant solution to this kind of problems, which are also known as eigenvalue problems for a matrix. The solution is a series of eigenvalues and eigenvectors. The principal components can be directly constructed as linear combination of the original variables using the coordinates of the eigenvectors. By construction the variance in the principal components is maximized. The first principal component is the most interesting one as it captures the largest part of variation in the original variables. The following principal component will ever capture a smaller part of this variation. The principal components are mutual orthogonal, as this is a characteristic of the eigenvectors, which is useful for representations as they can be used as the axes in 2D or three-dimensional representations (**Fig. 2**).

For practical use of PCA, there are some additional considerations to be made. The construction of the principal components is influenced by the overall level and the scale of the measurements for the original variables. Before calculating a PCA, the data should be normalized data in the sense of being centered and scaled. This is especially important to realize if PCA analyses of different data sets are compared.

1. In case of a PCA calculated on the genes, the PCA plot is centered around the origin by subtracting the average of the array profile from all expression values on the array profile, this is recommended for general purposes.
2. If the expression values on the array profile are divided by the variance of the array profile, each array profile is considered equally important in the analysis. This can be interesting to correct for array profiles that are of unequal intensity. If no correction is applied in such a case, the information in the array profiles with lower expression values will be completely masked by the information in the array profiles with high expression values.
3. Subtracting the average of the gene profile from all expression values on the gene profile results in an expression matrix, where the sum of expression values on each gene profile equals zero.
4. An expression matrix for which the intensity is normalized for all gene profiles is obtained by dividing the expression values on the gene profile by the variance of the gene profile. This has the effect that genes with different overall reaction fall together as long as the relative reactions of the profiles are the same.

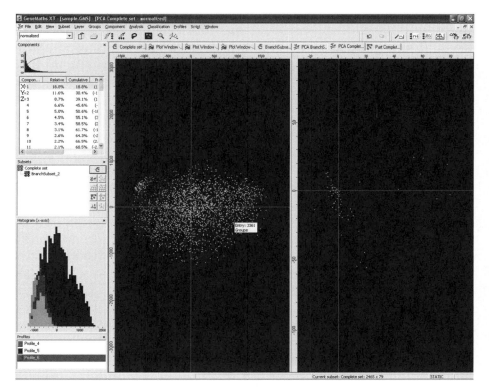

Fig. 2. The two central panels show typical results from a PCA analysis. The clouds represent the projections of the gene profiles (left) and array profiles (right) on the first two principle components.

6.2.2. Results From a PCA

The best known results from a PCA are projection scatter plots for the original data set. These plots are constructed by projecting the original profiles; these can be either gene profiles or array profiles, on the principal components. For example, in case of a PCA calculated on the genes, a projection of the gene profiles on the first two or three principal components can reveal some structure present in the data. A projection of the array profiles on the principal components reveals the contribution of each of the profiles to the components. There is a good interpretation of such a scatter plot: the further a gene away from the center of the plot or alternatively the larger its projection on one of the components, the larger its contribution to the variation accounted for by that component. A PCA calculated on the arrays used the same principal components in its representation and the interpretation as explained earlier can be translated by permutation of the words gene to an array.

One can also plot a PCA spectrum. For an example, the squares of the eigenvalues used to construct the principal components are proportional to the variance explained by the component. Plotting these quantities in a bar graph shows at a glance how quickly the importance of subsequent principal components decreases. Moreover, a cumulative distribution of these quantities is a helpful tool to see how many components are needed to capture the variation in the data up to an acceptable level. For a more detailed description of the PCA technique *(6)* and for applications (*see* **refs.** *12* and *13*).

6.2.3. Discriminant Analysis

This method is often mentioned in relation with PCA because it is mathematically very similar. On the other hand the input for the analysis is quite different and it is not really a grouping method, but it does reduce the complexity of the data set by means of a linear transformation. A discriminant analysis starts from a data set on which a number of predefined groups are indicated and constructs such a space of new parameters in which these groups are maximally separated. These new parameters are called discriminant profiles and just like principle components, they are linear combinations of the original parameters. Also, the representation of the results, a scatter plot in which the discriminant profiles as axes, closely resembles a PCA plot.

A discriminant analysis can also be used to answer the question whether the groups in the arrays can be characterized mainly looking at a limited number of genes (or vice versa). The coefficients for the genes in the discriminant profiles can be seen as the weights attributed to the genes that are important for maximizing the differences between the different array groups. As for the representation of the results, this is very similar to the example shown in Error! Reference source not found (**Fig. 2**).

6.3. Partitioning

A partitioning method divides the data set in a predefined number of groups in a fairly simple way. At the start, the data points are randomly assigned to the groups. Then the proximity between the individual data points and the groups is calculated using one of several criteria. Next, the data points are shuffled between the groups: a data point is assigned to the group it is closest to. In subsequent iterations, the distances of the data points to the new groups are calculated and they are again replaced to the group they are closest to. This process continues until all data points belong to the closest group. The challenge here is to estimate the number of groups in the data. This can be done based on results from other grouping techniques that do not require this parameter as input. Calculating a partitioning may well be an iterative process. The *k*-means clustering algorithm is a popular way of calculating a partitioning (**Fig. 3**).

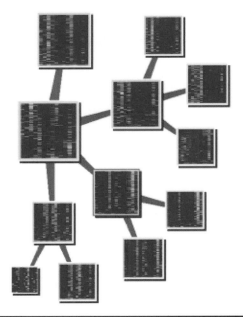

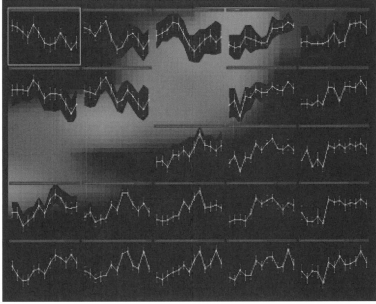

Fig. 3. Top: an example of a partitioning. The entries in the cells are represented by the corresponding rows from the similarity matrix. Bottom: an example of a self-organizing map. The cells are represesnted by an average profile, the shading around the profiles represent the internal spread between the profiles in the cell.

6.4. Self-Organizing Map

A is a method used to map a high dimensional feature space onto a low dimensional one (generally a 2D one), with a number of predefined cells in X and Y direction. The map is arranged such that neighboring cells represent groups of very similar profiles, while well separated points represent groups of very different profiles. If a SOM is constructed based on, for example, genes, the gene profiles are to be used as input data, called sample vectors. The nodes of these cells are weight vectors of the same dimension as the sample vectors.

The initialization of a SOM assigns randomly values to the weight vectors. The map is organized through a number of iterations of the learning cycle. Each iteration in the learning cycle focuses on one sample vector and to find the weight vector that is located closest to a chosen sample vector. For this step the Euclidean distance is used. It is also necessary to obtain the weight vectors that are considered as neighbors for the chosen sample vector. The weight vectors close to the sample vector will be moved even closer, the further away the weight vectors, the less they will be adapted. The moving of the weight vectors is scaled such that at the beginning of the learning cycle the effects are largest and then decreasing, while the learning cycle evolves. When this learning process guided by one sample vector is finished, another one will be used to train the map in the next learning cycle. The complete learning process starts with a global order and proceeds with ever more detailed order. A sufficient number of iterations should be considered in order to obtain a representative map.

The similarities are indicated by means of a gray shaded map on which the sample vectors are indicated as points. If the average distance between cells is high, the surrounding neighbors are very different and this is indicated with a light color (white in the limit). The places where the similarity between the surrounding neighbors is high are indicated with a dark color (black in the limit). This gray scale map makes a SOM easy to interpret, if weight vectors are close together and they are connected by a dark gray or black background, then they are clusters of similar expression profiles. If on the other hands they are separated by a white or light gray region, then they are clusters of different expression profiles.

Also for this method, the estimation of the number of weight vectors in the map is the most difficult part. If there is a large scatter on the profiles within the groups, this indicates that not enough nodes have been used. If not all groups show fundamentally different average profiles, this indicates that too many nodes have been used. For principles and applications *see* **refs. *14–16***.

7. Classification Tools or Supervised Grouping Techniques

If a data set contains a set of profiles clearly attributed to certain classes; these profiles can be used to construct a classifier that is able to determine the

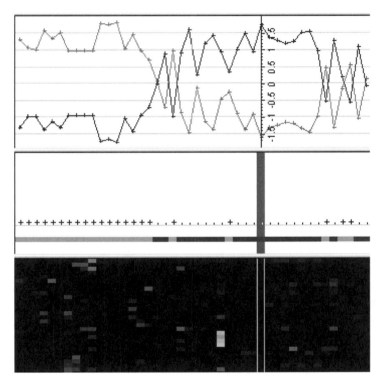

Fig. 4. The result of the training of a classifier of the execution of a classifier is a set of profiles. Each of these profiles represents a class used by the classifier and pictures the probability that a certain gene or array profile belong to that class.

class membership of a set of profiles for which the class is unknown. Classifiers or supervised grouping techniques use knowledge on the data set that is not contained in the expression matrix but, for example, in a text field to learn characteristics for the profiles in known classes. A classifier that works well on the training set is constructed and that is assumed to classify other but similar profiles correctly. The result of training or a classification can be expressed as a set of profiles, one for each class, giving a measure for the probability that the corresponding profile belong tot that class. Some classification techniques rely on rather simple concepts, like for example, the *k*-nearest neighbor classifier, whereas others are related to more complex specific research domains, for example, neural networks and support vector machines. These techniques may indeed give better results, but the complexities of the parameters they require make them sometimes hard to use appropriately (**Fig. 4**).

As the number of profiles involved in the classification may become very large, it is more efficient to use only those profiles that capture the largest part of the variation in the data set. Performing a PCA and subsequently using the relevant principal components for the classification can achieve this dimensional reduction. This step is compatible with all classifiers. It is really rewarding to perform this, as the results for the classifier are clearly more satisfying.

7.1. K-Nearest Neighbor Classifier

The principle behind this classifier is very easy to understand. For each profile in the training set, its K-nearest neighbors are searched for. The class the profile will be placed in is the class that is most common among its K-nearest neighbors. This technique obviously needs a way to estimate what the nearest neighbors to a profile are; this is generally done by means of the Euclidean distance. The number of neighbors K is an important user defined parameter.

7.2. Neural Network

A neural network is a means of calculating a function of which one does not have a clear description, but of which many examples with known input and output are present. In the context of classifying genes, the input for each example is a numerical gene profile, and the output is the name of a group to which the profile belongs. The neural network is trained with the examples and can be used to perform a classification of other data. The number of nodes in the input layer equals the dimension of the gene profiles; the number of nodes in the output layer equals the number of identified groups. The number of nodes in the additionally used hidden layer(s) depends on the nature and complexity of the data set and identification system.

A known example is entered in the neural network and the output is calculated. Initially this output will not be the expected output. The weights between every pair of consecutive neurons are then adjusted slightly, so that the output gets closer to the correct output. This training process is typically repeated a few thousand times, each time with another known example. After a sufficient number of iterations the calculated outputs will be close to the correct ones, provided that the number of layers and number of nodes per hidden layer is chosen correctly. A higher number of layers and/or neurons imply that the training and calculation takes longer, while a better result can be expected. There is also a danger of overtraining: too many layers and/or neurons used in the training will give very good results for the example profiles, but not at all for other inputs. This can be checked using a validation set, this is a set of example profiles that is not used for training, but it is used to check how well the neural network performs on this set. If the performance is significantly worse than for the training set, one knows that there are too many layers and/or neurons.

An important requirement for successfully applying neural networks is that the example data set is sufficiently large and that many examples are present for each class. The number of required entries depends on the heterogeneity of the class: the more heterogeneous a class, the more entries that will be needed to create a reliable neural network.

7.3. Support Vector Machine

This technique can be explained assuming there are only two classes for the data. A training set is required: a number of profiles that are clearly member of one of the two classes. The number of profiles in the training set should be in balance with the total number of profiles that has to be classified.

The profiles in the training set represent the vectors in input space. The idea is to transform these vectors in input space to a higher dimensional space (called feature space), where a hyperplane can be found that provides a maximal separation between the two classes. Finding the best separating hyperplane is equivalent to solving a quadratic programming problem. The transformation to feature space is done using a so-called kernel function. For simple problems, a linear kernel function can be used. For more complex problems one can use a polynomial function of a certain degree. Even more sophisticated kernels are radial basis function that. In this case each cluster is separated from the other members by placing a Gaussian over each support vector in the training set. It can be used when the classes are expected to form one or more clusters. The exact input of the kernel is calculated from the training set. For more information is available on the method and applications *(11,17,18)*.

8. Conclusions

There is clearly more than one way from raw microarray data to be analyzed. Every step in the reduction and analysis process is important and one should have considerable knowledge of the underlying principles of the applied tools, in order to understand the results or to develop a critical attitude toward them. Moreover, in publications, it is important to detail the different steps the researcher implemented, with clear communication of the relevant parameters utilized in order to allow the reader to fully understand how the results were achieved.

Acknowledgments

Fahd Al-Mulla is supported by Kuwait University Shared Facility Grant GM/0101.

References

1. Troyanska, O., Cantor, M., Sherlock, G., et al. (2001) Missing value estimation methods for DNA microarrays. *Bioinformatics* **17,** 250–525.
2. Yang, Y. H., Dutoit, S. P., Luu, P., and Speed, T. P. (2001) Normalization for cDNA Microarray Data. Technical Report 589, Department of Statistics, UC Berkeley.

3. Yang, Y. H., Dutoit, S., Luu, P., et al. (2002) Normalization for cDNA microarray data: a robust composite method addressing single and multiple slide systematic variation. *Nucleic Acids Res.* **30,** E15.

4. Quackenbush, J. (2001) Computational analysis of microarray data. *Nat. Rev. Genet.* **2,** 418–427.

5. Quackenbush, J. (2002) Microarray data normalization and transformation. *Nature Genetics Suppl.* **32,** 496–501.

6. Dobson, J. D. (1992) Applied multivariate data analysis, vol. II: *Categorical and Multivariate Methods.* Berlin: Springer—Verlag, p 731.

7. Dudoit, S., Shaffer, J. P., and Boldrick, J. C. (2003) Multiple hypothesis testing in microarray experiments. *Statist. Sci.* **18,** 71–103.

8. Jain, A. K., Murty, M. N., and Flynn, P. J. (1999) Data clustering: a review. *ACM Comput. Surveys* **31,** 264–323.

9. Eisen, M. B., Spellman, P. T., Brown, P. O., and Botstein, D. (1998) Cluster analysis and display of genome-wide expression patterns. *Proc. Natl. Acad. Sci. USA* **95,** 14,863–14,868.

10. Ramaswamy, S., Tamayo, P., Rifkin, R., et al. (2001) Multiclass cancer diagnosis using tumor gene expression signatures. *Proc. Natl. Acad. Sci. USA* **98,** 15,149–15,154.

11. Alizadeh, A., Eisen, M., Davis, R. E., et al. (2000) Distinct types of diffuse large B-cell lymphoma identified by gene expression profiling. *Nature* **403,** 503.

12. Raychaudhuri, S., Stuart, J. M., and Altman, R. B. (2000) *Principal Components Analysis to Summarize Microarray Experiments: Application to Sporulation Time Series. Pacific Symposium on Biocomputing,* Honolulu, Hawaii, pp. 452–463.

13. Yeung, K. W. and Ruzzo, W. L. (2001) An empirical study on principal component analysis for clustering gene expression data. *Bioinformatics* **17,** 763–774.

14. Tamayo, P., Slonim, D., Mesirov, J., et al. (1999) Interpreting patterns of gene expression with self-organizing maps: Methods and application to hematopoietic differentiation. *Proc. Natl. Acad. Sci.* **96,** 2907–2912.

15. Golub, T. R., Slonim, D. K., Tamayo, P., et al. (1999) Molecular classification of cancer: class discovery and class prediction by gene expression monitoring. *Science* **286,** 531–537.

16. Kohonen, T. (2001) *Self-Organizing Maps.* Springer, Berlin, Germany.

17. Burges, C. (1998) A tutorial for support vector machines for pattern recognition. *Data Mining Knowledge Discov.* **2,** 121–167.

18. Hearst, M. A. (1998) Trends and controversies: support vector machines. *IEEE Intell. Syst.* **13,** 18–28.

24

Predicting DNA Duplex Stability on Oligonucleotide Arrays

Arnold Vainrub, Norha Deluge, Xiaolin Zhang, Xiaochuan Zhou, and Xiaolian Gao

Summary

DNA duplex stability on oligonucleotide microarray was calculated using recently developed electrostatic theory of on-array hybridization thermodynamics. In this method, the first step is to finding the enthalpy and entropy of duplex formation in solution. This standard calculation was done with nearest-neighbor scheme and on-line software. Next the defined parameters and the array's single characteristic, the surface density of probes, are used to predict on-array duplex melting behavior. Reasonable accords of calculated and experimental melting curves for *in situ* synthesized microfluidic array were observed. The proposed method could be useful in microarray design and hybridization optimization. However, lack of melting curve measurements for different microarray platforms makes more experiments desirable to determine the method's accuracy.

Key Words: Hybridization thermodynamics; melting curve; oligonucleotide microarray.

1. Introduction

In this chapter, the use of a recently developed electrostatic theory of DNA hybridization with oligonucleotide probes tethered to the surface and compare the theoretical results with new microfluidic array experiments presented here were demonstrated. Two similar theoretical approaches have been developed by Vainrub (*1–3*) and afterwards by Halperin (*4,5*), the first is followed, which uses the terms and notations adopted in experimental microarray literature. The goals are to introduce the theory to the microarray community and present it in a form convenient for practical use. Therefore, the presentation of the theory is simple and omits the derivations that can be found elsewhere (*2,3*). Rather on providing a set of relations for calculation of the theoretical values for melting temperature, the width of the melting curve, and the degree of hybridization to enable comparison with experimental results were focused. In particular, how to calculate and plot the melting curves were described. These applications

From: *Methods in Molecular Biology, vol. 382: Microarrays: Second Edition: Volume 2*
Edited by: J. B. Rampal © Humana Press Inc., Totowa, NJ

of the theory are illustrated for the presented melting experiments with *in situ* synthesized oligonucleotide arrays.

Experimentally, a DNA double helix tethered by one strand to the surface of an array is less thermally stable *(6–8)* and has extended temperature range of denaturation (melting) *(7,9)* compared with the duplex free in solution. These differences are illustrated in **Fig. 1** showing substantial discrepancy between on-array (our experiment) and solution (theoretical) melting curves. In contrast, the electrostatic theory (solid line in **Fig. 1**) is in better agreement with experiment and accounts for the low temperature shift and broadening of the melting curve. Hence, the electrostatic theory can be used for estimates of on-array hybridization thermodynamic parameters. However, the theory includes the dominant electrostatic interaction between the hybridizing DNA and an array of the surface tethered oligonucleotide probes, but ignores known effect of the linker molecule *(10,11)* and also the effect of surface charge considered in *(12)*. Therefore, this first theory of on-array hybridization thermodynamics *(2,3)* is anticipated to be better approximation for oligonucleotide array designs using long linker molecule and low charged surface.

2. Thermodynamics of DNA Hybridization–Denaturation in Solution

The equilibrium of oligonucleotide hybridization in solution phase can be predicted using a well-established empirical nearest-neighbor model *(13–15)*. Hybridization of two complementary single-stranded oligonucleotides (probe *P* and target *T*) leads to formation of a double helix (duplex) *D* and could be presented as a reversible reaction.

$$T + P \rightleftharpoons D \tag{1}$$

Thermodynamic equilibrium in this reaction at temperature T is determined by the Gibbs free energy of duplex formation

$$\Delta G = \Delta H - T\Delta S. \tag{2}$$

Here ΔH and ΔS are the enthalpy and entropy of the duplex formation. The entropy decreases ($\Delta S < 0$) on double helix formation because the probe and target adopt a helical structure, and thus reduce the number of the possible configurations *(16)*. If the initial concentration of the target is C_0, the thermodynamic equilibrium is described by the well-known relation:

$$C_0 = \frac{\theta}{1-\theta} \exp\left(\frac{\Delta H_0 - T\Delta S_0}{RT}\right) \tag{3}$$

where the extent of hybridization θ ($0 < \theta < 1$) is the fraction of the total number of probes that form the duplexes. Keeping in mind for further comparison,

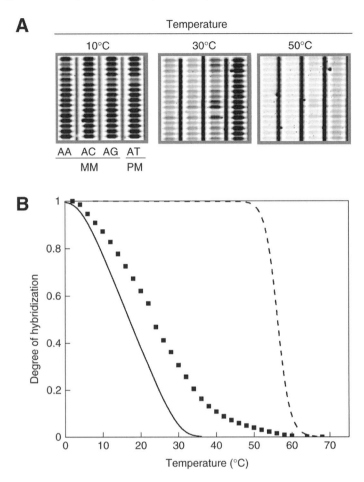

Fig. 1. (**A**) Images taken using a cooled CCD camera from temperature-dependent melting experiments measured in real-time monitoring of Cy3 (532 nm) fluorescence signals of the hybridization targets. The first three columns contain probes which have a central mismatch pair (MM) when hybridizing with the target (**Table 1**); from the first to the third row, MM = AA, AC, or AG; the last column contains probes which form perfect complementary (PM) duplexes. The 16 probes in each column vary in the residues nearest to the MM or PM base pairs; the probes in each row have the same nearest neighbors. The three images were selected from a series of temperature melting data (2–68°C) to illustrate the signal transition as a function of temperature. (**B**) Experimental and theoretical melting curves for D_1^{PM} oligonucleotide duplex. The electrostatic theory (solid line) is in significantly better agreement with array-based experiment (points) compared with prediction of the nearest-neighbor solution model (dashed line). *See* **Table 1** for the oligonucleotide sequence and the hybridization conditions.

this equation is written for typical microarray case when the targets in solution are in excess to surface immobilized probes and thus the concentration of non-hybridized targets practically does not depend on the hybridization extent θ. Standard characteristics of DNA hybridization are the melting temperature T_M and the width of the melting curve W. T_M is defined as the temperature when half of the probes form duplexes ($\theta = 0.5$). From **Eq. 3** we get

$$T_M = \frac{\Delta H_0}{\Delta S_0 + R \ln C_0} \tag{4}$$

Melting width W is defined by the slope of the melting curve at the temperature T_M

$$W = -\left(\frac{dT}{d\theta}\right)_{T = T_M} = -\frac{4R\Delta H_0}{(\Delta S_0 + R \ln C_0)^2} = -\frac{4R}{\Delta H_0}T_M^2 \tag{5}$$

Comprehensive experimental studies of hybridization in solution established an empirical nearest-neighbors model *(13–15)* that allows prediction of the enthalpy ΔH_0 and entropy ΔS_0 based on the oligonucleotide sequence. Several convenient programs are available on-line (used melting at http://bioweb.pasteur.fr/seqanal/interfaces/melting.html *[17]*) and make these calculations simple. The user just specifies the probe and target sequences that can either be perfectly matched or contain one mismatch, the target concentration, and the content of NaCl salt in hybridization solution. The program's outputs are enthalpy ΔH_0 and entropy ΔS_0, and the melting temperature T_M of the duplex.

In the experimental data used for model analysis herein, a set of 640 DNA oligonucleotides of 19 residues were synthesized on a microfluidic chip; these probes were hybridized with a set of 64 solution oligonucleotides containing 5′-Cy3 fluorescence dye labeling and the temperature-dependence curves were recorded *(18)*. **Table 1** lists example probe and target sequences of the interest in this discussion with their perfect matched (PM) and single mismatched (MM) in the central of duplexes. The computed ΔH_0 and ΔS_0 of these sequences in a 1 M NaCl hybridization buffer are shown in **Table 2**. In fact, in our experiment a denaturating agent, formamide (25%, v/v) was also present.

The known formamide effect is to decrease the melting temperature by 0.61°C per a volume percent of formamide without change of the melting curve shape *(19)*. Hence, at formamide concentration ρ (in % v/v) the duplex formation enthalpy ΔH_{0d} and entropy ΔS_{0d} could be approximated as

$$\begin{aligned} \Delta H_{0d} &= \Delta H_0 - 0.61\ \rho \Delta S_0 \\ \Delta S_{0d} &= \Delta S_0 \end{aligned} \tag{6}$$

Table 1
Oligonucleotide Probe and Target Sequences

Target	Probe	Notation of the duplex
5'-ACAAGGACCATGACAACGA-3'	5'-TCGTTGTCATGGTCCTTGT-3'	D_1 PM
	5'-TCGTTGTCAAGGTCCTTGT-3'	D_1 MM
5'-ATACCGATCCTGATACGGA-3'	5'-TCCGTATCAGGATCGGTAT-3'	D_2 PM
	5'-TCCGTATCACGATCGGTAT-3'	D_2 MM

Table 2
Nearest-Neighbor Model Enthalpy and Entropy of Duplex Formation in 1 *M* NaCl Solution (ΔH_0, ΔS_0) and in the Same Solution With 25% (v/v) Formamide (ΔH_{0d}, ΔS_{0d})

Duplex	ΔH_0 (kJ/mol)	$\Delta S_0 = \Delta S_{0d}$ (kJ/molK)	ΔH_{0d} (kJ/mol)
D_1 PM	−585.2	−1.542	−561.7
D_1 MM	−525.8	−1.393	−504.6
D_2 PM	−559.7	−1.479	−537.1
D_2 MM	−512.5	−1.395	−491.2

3. Thermodynamics of DNA Hybridization on an Oligonucleotide Array

3.1. Outline of the Electrostatic Theory

Single-stranded DNA in solution is a highly negatively charged molecule bearing one electron charge per each backbone phosphate group. The electrostatic repulsion between single strands in DNA double helix is known to destabilize the DNA double helix. The repulsion becomes weaker when small ions (typically from added NaCl) are present in solution and screen the electrostatic forces increasing the duplex stability. This has been confirmed by numerous experiments demonstrating the melting temperature increase from 12 to 16.6°C with 10-fold increase in NaCl concentration *(20)*.

The theory of on-array DNA hybridization thermodynamics *(2,3)* is based on the intuitively clear idea that the electrostatic repulsion barrier is stronger on-array compared with solution phase hybridization. Indeed, under array application conditions an extra repulsion of hybridizing target from the arrayed probes that surround the hybridizing probe occurs and adds to the repulsion when the target approaches the probe. This is measured by decrease in the T_M of the probe-target pair. Because additional on-array repulsion grows with surface density of probes, so does the drop in melting temperature. The second effect, the broadening of the melting curve, is also owing to the on-surface electrostatic repulsion. The charge in probe layer on the surface increases in each next hybridization event by the charge of hybridizing target. As a result, the electrostatic repulsion between the hybridized target and probe layer becomes stronger in the course of hybridization. This extends melting curve toward low temperature and thus causes melting curve broadening.

3.2. Calculation of the Melting Temperature and Width

According to electrostatic theory on-array hybridization equilibrium is described by the relation *(3)*

$$C_0 = \frac{\theta}{1-\theta} \exp\left(\frac{\Delta H_0 - T\Delta S_0}{RT}\right) \exp\left(\frac{wN_p Z_P (Z_P + \theta Z_T)}{RT}\right) \tag{7}$$

The array surface effects are described in this equation by the last exponential in the right side. Here N_p is the surface density of probes, Z_P and Z_T are the lengths of probe and target expressed in a number of bases, and w is the electrostatic interaction parameter that depends on the NaCl concentration in the hybridization solution. At low surface density N_p (no close neighbors around the hybridizing probe) the conditions become solution-like and this relation reduces to **Eq. 3** for solution hybridization.

From **Eq. 7** we derive formulas for the on-array melting temperature *(2)*

$$T_M^{array} = \frac{\Delta H_0 + wN_p Z_P \left(Z_P + \frac{1}{2} Z_T\right)}{\Delta S_0 + R\ln C_0} \tag{8}$$

and on-array melting width

$$W^{array} = -\frac{4R\Delta H_0}{(\Delta S_0 + R\ln C_0)^2} - \frac{wN_p Z_P^2}{\Delta S_0 + R\ln C_0} - \frac{wN_p Z_P \left(Z_P + \frac{1}{2} Z_T\right) R}{4(\Delta S_0 + R\ln C_0)^2} \tag{9}$$

The difference between on-array **(Eq. 8)** and in solution **(Eq. 4)** melting temperatures is

$$\Delta T_M^{array} = \frac{wN_p Z_P \left(Z_P + \frac{1}{2} Z_T\right)}{\Delta S_0 + R\ln C_0} \tag{10}$$

The broadening of on-array melting curve is

$$\Delta W^{array} = -\frac{wN_p Z_P Z_T}{\Delta H_0} T_{M0} + \frac{\Delta T_M^{array}}{T_{M0}} W_0 \tag{11}$$

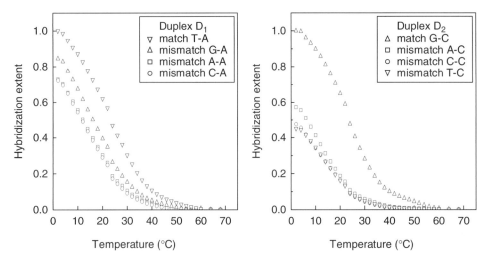

Fig. 2. Experimental on-array melting curves of *D*1 and *D*2 duplexes. In each case the four curves correspond to the perfect match and three different mismatches in the central base position.

3.3. Calculation and Plotting the Melting Curve

The on-array melting curve function θ(T) cannot be found in analytical form from transcendental **Eq. 7**, but a simple procedure allows to get it numerically and as a graph. We rewrite **Eq. 7** as

$$T = \frac{\Delta H_0 + w N_P Z_P (Z_P + \theta Z_T)}{\Delta S_0 + R \ln[C_0 (1 - \theta) / \theta]} \tag{12}$$

to express the temperature T as a function of the degree of hybridization θ. Then we make a function plot $T(\theta)$ using a commercial plotting software, Origin 6.0 (OriginLab Corporation, Northampton, MA). Finally, applying standard in plotting software operation to exchange *x*- and *y*-axes, produce the melting curve plot. The generated melting curves will be discussed in **Subheading 3.5.**

3.4. Experimental Melting Curves

Melting experiments were performed with oligonucleotide microarray based on a microfluidic chip with 3968 cells. Oligonucleotide probes of 19-mer were synthesized *in situ* using photogenerated acids method and digital maskless illumination patterns *(18,21–23)*. $(dT)_{10}$ oligonucleotide was used as a spacer between the probe and coupled to the surface silane linker. The microarray contains 640 probe sequences (each probe repeated six times) designed to study comprehensively the effect of single base mismatches located near 5′-end, center, and near 3'-end of the probe. For these three positions all possible perfect match and single mismatch combinations were assessed by an equimolar mixture of 64 fluorescent dye-labeled target oligonucleotides (19-mers). Sequence design avoided cross-hybridization and internal secondary structure of probes and targets. The hybridization was equilibrated at temperature 2°C, subsequently the temperature was raised in 2°C steps up to 68°C, and the equilibrium fluorescence intensity was measured at each temperature point. The control experiments showed that in this particular case of short 19-mer targets, after the initial equilibration was reached at low temperature, the equilibrium was established within minutes after each 4°C change in temperature.

Figure 2 shows the sets of the melting curves measured for two targets described in **Table 1**, each hybridizes to a perfect match probe and three different single mismatch probes.

3.5. Comparison of Theory With Experiment

The melting curves were calculated and plotted from **Eq. 12** as described in **Subheading 3.3.** Used parameters corresponding to the experimental hybridization conditions and the microarray design. The target concentration is $C_0 = 3$ nM, the nearest-neighbor model values ΔH_{0d} and ΔS_{0d} listed in **Table 2** are found for hybridization buffer with 1 M NaCl and corrected for 25% (v/v) formamide. The electrostatic interaction parameter for 1 M NaCl concentration was previously evaluated as $w = 4 \times 10^{-16}$ Jm²/mol *(2,3)*. For 19-mer probes and targets, $Z_p = 19$ $Z_T = 29$ because of (dT)$_{10}$ spacer. The surface density of probes N_p was not known. However, by trying several N_p values we found that accord of theory with all the four experimental melting curves occurs at unique N_p. Obtained value $N_p = 1.5 \times 10^{17}$ probes/m² means a surface area of 6.7 nm² per probe (compared with 3.1 nm² cross-section area of DNA double β-helix) in accord with typical surface probe density for *in situ* synthesis *(7)*. **Figure 3** shows general agreement between calculated and experimental melting curves for D_1^{MM}, D_2^{PM}, and D_2^{MM} duplexes (**Table 1**). Some systematic discrepancy appears at higher temperatures, where the experimental melting curves in **Fig. 3** exhibit occurrence of certain duplexes that are stable up to the melting temperature in solution (*see* **Fig. 1**). This feature as a result of large statistical fluctuations in a local probe density was interpreted. Indeed, in case of no other surface probes in vicinity of the duplex, the theory predicts its melting temperature to be the same as in solution.

4. Conclusion

We have demonstrated simple two-step calculation of DNA duplex melting behavior on oligonucleotide array. First, standard nearest-neighbor method is applied to evaluate the duplex formation enthalpy and entropy in solution environment. To use the method for a frequent case when formamide is added as a

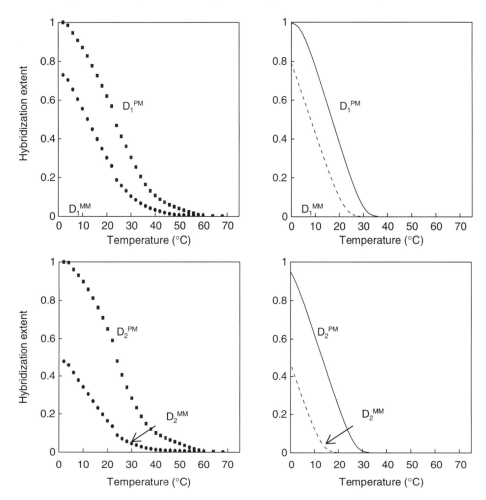

Fig. 3. Experimental (left column) and theoretical (right column) on-array melting curves. The duplexes are defined in **Table 1**.

denaturating agent, we suggest calculating the corresponding increase in duplex enthalpy. Second, a recent theory to account for on-array electrostatic effects was applied. A single array parameter, the surface density of the probes, is involved in these calculations. On-array melting temperature and width of the melting curve are estimated from simple analytical expressions. The procedure of the calculation for the prediction of the melting curve is accomplished using standard scientific graphic software.

This melting prediction method was illustrated by its application to the thermodynamic experiment with a particular *in situ* synthesized microarray design. It would be interesting to test other microarray platforms to verify the applicability and accuracy of this prediction method under the various physical and chemical experimental conditions.

Acknowledgments

This research is supported by grants by National Institutes of Health and the R. A. Welch Foundation (E-1270). The authors thank Professor Eugogan Gulari (University of Michigan) and Dr. Onnop Srivannavit (Atactic Technologies) for chip fabrication and Professor Richard Willson and Mr. Bill Jackson (both from University of Houston) for assisting automated data collection.

References

1. Vainrub, A. and Pettitt, B. M. (2000) Thermodynamics of association to a molecule immobilized in an electric double layer. *Chem. Phys. Lett.* **323,** 160–166.
2. Vainrub, A. and Pettitt, B. M. (2002) Coulomb blockage of hybridization in two-dimensional DNA arrays. *Phys. Rev. E.* **66,** art. no. 041905.
3. Vainrub, A. and Pettitt, B. M. (2003) Sensitive quantitative nucleic acid detection using oligonucleotide microarrays. *J. Am. Chem. Soc.* **125,** 7798–7799.
4. Halperin, A., Buhot, A., and Zhulina, E. B. (2004) Sensitivity, specificity, and the hybridization isotherms of DNA chips. *Biophys. J.* **86,** 718–730.
5. Halperin, A., Buhot, A., and Zhulina, E. B. (2005) Brush effects on DNA chips: thermodynamics, kinetics, and design guidelines. *Biophys. J.* **89,** 796–811.
6. Yu, C. J., Wan, Y. J., Yowanto, H., et al. (2001) Electronic detection of single-base mismatches in DNA with ferrocene-modified probes. *J. Am. Chem. Soc.* **123,** 11,155–11,161.
7. Forman, J. E., Walton, I. D., Stern, D., Rava, R. P., and Trulson, M. O. (1998) in *Molecular Modeling of Nucleic Acids,* (Leontis, N. B., and SantaLucia, J., eds.), vol. 682, Am. Chem. Soc., Washington, DC, pp. 206–228.
8. Lu, M. C., Hall, J. G., Shortreed, M. R., et al. (2002) Structure-specific DNA cleavage on surfaces. *J. Am. Chem. Soc.* **124,** 7924–7931.
9. Watterson, J. H., Piunno, P. A. E., Wust, C. C., and Krull, U. J. (2000) Effects of oligonucleotide immobilization density on selectivity of quantitative transduction of hybridization of immobilized DNA. *Langmuir* **16,** 4984–4992.
10. Guo, Z., Guilfoyle, R. A., Thiel, A. J., Wang, R. F., and Smith, L. M. (1994) Direct fluorescence analysis of genetic polymorphisms by hybridization with oligonucleotide arrays on glass supports. *Nucleic Acids Res.* **22,** 5456–5465.
11. Shchepinov, M. S., CaseGreen, S. C., and Southern, E. M. (1997) Steric factors influencing hybridization of nucleic acids to oligonucleotide arrays. *Nucleic Acids Res.* **25,** 1155–1161.
12. Vainrub, A. and Pettitt, B. M. (2003) Surface electrostatic effects in oligonucleotide microarrays: control and optimization of binding thermodynamics. *Biopolymers* **68,** 265–270.

13. SantaLucia, J., Allawi, H. T., and Seneviratne, A. (1996) Improved nearest-neighbor parameters for predicting DNA duplex stability. *Biochemistry* **35,** 3555–3562.

14. Peyret, N., Seneviratne, P. A., Allawi, H. T., and SantaLucia, J. Jr. (1999) Nearest-neighbor thermodynamics and NMR of DNA sequences with internal A center dot A, C center dot C, G center dot G, and T center dot T mismatches. *Biochemistry* **38,** 3468–3477.

15. Breslauer, K. J., Frank, R., Blocker, H., and Marky, L. A. (1986) Predicting DNA duplex stability from the base sequence. *Proc. Natl. Acad. Sci. USA* **83,** 3746–3750.

16. Cantor, C. R. and Schimmel, P. R. (1980) *Biophysical Chemistry*, Freeman, San Francisco, CA, pp. 1183–1264.

17. Le Novere, N. (2001) MELTING, computing the melting temperature of nucleic acid duplex. *Bioinformatics* **17,** 1226–1227.

18. Vaganay (Deluge), N. (2005) Comprehensive and high-throughput thermodynamic characterization of DNA/DNA and LNA-DNA duplexes on microarrays. PhD Thesis, University of Houston.

19. Blake, R. D. and Delcourt, S. G. (1996) Thermodynamic effects of formamide on DNA stability. *Nucleic Acids Res.* **24,** 2095–2103.

20. Wetmur, J. G. (1991) DNA probes—applications of the principles of nucleic-acid hybridization. *Crit. Rev. Biochem. Mol. Biol.* **26,** 227–259.

21. Gao, X. L., Yu, P. L., LeProust, E., Sonigo, L., Pellois, J. P., and Zhang, H. (1998) Oligonucleotide synthesis using solution photogenerated acids. *J. Am. Chem. Soc.* **120,** 12,698–12,699.

22. Gao, X. L., LeProust, E., Zhang, H., et al. (2001) A flexible light-directed DNA chip synthesis gated by deprotection using solution photogenerated acids. *Nucleic Acids Res.* **29,** 4744–4750.

23. Gao, X. L., Gulari, E., and Zhou, X. C. (2004) In situ synthesis of oligonucleotide microarrays. *Biopolymers* **73,** 579–596.

25

Bioinformatics

Microarray Data Clustering and Functional Classification

Hsueh-Fen Juan and Hsuan-Cheng Huang

Summary

The human genome project has opened up a new page in scientific history. To this end, a variety of techniques such as microarray has evolved to monitor the transcript abundance for all of the organism's genes rapidly and efficiently. Behind the massive numbers produced by these techniques, which amount to hundreds of data points for thousands or tens of thousands of genes, there hides an immense amount of biological information. The importance of microarray data analysis lies in presenting functional annotations and classifications. The process of the functional classifications is conducted as follows. The first step is to cluster gene expression data. Cluster 3.0 and Java Treeview are widely used open-source programs to group together genes with similar pattern of expressions, and to provide a computational and graphical environment for analyzing data from DNA microarray experiments, or other genomic datasets. Clustered genes can later be decoded by Bulk Gene Searching Systems in Java (BGSSJ). BGSSJ is an XML-based Java application that systemizes lists of interesting genes and proteins for biological interpretation in the context of the gene ontology. Gene ontology gathers information for molecular function, biological processes, and cellular components with a number of different organisms. In this chapter, in terms of how to use Cluster 3.0 and Java Treeview for microarray data clustering, and BGSSJ for functional classification are explained in detail.

Key Words: Bioinformatics; clustering; functional classification; gene ontology; microarray; gene expression.

1. Introduction

The microarray technology is an extremely powerful approach to accurately measure changes in global mRNA expression levels *(1,2)*. It can measure gene expressions of tens of thousands of discrete sequences in a single array *(3)*. This technique has been used for discovery of novel genes *(4)*, determination of gene functions, drug evaluation, pathway dissections, and classification of clinical samples *(3,5)*. The microarray technique is a high-throughput method

From: *Methods in Molecular Biology, vol. 382: Microarrays: Second Edition: Volume 2*
Edited by: J. B. Rampal © Humana Press Inc., Totowa, NJ

for determining gene expressions in the study of functional genomics. Searching gene and protein functions is very helpful in investigating the relationship between these significant gene expressions, biological categories, and experimental clusters.

Clustering, a natural basis for organizing gene expression data is to group together genes with similar patterns of expression *(6)*. Because of the large number of genes and the complexity of biological networks, clustering is a useful exploratory technique for the analysis of gene expression data. This clustering information helps researchers to better understand biological processes *(7)*, to investigate biological mechanisms *(8)*, and to find cancer diagnostics *(9–11)*. Cluster 3.0, together with Java Treeview, is an open-source clustering software tool that provides a computational and graphical environment for analyzing data from DNA microarray experiments, or other genomic datasets *(12,13)*. Cluster 3.0 can organize and analyze the data in a number of different ways and Java Treeview allows the organized data to be visualized and browsed.

The ultimate purpose of the gene expression experiments is to produce biological knowledge. Microarray data have to be interpreted before producing biological knowledge. The interpretation of these microarray data can be facilitated by well-presented functional annotations or classifications, which provide an overview of the dominate functions for differentially expressed or clustered genes. The search and match analysis of the voluminous data from microarray without adequate automation searching tools is very time-consuming *(14)*. Even if the information about the genes is complete and accurate, the mapping of list of tens or hundreds of differentially regulated genes to biological functions, molecular functions, and cellular components is not a trial matter *(15)*. Therefore, for the purpose of functional classification, we have developed Bulk Gene Searching Systems in Java (BGSSJ) (http://bgssj.source-forge.net/), an XML-based Java application that organizes lists of interesting genes or proteins for biological interpretation in the context of the gene ontology (http://www.geneontology.org/), which organizes information for molecular function, biological processes, and cellular components for a number of different organisms. The application allows for easy and interactive querying using different gene identifiers (GenBank ID, UniGene, Swiss-Prot, and gene symbol). It also generates a summary page with lists of the frequencies of Gene Ontology annotations for each functional category (cluster), and separate pages with lists of annotations for each gene in a cluster, and provides both quantitative and statistical output files. The visualization browser allows users to navigate the cluster hierarchy displayed in a tree-like structure and explore the associated genes of each cluster through a user-friendly interface. BGSSJ will save time and enhance the ability to analyze gene expression data.

In this chapter, how to use Cluster 3.0 and Java Treeview for microarray data clustering, and BGSSJ for functional classification are described in detail.

2. Materials

1. Software tools: all three software tools used for this analysis are open-source cross-platform software utilizing Java technology. One can download and install Java software (the Java Runtime Environment) from http://www.java.com/. Cluster 3.0 is a clustering software tool available for Windows, Mac OS X, and Linux/Unix can be downloaded and installed. Java Treeview is a program that allows interactive graphical analysis of the results from Cluster 3.0 or other clustering tools. It reads in matching "*.cdt"(clustered data table) and "*.gtr," "*.atr," "*.kgg," or "*.kag" files produced by Cluster 3.0 BGSSJ is an XML-based Java program for functional classification with groups of genes. Required information and formats about your microarray data to run these programs are listed next.
2. Microarray data: you will need microarray data set including accession numbers in GeneBank for each gene or gene names, time series, and gene expression intensities.
3. Input data format for Cluster 3.0: a tab-delimited text files containing microarray data is required as the input data for Cluster 3.0. Such tab-delimited text files can be created and exported in any standard spreadsheet program, such as Microsoft Excel.
4. Input data format for Java Treeview: Java Treeview reads in a tab-delimited text file called a "clustered data file" or CDT file (and the accompanied "*.gtr," "*.atr," "*.kgg," or "*.kag" files) produced by Cluster 3.0 or other programs.
5. An input data format for BGSSJ: GenBank accession numbers (e.g., AA026644), UniGene cluster ID (e.g., Hs.100002), gene symbols (e.g., YMEL1), UniProt/Swiss-Prot accession numbers (e.g., P05121) or entry names (e.g., PAI1_HUMAN) can be used as the input data for BGSSJ.

3. Methods

In this chapter, a dataset from human myeloid leukemia cells induced by 12-*O*-tetradecanoylphorbol-13-acetate (TPA) *(16)* is used to illustrate the methods for the clustering and functional classification. This array contained approx 4200 expressed sequence tags corresponding to known genes in the GenBank database. The array also contained 192 spots of total genomic DNA and 168 housekeeping genes. Finally, the data set consisted of 4324 genes in HL-60 and HL-525 cells, which were both induced by TPA. The clustering was performed at first, in the genes of TPA-induced HL-60 cells and then functional classification.

ACC #	HL60_15min	HL60_30min	HL60_1hr	HL60_2hr	HL60_3hr
H02243	−0.774727	−0.173953	−0.371564	−0.553885	−1.075
T62048	−0.615186	−0.0582689	−0.797507	−0.470004	−1.40364
AA055350	0.322083	0.553885	0.751416	0.751416	0.0582689
R89082	0.900161	0.904218	0.996949	0.451076	0
AA496691	0.482426	0.871293	0.8671	−0.139762	−0.0861777
W58092	0.239017	0.198851	−0.0861777	−0.0953102	−0.916291
AA455043	−0.285179	0.122218	−0.231112	−0.636577	−1.10526
AA464748	−0.165514	0.182322	0.262364	−0.765468	−0.71295
AA521232	0.0487902	0.350657	0.223144	0.24686	0.625938
⋮	⋮				⋮

Fig. 1. An example input file for Cluster 3.0. A tab-delimited text file containing microarray data is required as the input data to Cluster 3.0. Such tab-delimited text files can be created and exported in any standard spreadsheet program, such as Microsoft Excel.

3.1. Microarray Data Conversion to Text File

1. Prepare the microarray data for clustering analysis and functional classification.
2. If necessary, convert the data file to a tab-delimited text file, like the example shown in **Fig. 1**.

3.2. Hierarchical Clustering of the Microarray Data

1. First, install a Java runtime environment, which is downloadable from http://www.java.com/.
2. Download Cluster 3.0 in http://bonsai.ims.u-tokyo.ac.jp/~mdehoon/software/cluster/ and install it.
3. Open Cluster 3.0. Its graphical user interface is shown in **Fig. 2**.
4. Load the data into Cluster by selecting the menu bar item "File" → "Open Data" and it will give you information about the loaded data file.
5. Choose *Hierarchical Clustering* tab (*see* **Note 1**) for hierarchical clustering.
6. In "Genes," click the checkbox labeled "cluster" (*see* **Note 2**).
7. In the "Clustering method," click "average linkage" and the program will start to run (*see* **Note 3**).
8. Cluster 3.0 writes up to three output files for each hierarchical clustering run. The three output files are "JobName.cdt," "JobName.gtr," and "JobName.atr" (*see* **Note 4**).

3.3. Dendrogram of Clustering Using Java Treeview

1. Install Java Treeview, which is available at http://jtreeview.sourceforge.net/.
2. Open Java Treeview.
3. Open the ".cdt" file generated by Cluster 3.0 and the dendrogram output result is shown in **Fig. 3**.
4. Click a sub-dendrogram, and then a zoom-in of the selected cluster and its gene information will be shown in the right column.

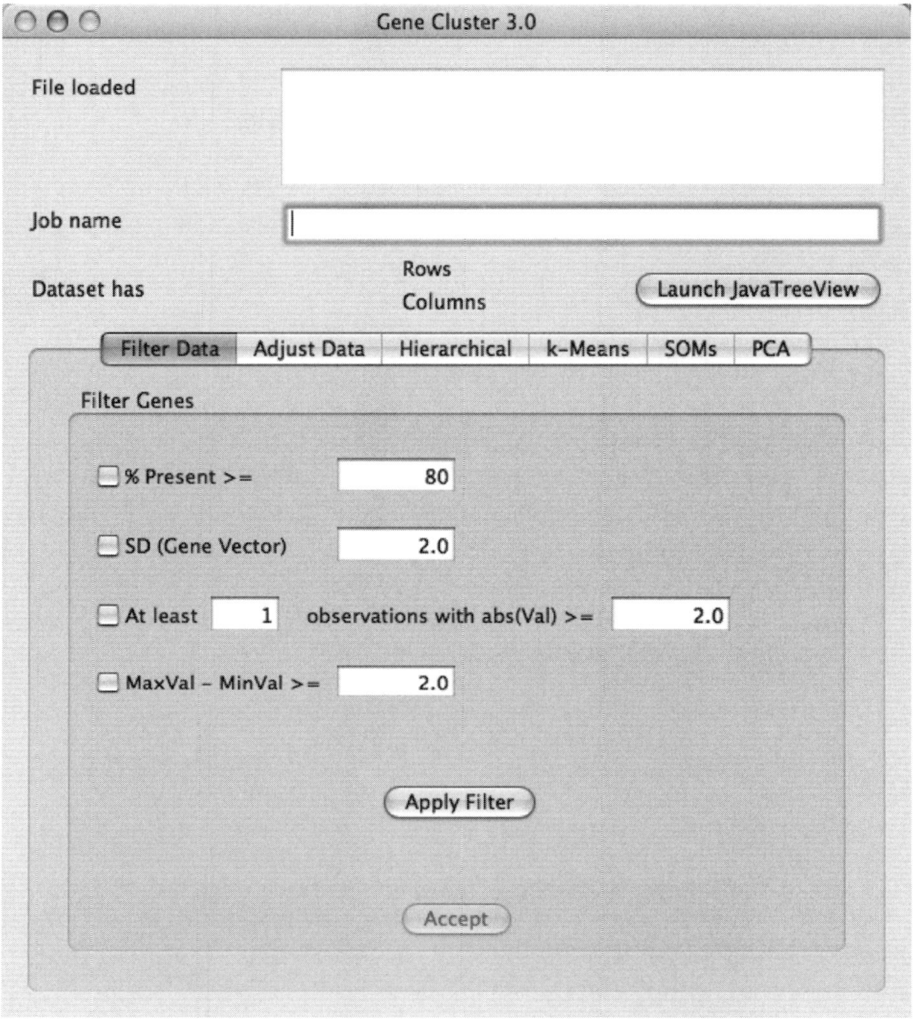

Fig. 2. A screenshot of the Cluster 3.0 graphical user interface. Cluster 3.0 is available at http://bonsai.ims.u-tokyo.ac.jp/~mdehoon/software/cluster/. It provides four clustering algorithms including hierarchical clustering, k-means clustering, self-organizing maps and principal component analysis.

5. The gene list of the sub-dendrogram you just selected can be exported by selecting the menu bar "Export" → "Save List" and will be used as the input for BGSSJ to search for their functions.

3.4. Functional Classification Using BGSSJ

1. Download BGSSJ, which is available in http://bgssj.sourceforge.net/ (*see* **Note 5**) and install it. The BGSSJ program interface is shown in **Fig. 4**.

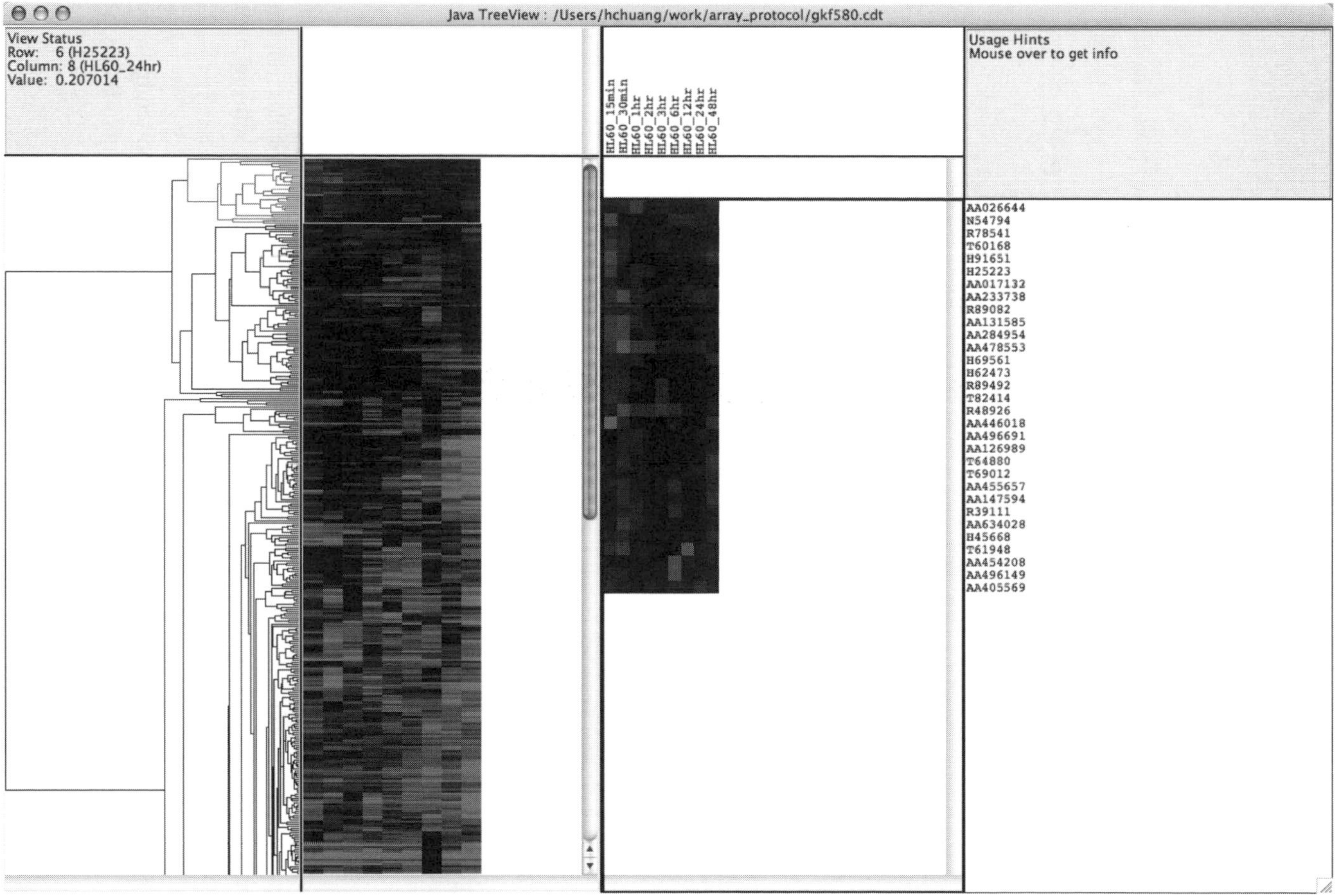
Java TreeView : /Users/hchuang/work/array_protocol/gkf580.cdt
View Status
Row: 6 (H25223)
Column: 8 (HL60_24hr)
Value: 0.207014
HL60_15min
HL60_30min
HL60_1hr
HL60_2hr
HL60_3hr
HL60_6hr
HL60_12hr
HL60_24hr
HL60_48hr
Usage Hints
Mouse over to get info
AA026644
N54794
R78541
T60168
H91651
H25223
AA017132
AA233738
R89082
AA131585
AA284954
AA478553
H69561
H62473
R89492
T82414
R48926
AA446018
AA496691
AA126989
T64880
T69012
AA455657
AA147594
R39111
AA634028
H45668
T61948
AA454208
AA496149
AA405569

2. Input the GenBank accession numbers (e.g., AA026644, N54794, T60168.), which are selected from the clustering analysis. Other kind of inputs like UniGene cluster ID (e.g., Hs.100002), gene symbols (e.g., YMEL1), UniProt/Swiss-Prot accession numbers (e.g., P05121) or entry names (e.g., PAI1_HUMAN) are also accessible by BGSSJ. Multiple accession numbers or symbols, separated by space or linefeed, can be entered and searched simultaneously.

3. Choose database (*see* **Note 6**), Data Type (*see* **Note 7**) and click "OK," then you will get the result for functional list as shown in **Fig. 4**. In "Name," it shows the functional classification of genes. It also shows the statistical values for your input data in "Times" and the percentage in "Percent" (*see* **Note 8**).

4. Click "Tree View" to display the hierarchical tree view about "cellular component," "molecular function," and "biological process" (*see* **Note 9**). Each function category can be expanded to show its child sub-categories. Number of the input genes belonging to each function category is displayed in parentheses alongside. The output result is shown in **Fig. 5**.

5. Click an interested function category or "Gene Information" to display the IDs, symbols and names of those genes belonging to a function category selected in the "Tree View" panel. The output result is shown in **Fig. 6**.

4. Notes

1. Cluster 3.0 provides four clustering algorithms including hierarchical (pairwise single-, average-, maximum-, and centroid-linkage) clustering, *k*-means clustering, self-organizing maps and principal component analysis (PCA). Hierarchical clustering organizes genes in a tree-like structure, based on their similarity. In *k*-means clustering, genes are organized into *k* clusters, where the number of clusters *k* needs to be chosen in advance. Self-organizing maps create clusters of genes on a two-dimensional rectangular grid, where neighboring clusters are similar. In PCA, clusters are organized based on the principal component axes of the distance matrix. The "Hierarchical Clustering" tab allows user to perform hierarchical clustering on the data. The "Filter Data" tab allows user to remove genes that do not have certain desired properties from the dataset. From the "Adjust Data" tab, one can perform a number of operations that alter the underlying data in the imported table.

2. To analyze the clustering among different arrays, click the checkbox in "Arrays." If you do not click "Calculate weight," all of the observations for a given item are treated equally by default.

3. Cluster 3.0 provides four methods to build the clustering hierarchy including centroid

---

Fig. 3. (*Opposite page*)   Visualization of the hierarchical clustering results using Java Treeview. Java Treeview is available at http://jtreeview.sourceforge.net/. The left column is the dendrogram of clustering; the middle column is a "heat map" representing the expression data and then a zoom-in of the selected cluster; the right column is the gene information.

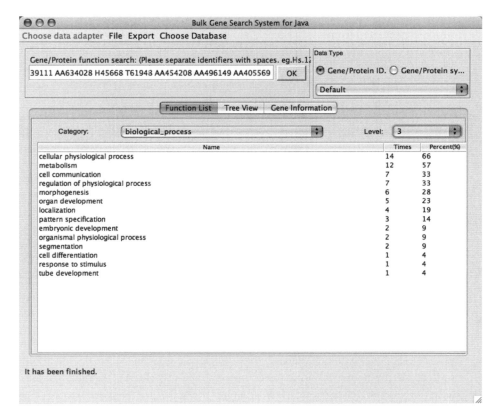

Fig. 4. A screenshot of the BGSSJ graphical user interface. BGSST is available at http://bgssj.sourceforge.net/. The input data are a list of gene (protein) IDs or symbols. Functional classification of the input genes sorted by the number of genes in each function category at the level 3 of "biological process" is shown here.

linkage, single linkage, complete linkage, and average linkage. If you click "Centroid linkage," a vector is assigned to each pseudo-item, and this vector is used to compute the distances between this pseudo-item and all remaining items or pseudo-items using the same similarity metric as used to calculate the initial similarity matrix. In "Single linkage," the distance between two items x and y is the minimum of all pairwise distances between items contained in x and y. In "Complete linkage," the distance between two items x and y is the maximum of all pairwise distances between items contained in x and y. In "Average linkage," the distance between two items x and y is the mean of all pairwise distances between items contained in x and y.

4. The root filename of each file is whatever text you enter into the Job Name dialog box. When you load a file, Job Name is set to the root filename of the input file. The ".cdt" file contains the original data with the rows and columns reordered based on the clustering result.

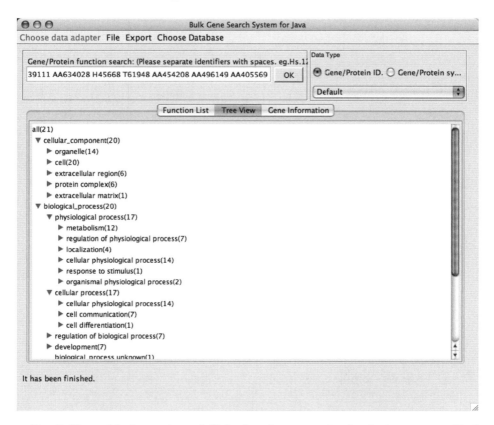

Fig. 5. Hierarchical tree view of all the function categories for the input genes. Each function category can be shrunk to hide or expanded to show its child sub-categories. The number of input genes belonging to each function category is displayed in parentheses alongside.

5. Install Java 1.3 or higher version in your computer. BGSSJ can be run on Windows 98/NT/2000/XP, Mac OS X, Solaris, Linux and FreeBSD. However, it may not work on Mac OS 9, because there is no Java 1.3 support for this platform.

6. Five database are available including "NCBI + GO," "GOA_Human + GO," "GOA_Mouse + GO," "GOA_Rat + GO", and "GOA_UniProt." The default is "NCBI + GO."

7. The Data Type has two options, one is "Gene/Protein ID" (in the cases of GenBank accession numbers, UniGene cluster IDs or UniProt/Swiss-Prot accession numbers) and the other is "Gene/Protein Symbols" (for gene symbols and UniProt/Swiss-Prot entry names).

8. The "Category" has four choices including "all," "biological process," "cellular component," and "molecular function." If you choose "all," it will show the result in all annotations consisting of biological process, cellular component, and molecular function. The "Level" means the GO hierarchy level. Deeper

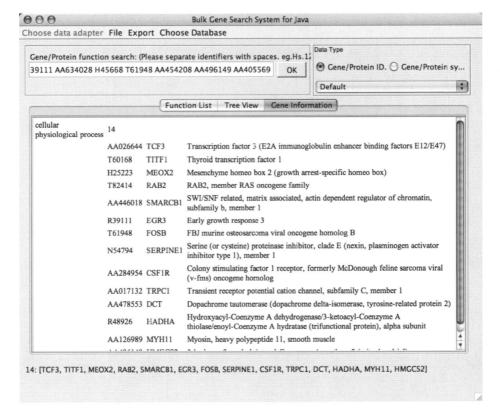

Fig. 6. A list of the genes belonging to some selected function category. When a user clicks an interested function category or the "Gene information" tab, a list of the IDs, symbols and names of those genes will be displayed.

terms in the GO hierarchy are more precise *(17)*. Obviously, the number of genes with annotations decreases at deeper GO levels. GO level 3 constitutes a good compromise between information quality and number of genes annotated at this level *(18)*.

9. BGSSJ is a biological interpretational program based on *Gene Ontology* (GO) database. The three organizing principles of GO are molecular function, biological process and cellular component. Molecular function, biological process and cellular component refer to the tasks performed by individual gene products, broad biological goals, subcellular structures, locations, and macromolecular complexes, respectively.

Acknowledgments

The authors would like to thank Jia-Je Li for development of BGSSJ and thank Nancy Lin and Jen-Tsung Hsiang for proofreading the manuscript. This work was supported by National Science Council of Taiwan (NSC 93-3112-B-002-042 and NSC 93-2311-B-010-017).

References

1. Yue, H., Eastman, P. S., Wang, B. B., et al. (2001) An evaluation of the performance of cDNA microarrays for detecting changes in global mRNA expression. *Nucleic Acids Res.* **29,** E41.
2. Ideker, T., Thorsson, V., Ranish, J. A., et al. (2001) Integrated genomic and proteomic analyses of a systematically perturbed metabolic network. *Science* **292,** 929–934.
3. Hughes, T. R., Marton, M. J., Jones, A. R., et al. (2000) Functional discovery via a compendium of expression profiles. *Cell* **102,** 109–126.
4. Seo, J., Kim, M., and Kim, J. (2000) Identification of novel genes differentially expressed in PMA-induced HL-60 cells using cDNA microarrays. *Mol. Cells* **10,** 733–739.
5. Naour, F. L., Hohenkirk, L., Grolleau, A., et al. (2001) Profiling changes in gene expression during differentiation and maturation of monocyte-derived dendritic cells using both oligonucleotide microarrays and proteomics. *J. Biol. Chem.* **276,** 17,920–17,931.
6. Eisen, M. B., Spellman, P. T., Brown, P. O., and Botstein, D. (1998) Cluster analysis and display of genome-wide expression patterns. *Proc. Natl. Acad. Sci. USA* **95,** 14,863–14,868.
7. Luo, X., Ding, L., Xu, J., Williams, R. S., and Chegini, N. (2005) Leiomyoma and myometrial gene expression profiles and their responses to gonadotropin-releasing hormone analog therapy. *Endocrinology* **146,** 1074–1096.
8. Sadlier, D. M., Connolly, S. B., Kieran, N. E., et al. (2004) Sequential extracellular matrix-focused and baited-global cluster analysis of serial transcriptomic profiles identifies candidate modulators of renal tubulointerstitial fibrosis in murine adriamycin-induced nephropathy. *J. Biol. Chem.* **279,** 29,670–29,680.
9. Ptitsyn, A. (2004) Class discovery analysis of the lung cancer gene expression data. *DNA Cell Biol.* **23,** 715–721.
10. Bernard, P. S. and Wittwer, C. T. (2002) Real-time PCR technology for cancer diagnostics. *Clin. Chem.* **48,** 1178–1185.
11. Ramaswamy, S., Tamayo, P., Rifkin, R., et al. (2001) Multiclass cancer diagnosis using tumor gene expression signatures. *Proc. Natl. Acad. Sci. USA* **98,** 15,149–15,154.
12. De Hoon, M. J. L., Imoto, S., Nolan, J., and Miyano, S. (2004) Open source clustering software. *Bioinformatics* **20,** 1453–1454.
13. Saldanha, A. J. (2004) Java Treeview—extensible visualization of microarray data. *Bioinformatics* **20,** 3246–3248.
14. Juan, H. F., Lin, J. Y., Chang, W. H., et al. (2002) Biomic study of human myeloid leukemia cells differentiation to macrophages using DNA array, proteomic, and bioinformatic analytical methods. *Electrophoresis* **23,** 2490–2504.
15. Draghici, S. (2003) Functional analysis and biological interpretation of microarray data, in *Data Analysis Tools for DNA Microarrays*, CRC, Boca Raton, FL, pp. 363–382.
16. Zheng, X., Ravatn, R., Lin, Y., et al. (2002) Gene expression of TPA induced differentiation in HL-60 cells by DNA microarray analysis. *Nucleic Acids Res.* **30,** 4489–4499.

17. Al-Shahrour, F., Diaz-Uriarte, R., and Dopazo, J. (2004) FatiGO: a web tool for finding significant associations of Gene Ontology terms with groups of genes. *Bioinformatics* **20,** 578–580.
18. Mateos, A., Herrero, J., Tamames, J., and Dopazo, J. (2002) Supervised neural networks for clustering conditions in DNA array data after reducing noise by clustering gene expression profiles, in *Methods of Microarray Data Analysis II,* (Lin, S. and Johnson, K., eds.), Kluwer Academic Publishers, Boston, MA, pp. 91–103.

26

In Silico Gene Selection for Custom Oligonucleotide Microarray Design

Conor W. Sipe, Vijay R. Dondeti, and Margaret S. Saha

Summary

A method for systematically selecting the large number of sequences needed to custom design an oligonucleotide microarray was presented. This approach uses a Perl script to query sequence databases with gene lists obtained from previously designed (and publicly available) microarrays. Homologous sequences passing a user-defined threshold are returned and stored in a candidate gene database. Using this versatile technique, microarrays can be designed for any organism having sequence data. In addition, the ability to select specific input gene lists allows the design of microarrays tailored to address questions pertaining to specific pathways or processes. Given recent concerns about the accuracy of annotation in public sequence databases, it is also necessary to confirm the correct orientation of candidate sequences. This step is performed by a second Perl script that extracts protein similarity information from individual Unigene records, checks for consistency of features, and adds this information to the candidate gene database. Discrepancies between the orientations determined using protein similarities and that predicted by a given sequence's assigned orientation are readily apparent by querying the candidate gene database.

Key Words: BLAST; gene selection; micro array; oligonucleotide; Perl; sequence orientation.

1. Introduction

Long oligonucleotide arrays are routinely used to assess gene expression on a large scale. In comparison with cDNA microarrays, this platform provides the advantages of having a uniform length and composition of probe sequences. Although there are numerous commercial arrays and oligonucleotide sets available for some organisms, the majority of these have been developed with the major model systems in mind (e.g., human, mouse, *Caenorhabditis elegans*, yeast, *Arabidopsis*). For many important organisms, commercial microarrays simply do not exist, despite having substantial amounts of sequence data. Moreover, even for major model organisms, some experimental protocols would greatly benefit

From: *Methods in Molecular Biology, vol. 382: Microarrays: Second Edition: Volume 2*
Edited by: J. B. Rampal © Humana Press Inc., Totowa, NJ

from the ability to design specialized microarrays tailored to a given question. With the availability of custom array manufacturing services, it has recently become feasible to custom design high-density arrays for such circumstances.

The amount of sequence data presently available for many organisms is such that a manual selection of the genes included on a chip is impractical at best. This logistical problem will intensify as public sequence databases continue to grow. Unfortunately, there have been few systematic methods proposed in the literature to aid in the selection of the genes included on high-density arrays (*see* **Note 1** for a brief discussion of cDNA array gene selection methods) *(1,2)*. Here, a protocol for *in silico* gene selection that relies on using publicly available microarray information to interrogate existing sequence databases to identify a set of homologous genes in the organism of interest were presented. It is particularly suitable for designing microarrays to investigate processes that are conserved among species, and can be tailored to any chip density. The rapid progress of various sequencing initiatives, as well as the dramatic increase in available microarray data, has served to increase the power of this selection method.

2. Materials

A PC computer with internet access running Windows 2000/XP with Microsoft Access installed (*see* **Note 2**) is required. The hard drive should have at least 200-mb free after installing the necessary sequence databases (which can be up to several hundred megabytes) and software. The computer also needs to have the Perl executable files and the necessary Perl modules installed. The scripts used in this procedure were written and tested using ActivePerl (available free from www.activestate.com), with the Bioperl (available from www.bioperl.org) and DBD-ODBC modules installed (*see* **Note 3** for detailed instructions on installing these modules). It is also helpful to have Microsoft Excel and an advanced text editor (e.g., EditPlus) installed to manipulate data files. Users inexperienced with the Windows operating system may wish to consult a basic windows tutorial (e.g., http://www.learnthat.com/courses/computer/windows/) before beginning the procedure.

3. Methods

3.1. Local Installation of BLAST Server

1. Download the latest version of the basic local alignment search tool (BLAST) software (available at ftp://ftp.ncbi. nlm.nih.gov/blast/executables/) into an easily accessible folder on your computer (e.g., C:\microarray\).
2. Double-click on the file to uncompress the BLAST executables and data directory. Before BLAST can be run locally, the relevant databases must be installed and formatted (*see* **Subheading 3.2.**).

3.2. Installation of a Local Sequence Database for Querying

1. Download the latest version of the Unigene unique sequence database for the species for which you are designing the array (available from ftp://ftp.ncbi.nih. gov/repository/UniGene/). Unigene unique databases contain the highest quality representative sequence from each Unigene cluster, and thus are a nonredundant collection of known transcripts. At the time of writing, there are >50 organisms cataloged in the Unigene database, including the major model organisms (*see* **Note 4** for a discussion of alternate sequence database sources).

2. Double-click on the file to uncompress the downloaded file into the directory where you installed the BLAST software. The resulting file should be in the form Xx.seq.uniq, where Xx is the NCBI abbreviation for the species.

3. Open a command prompt window (*see* **Note 5**) and navigate to the folder containing the downloaded BLAST software and Unigene database.

4. Type the following on the command line "formatdb –i Xx.seq.uniq –p F –o T" and press Enter (where Xx is the abbreviation from **step 2**). This program constructs the index files (which should appear in the directory after a several seconds depending on the speed of the computer) needed to run a BLAST search against the database (*see* **Note 6**).

5. Download the corresponding Unigene data file (named Xx.data) and uncompress the file into the same directory where you placed the Xx.seq.uniq file.

3.3. Gene Selection

1. Choose the arrays from which homologous sequences will be selected for inclusion on the microarray (*see* **Note 7** for more information on array selection). These arrays can be both general and more specialized, and should be tailored to the process under investigation, as they will ultimately determine the nature of the genes included in the candidate database.

2. Cut-and-paste the probe accession numbers and any description of the sequence provided by the array source into a text file with the name *Description.txt, where * is a unique prefix corresponding to a given array (*see* **Note 8** and **step 8** next). This information is obtained from the spreadsheets or tab-delimited text files distributed with the microarrays chosen in **step 1**. This description file should contain one row for each probe sequence, with each row consisting of an accession number followed by a tab character and then the description (can include white spaces) followed by a new line character.

3. In another text file (named *Acc.txt, where * is the unique prefix corresponding to a given array), store a list of the accession numbers only from the input array. This file should contain a single row for each accession number, and will be used to retrieve the sequences from GenBank. Duplicate entries in this list do not pose a problem.

4. Repeat **steps 2** and **3** to create description and accession files for each input array.

5. Open a web browser and navigate to http://www.ncbi.nlm.nih.gov/entrez/ batchentrez.cgi?db=Nucleotide to access the NCBI Batch Entrez tool. Click the "Browse" button located near the top of the screen and select the file you created

in **step 3** containing the accession numbers from an input array. Click the "Retrieve" button to return a list of input sequence entries from the given array (*see* **Note 9**).

6. In the Display drop-down list, choose "FASTA" to return the sequences in FASTA format. Ensure that "all to file" is chosen in the "Send" drop-down list, and click the Send button to save the returned FASTA sequences in the working directory. Name the file *Seq.txt, where * is the prefix unique to a given array.

7. Repeat **steps 5** and **6** to create sequence files for each input array.

8. Create another text file, named prefixes.txt that contains a list of the unique prefixes used to identify each input array. This file should contain a single row for each prefix followed by a new line character. Note that each input array needs a unique prefix.

9. BLAST the group of sequences from a given input array against the query database. Navigate to the directory containing the BLAST software. At the command prompt, type "blastall –p blastn –d Xx.seq.uniq –i *Acc.txt –o *Blast.txt –v 1 –b 1" and press enter (where Xx is the NCBI abbreviation of the query database species, and * is the prefix corresponding to the given array) (*see* **Note 10**). Upon completion, a text output file containing the results of this search will have been created in the working directory.

10. Repeat **step 9** to create BLAST result files for each input array.
    At this point, there should be four files associated with each input array: (1) *Acc.txt containing a list of the accession numbers a given array; (2) *Seq.txt containing the retrieved input sequences in FASTA format; (3) *Description.txt containing accession numbers and any description information provided with a source array; (4) *Blast.txt containing the results of a BLAST search against the query database (*see* **Note 8**).

11. Use a web browser to download the file containing the scripts necessary to build a database of candidate genes for inclusion on the new array (available at http://mssaha.people.wm.edu/methods-scripts.zip), and uncompress this file into the working directory.

12. The GeneSelectDBschema.mdb file is a Microsoft Access database that contains the necessary fields for use with these Perl Scripts. Create the database that will hold final gene selection data by making a copy of this file and renaming it GeneSelectDB.mdb.

13. In order to interface the Perl script with Microsoft Access functions, the database must be registered as a data source in the Windows control panel (*see* **Note 11**). From the Windows Control Panel, select Administrative Tools → Data Sources (ODBC).

14. Under the "User DSN" tab, click the "Add" button. From the list, select "Microsoft Access Driver" and click the "Finish" button. In the dialog box that appears, enter "GeneSelectDB" in the "Data Source Name:" box. In the Database box, press the "Select…" button and navigate to the working directory containing the GeneSelectDB.mdb file from **step 12**. Highlight this file and click "OK." Click "OK" again and the newly entered GeneSelectDB should appear in the list of registered data sources. To save these changes and return to Windows, click "OK" a final time.

A Perl script is now run that parses the BLAST results file from all input arrays, selects hits passing a set threshold value, and stores relevant information from these hits (query accession, hit accession, definition lines, E value, and so on) in the GeneSelect.mdb database. The final number of sequences placed in the candidate database is largely determined by a user-defined threshold value. The default threshold value is relatively stringent (E < 1e-15) to ensure a high degree of homology between input sequences and query database hits (*see* **Note 12** to modify this default threshold).

15. At the command prompt in the directory containing the four files for each input array and the GeneSelect.pl script, type "perl GeneSelect.pl" and press Enter. The script will prompt the user for the full path of the prefix file from **step 8** (for e.g., type "C:\microarray\user1\prefixes.txt" and press Enter). The script will also prompt for the full path of the *Blast.txt and *Description.txt files (for e.g., type "C:\microarray\user1\" and press Enter).

After a short time (up to several minutes depending on the number of input arrays and the speed of the computer), the script will display statistics about the number of sequences passing the threshold criteria, the number of nonredundant sequences passing the threshold criteria, and their average length in base pairs. This information is also placed in an output file, GeneSelectOutputInfo.txt, which is generated in the working directory.

16. Open the newly populated database in Microsoft Access by double-clicking on the GeneSelectDB.mdb file. Two tables in the database are now complete: (1) the "Genes" table contains a row for all input sequences where the database hit passes the set threshold criteria; (2) the "GenesNR" table is a view of the same sequences with redundant entries removed. *See* **Table 1** for a description of each field.

3.4. Determination of Correct Sequence Orientation

Expressed sequence tags (ESTs) usually represent the greatest proportion of chosen sequences utilizing the GeneSelect script. Given the recent concerns about possible annotation errors in public databases (e.g., NCBI) *(1,3,4)*, it is essential to determine the correct orientation of the selected sequences. Next, a second script first checks for consistency in the orientation of the sequence feature information (coding sequence and protein similarity) contained in individual Unigene records, and then compiles this information in the GeneSelectDB.mdb database (*see* **Note 13**).

1. Create a folder to hold downloaded individual Unigene records. It can be named arbitrarily (in our example, it could be C:\microarray\user1\unigene), but must contain two subdirectories named "HTML" and "TXT."
2. Close any databases that are open in Microsoft Access. Navigate to the directory containing the UniRecParser.pl script and at the command line, type "perl UniRecParser.pl" and press Enter to run the script.

Table 1
Explanation of the Fields in the Gene NR Table
in the Gene Select DB Database

Field	Description
SourceAccession	Accession number of the source gene.
SourceDefinitionLine	Definition line obtained from FASTA sequence of the source gene.
SourceDescription	Description of the source gene provided by array source.
Accession	Accession number of the homologous candidate gene.
ClusterID	Unigene cluster ID corresponding to the candidate gene.
SID	Unigene sequence identifier corresponding to the candidate gene.
DefinitionLine	Definition line obtained from the Unigene entry of the candidate gene.
Comment	Used to track the source microarray from which the sequence was obtained.
EValue	E-value resulting from the BLAST search.
Plus/Minus	Orientation of the alignment from the resulting BLAST report.

3. The script prompts the user with a choice of (1) downloading the pertinent Unigene records, (2) parsing and analyzing them, or (3) both operations. Type "3" and press Enter.
4. The user is then prompted for the full path of the directory you created in **step 1** (which contains the two subdirectories; for example, type "C:\microarray\user1\ unigene" and press Enter). The script also requests the full path of the Unigene data file (from **step 5** of **Subheading 3.2.**, for example, type "C:\microarray\user1\ Xx.data" and press Enter).
5. After a few moments, the script will connect to the NCBI Unigene database and download the individual Unigene records of the candidate sequences in the GenesNR table of the database. The speed of this step will vary depending on connection speed and the current load on the NCBI site, but a good approximation is 1 h for every 2000 UniGene records. A log of any errors that occur during this download is generated in the working directory (named DownloadErrorLog.txt). This text file should be checked after the script has finished downloading to ensure that all sequences have been downloaded properly.
6. The script will then parse and analyze these downloaded records and enter the pertinent information in the UnigeneRecordsSummary and ResultsSummary tables in the database. Depending on the number of sequences and the speed of the computer, this step will take minutes to several hours to run (during this time the hard drive should read and write data). The script has finished when the command prompt reappears.

Table 2
Explanation of Fields in the UnigeneRecords Summary Table in the GeneSelectDB Database

Field	Description
Accession	Accession number of the candidate gene sequence.
ProteinAccession	Accession number of the prot_sim feature (if applicable).
ProteinDescription	Description of the Unigene feature (if applicable).
Organism	Species from which the prot_sim feature is derived.
EValue	E-value resulting from a prot_sim feature alignment.
SeqType	Sequence type of the candidate gene sequence.
ProteinAlignment	Indicate the type of unigene feature for a given entry in the
CDS	table and the strand (Plus or Minus) on which it is found.
PolyASignal	

7. Open the GeneSelectDB.mdb database in Microsoft Access to view the results of the analysis. The information in tables can be sorted by the values in each column by clicking on the gray box at the head of the column (the mouse pointer will turn into a black downward arrow) to highlight the entire column. Click Records → Sort → Sort Ascending to rearrange the rows in the table according to the value in that column.

The UnigeneRecordsSummary table contains a list of all the sequence features the script was able to extract from individual Unigene records (*see* **Table 2** for an explanation of the table fields). Note that if a record contained no features, it will not be represented in this table.

The final table, ResultsSummary, summarizes the orientation information taken from individual Unigene records. Records that contained no features are represented in this table, though orientation information will be blank. To aid in visualization, a number of queries are embedded in the database to organize the data (*see* **Note 14**).

Explanations for each of the fields in this table can be found in **Table 3**, but the following deserve special attention:

1. *Consistency*—indicates whether the features within a sequence's individual Unigene record are consistent with each other with regard to orientation. A value of "NotConsistent" in this field (a relatively rare occurrence, as this is a curated database) may indicate a sequence peculiarity and deserves further inspection.
2. *Orientation*—if all Unigene features within a given record are consistent, this field indicates the strand on which they are found (Plus or Minus). This value is extracted from the CDS and/or prot_sim features found in a sequence's Unigene record by the Perl script. Note that this field may be blank, signifying that there were no features in the Unigene record or that the features were inconsistent with one another.

Table 3
Explanation of Fields in the Results Summary Table
in the Gene Select DB Database

Field	Description
Accession	Accession number of the candidate gene sequence.
Features	Denotes if sequence features were found in the individual unigene record.
NumberOfRecords	Number of Unigene sequence features extracted from the individual record.
Consistency	Indicates if all sequence features in the given record were consistent with respect to orientation.
Orientation	If consistent, the orientation of the Unigene features.
SeqType	Sequence type of the candidate gene sequence.
PlusMinus	Nature of the original source and candidate sequence alignment.
PolyASignal	Denotes whether a poly-A signal was extracted from the Unigene record.

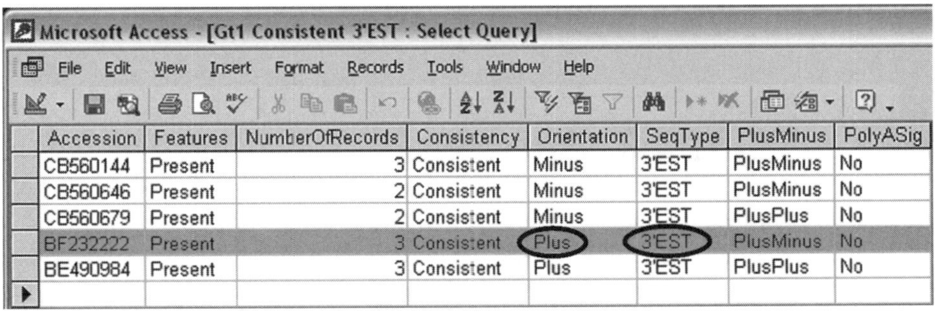

Fig. 1. Database view of incorrectly annotated sequence. The 3′ EST BF232222 (highlighted in gray) consistently aligns to 3′ Unigene features in the sense orientation, indicating this sequence is annotated incorrectly.

3. *SeqType*—the candidate sequence's type (5′ EST, 3′ EST, or mRNA). According to sequence submission guidelines, 5′ ESTs and mRNAs are conventionally submitted to the NCBI database in the sense orientation (i.e., the Plus strand represents the coding sequence) and 3′ ESTs in the antisense orientation (i.e., the Minus strand corresponds to the reverse complement of the coding sequence).

Therefore, discrepancies between the consensus orientations derived from individual Unigene records (based on conceptual translations) and the orientation predicted by a sequence's type indicate incorrect annotation of the candidate sequence (*see* **Fig. 1**). For example, according to convention, a 3′ EST should be in the antisense orientation; thus, if the Orientation column registers a "Plus" value for this EST, it is likely the sequence has been incorrectly annotated as a 3′ read.

4. Notes

1. Numerous investigators have utilized cDNA microarrays, and have for the most part relied on randomly generated EST libraries to dictate the number and makeup of probe sequences included on a given chip *(5)*. Although subtractive methods can be applied to enrich EST libraries to select for specific transcripts, the final gene makeup remains relatively indiscriminate. Moreover, the current level of sequence annotation for many EST collections used to spot arrays is unsatisfactory, as numerous genes have not been assigned an identity based on homology *(6)*.

2. The current distribution of the Perl scripts (available at http://mssaha.people. wm.edu/methods-scripts.zip) implementing this gene selection method are written to interface with a Microsoft Access database. Minor modifications to the scripts would allow them to work with other common database systems, including MySQL or Oracle.

3. The required add-on modules to Perl can be installed directly from the command line using ActivePerl's ppm shell. From the command line, type "ppm3" and press Enter. After a moment, the shell command prompt will appear. Type "install bioperl" and press Enter. The required module will automatically be downloaded and installed. When the ppm3 command prompt returns, type the following command "install DBD-ODBC" and press Enter. All Perl components needed to run both scripts should now be installed. Type "quit" to exit the ppm3 shell and return to the Windows command line prompt.

4. The query database need not be a Unigene database, though this is certainly the best choice if available for a given organism. The script and database schema have been developed to extract Unigene-specific information for inclusion in the candidate gene database, but could be modified to work with a different query database by removing the Unigene-specific fields.

5. To open a command prompt window from within Microsoft Windows, click on the Start menu in the lower left side of the screen. Click on "Run…" and in the "Open:" box that appears type "cmd" and press Enter. The black command prompt screen should now open. If you are unfamiliar with the MS-DOS command prompt, a brief tutorial of basic commands is available online at http://www. bleepingcomputer.com/forums/Introduction_to_the_Windows_Command_Prompt-tut76.html

6. A BLAST-searchable database can be constructed from any collection of sequence files, provided they are in FASTA format (i.e., a single text file with individual FASTA records separated by a newline character). For instance, such a database could consist of all the genes from a particular species in the NCBI GenBank. To construct such a database, query the GenBank nucleotide database with "species-name[orgn]" to return all sequences whose source is from a given species. After the results are returned, choose "FASTA" in the Display drop-down list, "all to file" in the "Send" drop-down list, and click the Send button to save the returned

FASTA sequences into the working directory containing the BLAST software (this file can be named arbitrarily). To construct a searchable sequence database, type the following on the command line "formatdb –i FileName –p F –o T" and press Enter (where FileName is the filename you gave the list of sequences). In the same manner, a searchable database of sequencing runs or cDNA library sequences may be constructed.

7. Arrays or oligonucleotide sets from which to draw input sequences for the gene selection script are readily available on the internet. The majority of commercial suppliers of microarrays or oligo sets (e.g., Agilent, SuperArray, Clontech, Operon, Sigma-Genosys, and so on) make lists of genes included on their arrays available for download from their website as Microsoft Excel spreadsheets, HTML files, or tab-delimited text files. Gene list information can also be found for custom arrays (usually designed to answer questions pertaining to specific pathways or processes) reported in the primary literature, as most journals will make this available as supplementary material. Information about arrays and collections of clones distributed by government sources (e.g., National Institute of Aging mouse 15k clone set) is also accessible via the websites mentioned in most primary literature.

8. Because of the number of text files required to run the script, it is essential that good naming practices be followed to maintain organization. As there is no practical limit to the number of characters for a given filename, be as explicit as necessary to make identification of each input array possible. It is also important to include the dates of gene lists, BLAST searches, and database builds in filenames. This will avoid confusion when working with updated versions of these files in the future. Note that the description and BLAST result files must end with the appropriate suffix (e.g., filenameDescription.txt and filenameBlast.txt) for the script to execute properly.

9. If any errors are encountered during the sequence retrieval process, a screen will appear listing the records causing the problem. These errors are usually caused by formatting problems in the *Acc.txt file. Occasionally, the underscores in some accession numbers (e.g., NM_111111) will be changed to spaces when copied from a formatted document (e.g., Adobe PDF file) into a plain text file. Some array manufacturers also list their own proprietary sequence identification codes in gene lists alongside accession numbers; ensure that the genuine NCBI accession numbers are placed into the *Acc.txt file. Large genomic contigs (usually with the prefix NC_ or NT_ in their accession numbers) are also problematic in the Batch Entrez tool, so avoid including them in the list of accessions.

10. The arguments –v (number of database sequences for which to show one-line descriptions) and –b (number of database sequences for which to show alignments) are set equal to 1. This ensures that no more than one sequence, the best hit (having the lowest E-value), is returned for each query sequence.

11. If the database if not registered in Windows correctly, cryptic errors such as the following will occur (exact wording will vary with the computer system):
DBI connect (GeneSelectDB) failed: (Microsoft [ODBC Driver Manager]) Data source name not found and no default driver specified (SQL-IM002)(DBD:

db_login/SQLConnect err=-1) at GeneSelect.pl line 91"
Ensure that the database is registered correctly by following **steps 12** and **13** in **Subheading 3.3.**

12. To modify the threshold value from the default setting of E < 1e-15, open the GeneSelect.pl script in a text editor. Scroll down to line 48, which reads: "my $EValueThreshold" = "1e-015"

    The exponential in this expression determines the set threshold that candidate sequences must pass to be included in the database. A less stringent value, while increasing the number of candidate sequences placed in the database, will lower confidence in the homology of these sequences. Change the exponential in the above expression to the threshold value desired, keeping it in the same form. For example, to set a threshold value of 1 (E <1.0), the edited line would be as follows: "my $EValueThreshold" = "1e0"

    Save the edited file and run the script. Note that the script can be run on the same GeneSelectDB.mdb database multiple times. Sequences passing the relaxed threshold value will be appended to the database.

13. For more information regarding how the protein similarities found in Unigene records are constructed, see http://www.ncbi.nlm.nih.gov/UniGene/protest/ and the NCBI Unigene FAQ (http://www.ncbi.nlm.nih.gov/UniGene/FAQ.html). The UniRecParse.pl script determines proper sequence orientation by checking for consistency in orientation between the CDS and protein similarities contained in a given record. In the current implementation of the script, only protein similarities from mouse, rat, and human are considered, because these are generally the best annotated. Information on the orientation of the polyA signal is also recorded in the GeneSelectDB database, though we have determined that this feature alone does not serve as a reliable indicator of sequence orientation. Thus, it is not taken into account when checking for consistency within a record.

14. Queries embedded in the database can be accessed from the window initially displayed upon loading the GeneSelectDB database in Microsoft Access. In the gray Objects box located just to the left of the database tables, click on "Queries." You will be presented with a list of pre-set filters that can be applied to the ResultsSummary table; double-clicking on a filter will open up just the entries in the table that satisfy the query. These will divide the table into entries by whether or not Unigene features were present (Present, Absent), the number of Unigene features extracted (NumOfRecsEq1, NumOfRecGt1), consistency of Unigene Records (Consistent, NotConsistent), sequence type (Consistent 3′ EST, Consistent 5′ EST, Consistent mRNA), or combinations thereof (Gt1 Consistent 3′ EST, Gt1 5′ EST, and Gt1 mRNA). Abbreviations in the query names are: Eq 1, = 1; Gt1, >1.

References

1. Dondeti, V. R., Sipe, C. W., and Saha, M. S. (2004) In silico gene selection strategy for custom microarray design. *Biotechniques* **37,** 768–770.
2. Lorenz, M. G., Cortes, L. M., Lorenz, J. J., and Liu, E. T. (2003) Strategy for the design of custom cDNA microarrays. *Biotechniques* **34,** 1264–1270.

3. Wang, J. P., Lindsay, B. G., Leebens-Mack, J., et al. (2004) EST clustering error evaluation and correction. *Bioinformatics* **20,** 2973–2984.

4. Lambert, J. C., Testa, E., Cognat, V., et al. (2003) Relevance and limitations of public databases for microarray design: a critical approach to gene predictions. *Pharmacogenomics J.* **3,** 235–241.

5. Chen, Y. A., McKillen, D. J., Wu, S., et al. (2004) Optimal cDNA microarray design using expressed sequence tags for organisms with limited genomic information. *BMC Bioinforma.* **5,** 191.

6. Shin, Y., Kitayama, A., Koide, T., et al. (2005) Identification of neural genes using *Xenopus* DNA microarrays. *Dev. Dyn.* **233,** 248.

27

Integrated Analysis of Microarray Results

Olga G. Troyanskaya

Summary

Gene expression microarrays are becoming increasingly widespread, especially as a way to rapidly identify putative functions of unknown genes. Accurate microarray data analysis, however, still remains a challenge. The recent availability of multiple types of high-throughput functional genomic data can facilitate accurate and effective analysis of microarray experiments and thereby accelerate functional annotation of sequenced genomes. But genomic data often sacrifice specificity for scale, yielding very large quantities of relatively lower quality data than traditional experimental methods. Advanced analysis methods are thus necessary to make accurate functional interpretation of these large-scale datasets. This chapter outlines recently developed methods that integrate the analysis of microarray data with sequence, interaction, localization, and literature data and further outlines specific problems in currently available integrated analysis technologies.

Key Words: Bayesian networks; data integration; function prediction; gene expression; microarray analysis; functional genomics.

1. Introduction

At present, complete genomic sequences of several eukaryotic organisms, including the human genome *(1–6)*, are available; molecular biology has entered into a new era of systematic functional understanding of cellular processes. The sequences themselves provide a wealth of information, but functional annotation is a necessary step toward comprehensive description of genetic systems of cellular controls *(7–9)*. High-throughput functional technologies, such as genomic *(10,11)* and soon proteomic microarrays *(12–16)*, allow one to rapidly assess general functions and interactions of proteins in the cell. In addition to gene expression microarrays *(17)*, other high-throughput experimental methods are generating increasing amounts of data. In yeast *Saccharomyces cerevisiae*, the most well-studied eukaryotic organism that is

From: *Methods in Molecular Biology, vol. 381: Microarrays: Second Edition: Volume 2*
Edited by: J. B. Rampal © Humana Press Inc., Totowa, NJ

commonly used in computational and experimental genomic studies, these datasets include protein–protein interaction studies (affinity precipitation *[18]*, two-hybrid techniques *[19]*), synthetic rescue *(20)*, and lethality *(20,21)* experiments, and microarray analysis *(10,11)*.

This increase in functional data is also reflected in the rise of multiple functional databases, especially for yeast, including: the biomolecular interaction network database *(22)*, the database of interacting proteins *(23)*, the molecular interactions database *(24)*, the general repository for interaction datasets *(25)*, the MIPS comprehensive yeast genome database *(26)*, and the model organism database for yeast—*Saccharomyces* genome database *(27)*. Although, classical genetic and cell biology techniques continue to play an important role in the detailed understanding of cellular mechanisms, the combination of rapid generation and analysis of functional genomics data with targeted exploration by traditional methods will facilitate fast and accurate identification of causal genes and key pathways affected in cellular regulation, development, and in disease.

Therefore rapid functional annotation of the sequenced genomes and understanding of gene regulation is a key goal of these high-throughput data. Even in yeast, the well-studied eukaryote, 1481 of 5788 open reading frames are still unnamed, and functional annotation is unknown for 1865 open reading frames. High-throughput functional data, especially the large number of microarray datasets, are important for rapid functional annotation of these unknown genes, but it is important to recognize that high-throughput methods sacrifice specificity for scale in the quality to coverage tradeoff, yielding to many false-positives in the datasets *(8,28–32)*. Recent work has highlighted this problem, showing that different complementary deoxyribonucleic acid microarrays exhibit between 10 and 30% variation among corresponding microarray elements *(33)*. For gene function annotation and biological network analysis, an increase in accuracy is essential, even if it comes at the cost of some sensitivity *(30)*. This chapter presents an overview of computational methods that incorporate the abundant microarray data with other data sources for increased specificity in gene function prediction and in identification of biological networks. First outline resent progress in integrated analysis of heterogeneous data and then focus on a specific Bayesian network-based data integration methodology.

2. Methods
2.1. Challenges and Promises of Data Integration Methodologies

Integrated analysis of microarray data with other genomic data sources can increase prediction accuracy and provide a coherent view of functional information derived from diverse data types. As outlined in this chapter, integrated methods can be based on formal probabilistic reasoning and can generate predictions based upon heterogeneous data sources, and some are generalizable to new data sources as they become available. Although several promising probabilistic methods for integrated analysis have been developed, no truly general and robust method that can be routinely applied to noisy, heterogeneous data has yet been developed. Additionally, the majority of methods have been demonstrated only in baker's yeast, as multicellular organisms present a host of additional challenges for data integration and microarray analysis. It is also important to note that computational methods are always limited by the coverage and quality of experimental data they use. Public availability of high-quality high throughput datasets is, therefore, essential for rapid functional annotation. Furthermore experimental validation of computational predictions by traditional laboratory techniques is ideal for validation and for improvement of the computational methodology. Such validation can be accomplished through collaborations with biological researchers and through open publication of predictions in the form easily accessible to biologists.

A very promising direction in functional analysis of microarray data is integration of data from multiple organisms. Recently, several groups have started using coexpression information from homologous genes in several species to increase specificity of functional relationships identified from gene expression experiments. Such comparative genomics techniques, on their own or combined data integration methods described in this chapter, will undoubtedly contribute to functional annotation.

In general, development of accurate methods for integrated analysis of microarray data relies on labeled data for training and validation, for example, genes with known functions or known biological pathways. Such data, generated by traditional biological methods, are often scarce and for the most part represented in biological literature in the free-text format that cannot be readily used for automatic training or validation. One very effective solution to this problem is human curation, employed by several databases. However, curation is costly and thus currently limited, and many curated databases show biases toward specific biological processes that can influence integration results. Therefore, care must be taken with the "gold standard" data in both learning and evaluation.

2.2. Integrated Analysis of Microarray Data for Gene Function Prediction

Gene expression microarray datasets are the most commonly available functional genomic data owing to their relatively low cost and easily accessible technology. The largest public repository of gene expression data is NCBI's gene expression omnibus database *(17)*. These data can be used to identify groups of coexpressed genes, and such groups, through the principle of "guilt by association," can facilitate function prediction for unknown proteins and identification of regulatory elements. However, although gene coexpression data are an excellent tool for hypothesis generation, microarray data alone often lack the degree of specificity needed to make accurate biological conclusions. For such purposes, an increase in accuracy is needed, even if it comes at the cost of some sensitivity. This improvement in specificity can be achieved through incorporation of other data sources in an integrated analysis of gene expression data. These additional data sources include other high-throughput functional data (e.g., protein–protein interactions, genetic interaction data, and localization information), DNA and protein sequence data, published literature, and phylogenetic information.

Bioinformatics methods for effective integration of high throughput heterogeneous data can provide the improvement in specificity necessary for accurate gene function annotation and network analysis based on high-throughput data *(8,9,34,35)*. Although, the exact amount of overlap and correlation among functional datasets is unclear *(32,36–38)*, data integration has been shown to increase the accuracy of gene function prediction compared with a single high-throughput method *(31,34,39–43)*. Specifically, studies demonstrated that using more than one type of functional data for predictions increased accuracy *(31)* and that integrating more heterogeneous information increases the number of protein–protein interactions correctly identified *(42)*, leading to better prediction of function for unknown proteins. This potential of data integration recently led to development of several computational methods for integrated analysis of microarray data with other data sources.

2.2.1. Nonprobabilistic Integration

A simple scheme for increasing accuracy in function prediction based on heterogeneous data is to consider the intersection of interaction maps for different high throughput datasets *(44)*. Although, this scheme reduces the false-positives, it has the drawback that the lowest-sensitivity dataset will limit sensitivity of the entire analysis. As published large-scale interaction studies are not comprehensive even in model organisms, this strict sensitivity limitation is too restrictive for large-scale and general function prediction. Several other groups suggested approaches that provide increased sensitivity of function prediction from the intersection scheme previously listed. In the first study of this type, predicted a number of potential protein functions for *S. cerevisiae* based on a heuristic combination of different types of data *(34,39)*. In another early study, putative protein function assigned based on the number of interactions an unknown protein has with proteins from different functional categories *(40)*. These studies demonstrated the potential of integrated data analysis, but they combine the information from different sources in a heuristic fashion, where confidence levels for protein–protein links are defined on a case-by-case basis. This approach is successful in these studies and served as a clear proof of concept, but it may be hard to generalize to new datasets, data types, or other organisms because each approach is developed with specific data and application goal in mind and therefore, lacks a general scheme or representation.

A more general method was developed to introduce a rule-based method, in which heuristics are learned based on heterogeneous data sources and known functional predictions *(45)*. These heuristics are then applied to genes with unknown function to predict function. This study uses a modified C4.5 decision tree algorithm, and includes sequence, phenotype, expression, and predicted secondary structure data. In a different approach, combined interactions and expression data by creating a weighted graph of protein–protein interactions with the weight between two genes derived from coexpression values of these genes in one gene expression dataset *(46)*. They then used a variant of discrete-state Hopfield network to assign function for unknown proteins, based on known annotations in the gene ontology *(47)*.

2.2.2. Probabilistic Integration

Recently, several computational methods have been suggested that combine datasets in a confidence-dependent manner. The advantage of such statistical approaches is that they enable general data integration and can easily adapt to new data sources. In addition, because these methods are probabilistic, their outputs can be filtered by the confidence or probability cutoff to a desired level of sensitivity and specificity (estimated based on the cross-validation trials or a test data set).

In a general methodology based on support vector machines, Lanckriet et al. have combined interactions, expression, and sequence data by representing each input as a separate kernel. The weighted optimized combination of these kernels was then used to recognize membrane and ribosomal proteins *(48)*, as well as other general classes of proteins *(49)*. This method is general and can also readily provide information, encoded in the kernel weights, on the extent to which each data source contributes to the final prediction. One disadvantage of such discriminative approaches is that a separate classifier is generally built for each functional category, thereby making it possible to only predict general functional categories (e.g., metabolism) because of lack of training data for more specific functions.

Methodologies that first perform general data integration, creating a general graph of functional relationships, and then predict function based on such graph, can alleviate this problem. For example, Bayesian network-based method called multisource association of genes by integration of clusters (MAGIC) for general integrated analysis of gene expression microarrays in the context of functional genomic data *(35)*. MAGIC's key advantage is generality and adaptability because it can easily incorporate new data sources, datasets, and analysis methods. It incorporates expert knowledge in the prior probability parameters in the Bayesian framework or learns from available data, thus formally integrating relative accuracies of different experimental and computational techniques in the analysis and minimizing potential bias toward well-studied areas in its reasoning. In addition, Bayesian networks are generally robust to noise in prior probabilities and in training data. These characteristics of Bayesian networks yield high accuracy of gene function predictions, and the probabilistic nature of the system provides confidence levels for each output.

The MAGIC integration system takes as input groupings (or clusters) of genes based on each experimental data set (e.g., shared transcription factor binding sites, protein–protein interaction, or coexpression data). The system represents all input groupings as gene i –gene j pairs with corresponding scores s_{ij}. The score s_{ij} corresponds to the strength of each method's belief in the existence of a relationship between gene i and gene j. The score $s_{ij} > 0$ if gene i and gene j appear in the same cluster or grouping with strength s_{ij} or if they interact on the basis of an experimental method. This score can be binary (e.g., results of coimmunoprecipitation experiments), continuous or discrete (e.g., $-1 \leq s \leq 1$ for Pearson correlation). MAGIC's Bayesian network combines evidence from input groupings and generates a posterior belief for whether each gene i–gene j pair has a functional relationship. For each pair of genes, the network essentially asks the following question: "what is the probability, based on the evidence presented, that products of gene i and gene j interact or are involved in the same biological process?" MAGIC's Bayesian network structure was determined through consultation with experts in yeast genomics and microarray analysis, and the prior probabilities can be either based on expert consultation or learnt from example data (such as functional annotations of genes with the gene ontology biological process annotations).

Thus, MAGIC addresses the need for a generalizable method for comprehensive data integration, by combining microarray data with heterogeneous data types of various levels of accuracy in an algorithmic fashion. Other such approaches have recently become available, for predicting cocomplexed protein pairs with probabilistic decision trees based on expression and proteomics data *(50)*. Preliminary success of these methods in function prediction demonstrates the potential of sophisticated data integration algorithms. Further development of such computational methods combined with increased availability of large-scale functional genomics data in downloadable standardized formats should enable accurate prediction of function for most unknown yeast proteins. These predictions, followed by targeted laboratory experiments, may enable fast and relatively low cost annotation of whole genomes.

References

1. Goffeau, A., Barrell, B. G., Bussey, H., et al. (1996) Life with 6000 genes. *Science* **274,** 546–567.
2. Wood, V., Gwilliam, R., Rajandream, M. A., et al. (2002) The genome sequence of *Schizosaccharomyces pombe. Nature* **415,** 871–880.
3. Adams, M. D., Celniker, S. E., Holt, R. A., et al. (2000) The genome sequence of *Drosophila melanogaster. Science* **287,** 2185–2195.
4. *C. elegans* Sequencing Consortium. (1998) Genome sequence of the nematode *C. elegans*: a platform for investigating biology. *Science* **282,** 2012–2018.
5. Lander, E. S., Linton, L. M., Birren, B., et al. (2001) Initial sequencing and analysis of the human genome. *Nature* **409,** 860–921.
6. Venter, J. C., Adams, M. D., Myers, E. W., et al. (2001) The sequence of the human genome. *Science* **291,** 1304–1351.
7. Kitano, H. (2002) Looking beyond the details: a rise in system-oriented approaches in genetics and molecular biology. *Curr. Genet.* **41,** 1–10.
8. Steinmetz, L. M. and Deutschbauer, A. M. (2002) Gene function on a genomic scale. *J. Chromatogr. B Analyt. Technol. Biomed. Life Sci.* **782,** 151–163.
9. Ideker, T., Galitski, T., and Hood, L. (2001) A new approach to decoding life: systems biology. *Annu. Rev. Genomics Hum. Genet.* **2,** 343–372.
10. Lipshutz, R. J., Fodor, S. P., Gingeras, T. R., and Lockhart, D. J. (1999) High density synthetic oligonucleotide arrays. *Nat. Genet.* **21,** 20–24.
11. Schena, M., Shalon, D., Davis, R. W., and Brown, P. O. (1995) Quantitative monitoring of gene expression patterns with a complementary DNA microarray. *Science* **270,** 467–470.
12. Cahill, D. J. and Nordhoff, E. (2003) Protein arrays and their role in proteomics. *Adv. Biochem. Eng. Biotechnol.* **83,** 177–187.
13. Sydor, J. R. and Nock, S. (2003) Protein expression profiling arrays: tools for the multiplexed high-throughput analysis of proteins. *Proteome Sci.* **1,** 3.
14. Oleinikov, A. V., Gray, M. D., Zhao, J., Montgomery, D. D., Ghindilis, A. L., and Dill, K. (2003) Self-assembling protein arrays using electronic semiconductor microchips and in vitro translation. *J. Proteome Res.* **2,** 313–319.
15. Huang, R. P. (2003) Protein arrays, an excellent tool in biomedical research. *Front Biosci.* **8,** D559–D576.
16. Cutler, P. (2003) Protein arrays: the current state-of-the-art. *Proteomics* **3,** 3–18.
17. Edgar, R., Domrachev, M., and Lash, A. E. (2002) Gene Expression Omnibus: NCBI gene expression and hybridization array data repository. *Nucleic Acids Res.* **30,** 207–210.
18. Larsson, P. O. and Mosbach, K. (1979) Affinity precipitation of enzymes. *FEBS Lett.* **98,** 333–338.
19. Fields, S. and Song, O. (1989) A novel genetic system to detect protein–protein interactions. *Nature* **340,** 245–246.

20. Novick, P., Osmond, B. C., and Botstein, D. (1989) Suppressors of yeast actin mutations. *Genetics* **121,** 659–674.
21. Bender, A. and Pringle, J. R. (1991) Use of a screen for synthetic lethal and multicopy suppressee mutants to identify two new genes involved in morphogenesis in *Saccharomyces cerevisiae. Mol. Cell Biol.* **11,** 1295–1305.
22. Bader, G. D., Betel, D., and Hogue, C. W. (2003) BIND: the bimolecular interaction network database. *Nucleic Acids Res.* **31,** 248–250.
23. Xenarios, I., Salwinski, L., Duan, X. J., Higney, P., Kim, S. M., and Eisenberg, D. (2002) DIP, the database of interacting proteins: a research tool for studying cellular networks of protein interactions. *Nucleic Acids Res.* **30,** 303–305.
24. Zanzoni, A., Montecchi-Palazzi, L., Quondam, M., Ausiello, G., Helmer-Citterich, M., and Cesareni, G. (2002) MINT: a molecular INTeraction database. *FEBS Lett.* **513,** 135–140.
25. Breitkreutz, B. J., Stark, C., and Tyers, M. (2003) The GRID: the general repository for interaction datasets. *Genome Biol.* **4,** R23.
26. Mewes, H. W., Frishman, D., Guldener, U., et al. (2002) MIPS: a database for genomes and protein sequences. *Nucleic Acids Res.* **30,** 31–34.
27. Issel-Tarver, L., Christie, K. R., Dolinski, K., et al. (2002) *Saccharomyces* genome database. *Methods Enzymol.* **350,** 329–346.
28. Grunenfelder, B. and Winzeler, E. A. (2002) Treasures and traps in genome-wide data sets: case examples from yeast. *Nat. Rev. Genet.* **3,** 653–661.
29. Chen, Y. and Xu, D. (2003) Computational analyses of high-throughput protein-protein interaction data. *Curr. Protein Pept. Sci.* **4,** 159–181.
30. Bader, G. D., Heilbut, A., Andrews, B., Tyers, M., Hughes, T., and Boone, C. (2003) Functional genomics and proteomics: charting a multidimensional map of the yeast cell. *Trends Cell Biol.* **13,** 344–356.
31. von Mering, C., Krause, R., Snel, B., et al. (2002) Comparative assessment of large-scale data sets of protein-protein interactions. *Nature* **417,** 399–403.
32. Deane, C. M., Salwinski, L., Xenarios, I., and Eisenberg, D. (2002) Protein interactions: two methods for assessment of the reliability of high throughput observations. *Mol. Cell Proteomics* **1,** 349–356.
33. Yue, H., Eastman, P. S., Wang, B. B., et al. (2001) An evaluation of the performance of cDNA microarrays for detecting changes in global mRNA expression. *Nucleic Acids Res.* **29,** E41.
34. Marcotte, E. M., Pellegrini, M., Ng, H. L., Rice, D. W., Yeates, T. O., and Eisenberg, D. (1999) Detecting protein function and protein–protein interactions from genome sequences. *Science* **285,** 751–753.
35. Troyanskaya, O. G., Dolinski, K., Owen, A. B., Altman, R. B., and Botstein, D. (2003) A Bayesian framework for combining heterogeneous data sources for gene function prediction (in *Saccharomyces cerevisiae). Proc. Natl. Acad. Sci. USA* **100,** 8348–8353.
36. Edwards, A. M., Kus, B., Jansen, R., Greenbaum, D., Greenblatt, J., and Gerstein, M. (2002) Bridging structural biology and genomics: assessing protein interaction data with known complexes. *Trends Genet.* **18,** 529–536.

37. Kemmeren, P., van Berkum, N. L., Vilo, J., et al. (2002) Protein interaction verification and functional annotation by integrated analysis of genome-scale data. *Mol. Cell.* **9,** 1133–1143.
38. Werner-Washburne, M., Wylie, B., Boyack, K., et al. (2002) Comparative analysis of multiple genome-scale data sets. *Genome Res.* **12,** 1564–1573.
39. Marcotte, E. M., Pellegrini, M., Thompson, M. J., Yeates, T. O., and Eisenberg, D. (1999) A combined algorithm for genome-wide prediction of protein function. *Nature* **402,** 83–86.
40. Schwikowski, B., Uetz, P., Fields, S. (2000) A network of protein–protein interactions in yeast. *Nat. Biotechnol.* **18,** 1257–1261.
41. Bader, G. D. and Hogue, C. W. (2002) Analyzing yeast protein–protein interaction data obtained from different sources. *Nat. Biotechnol.* **20,** 991–997.
42. Gerstein, M., Lan, N., and Jansen, R. (2002) Proteomics. Integrating interactomes. *Science* **295,** 284–287.
43. Ge, H., Liu, Z., Church, G. M., and Vidal, M. (2001) Correlation between transcriptome and interactome mapping data from *Saccharomyces cerevisiae*. *Nat. Genet.* **29,** 482–486.
44. Tong, A. H., Drees, B., Nardelli, G., et al. (2002) A combined experimental and computational strategy to define protein interaction networks for peptide recognition modules. *Science* **295,** 321–324.
45. Clare, A. and King, R. D. (2003) Predicting gene function in *Saccharomyces cerevisiae*. *Bioinformatics* **19,** II42–II49.
46. Karaoz, U., Murali, T. M., Letovsky, S., et al. (2004) Whole-genome annotation by using evidence integration in functional-linkage networks. *Proc. Natl. Acad. Sci. USA* **101,** 2888–2893.
47. Ashburner, M., Ball, C. A., Blake, J. A., et al. (2000) Gene ontology: tool for the unification of biology. The Gene Ontology Consortium. *Nat. Genet.* **25,** 25–29.
48. Lanckriet, G. R., De Bie, T., Cristianini, N., Jordan, M. I., and Noble, W. S. (2004) A statistical framework for genomic data fusion. *Bioinformatics* **20,** 2626–2635.
49. Lanckriet, G. R., Deng, M., Cristianini, N., Jordan, M. I., and Noble, W. S. (2004) Kernel-based data fusion and its application to protein function prediction in yeast. *Pac. Symp. Biocomput.* 300–311.
50. Zhang, L. V., Wong, S. L., King, O. D., and Roth, F. P. (2004) Predicting co-complexed protein pairs using genomic and proteomic data integration. *BMC Bioinformatics* **5,** 38.

Index